Distributed Control Applications

Guidelines, Design Patterns, and Application Examples with the IEC 61499

Distributed Control Applications: Guidelines, Design Patterns, and Application Examples with the IEC 61499
Edited by Alois Zoitl and Thomas Strasser

Industrial Communication Technology Handbook, Second Edition
Edited by Richard Zurawski

Formal Methods in Manufacturing
Edited by Javier Campos, Carla Seatzu, and Xiaolan Xie

Embedded Systems Handbook, Second Edition
Edited by Richard Zurawski

Automotive Embedded Systems Handbook
Edited by Nicolas Navet and Françoise Simonot-Lion

Integration Technologies for Industrial Automated Systems
Edited by Richard Zurawski

Electronic Design Automation for Integrated Circuits Handbook
Edited by Luciano Lavagno, Grant Martin, and Lou Scheffer

Distributed Control Applications

Guidelines, Design Patterns, and Application Examples with the IEC 61499

EDITED BY
Alois Zoitl • Thomas Strasser

CRC Press
Taylor & Francis Group
Boca Raton London New York

CRC Press is an imprint of the
Taylor & Francis Group, an **informa** business

CRC Press
Taylor & Francis Group
6000 Broken Sound Parkway NW, Suite 300
Boca Raton, FL 33487-2742

First issued in paperback 2017

© 2016 by Taylor & Francis Group, LLC
CRC Press is an imprint of Taylor & Francis Group, an Informa business

No claim to original U.S. Government works

ISBN-13: 978-1-4822-5905-6 (hbk)
ISBN-13: 978-1-138-89295-8 (pbk)

Visit the Taylor & Francis Web site at
http://www.taylorandfrancis.com

and the CRC Press Web site at
http://www.crcpress.com

Contents

IV   Laboratory Automation Examples   395

18   Workspace Sharing Assembly Robots: Applying IEC 61499   397

*Matthias Plasch, Gerhard Ebenhofer, Michael Hofmann, Martijn
Rooker, Sharath Chandra Akkaladevi, and Andreas Pichler*

19   Hierarchically Structured Control Application for Pick and Place Station   423

Monika Wenger, Milan Vathoopan, Alois Zoitl, and Herbert Prähofer

Preface

Industrial automation and control systems in manufacturing, power, and energy systems, as well as the logistics sector are becoming increasingly complex. Usually in such systems a huge number of actuators and sensors are working together as a network of different types of devices. The latest trend in research and development shows that these devices are becoming more and more intelligent and networked. This means they can perform tasks autonomously and directly interact with each other.

In order to master the complexity of such highly interconnected and collaborative devices, advanced methods and concepts applied to the whole life-cycle of those devices and networks of devices (planning, engineering, operation, maintenance) are necessary, because the distributed control services have to provide advanced functions that include reconfigurabiltiy of software components, real-time execution (time-triggered, event-based), diagnostics, and other services. The IEC 61499 reference architecture was developed to provide a better way for developing and maintaining such automation systems and to better fulfill the above requirements.

The main aim of this book is to discuss the usage of the IEC 61499 reference architecture for distributed and reconfigurable control and its adoption by industry. It provides design patterns, application guidelines, and rules for designing distributed control applications based on the IEC 61499 reference model. Moreover, selected examples from various industrial domains as well as from laboratory environments are introduced and discussed. With this content, this book provides a good overview on application domains of IEC 61499 and helps to apply the IEC 61499 models for implementing current automation solutions.

Editors

Alois Zoitl earned a PhD in electrical engineering with a focus on dynamic reconfiguration of real-time constrained control applications and a master's degree in electrical engineering with a focus on distributed industrial automation systems from the Vienna Institute of Technology. He currently leads the Industrial Automation Research Group at fortiss GmbH in Munich. Previously, he headed the Distributed Intelligent Automation Group (Odo Struger Laboratory) at the Vienna University of Technology's Automation and Control Institute. He is active as a lecturer at the Technical University Munich.

Dr. Zoitl is the co-author of more than 100 publications (books, book chapters, conference papers, and journal articles) and is named as co-inventor on 4 patents in his research areas. In the course of his research, he has managed several industry-funded projects and coordinated and participated in a number of publicly funded Austrian and European projects. He is a founding member of the 4DIAC and OpENer open-source initiatives, a member of IEEE and the PLCopen user organization, a consultant for CAN in Automation, and a member of IEC SC65B/WG15 for the IEC 61499 distributed automation standard. He was named convenor of the group in May 2015.

Thomas Strasser earned a PhD in mechanical engineering with a focus on automation and control theory and a master's degree in industrial engineering with a focus on production and automation systems from the Vienna University of Technology.

For several years, he has been a senior scientist in the Energy Department of the AIT Austrian Institute of Technology. His main responsibilities involve strategic development of smart grid research projects and mentoring and advising junior scientist and PhD candidates. Before joining AIT, Dr. Strasser spent more than 6 years as a senior researcher investigating advanced and reconfigurable automation and control systems at PROFACTOR. He is active as a lecturer at the Vienna University of Technology and as guest professor at the University of Applied Sciences in Salzburg.

Dr. Strasser has co-authored more than 120 scientific publications (editorials, book chapters, conference papers, and journal articles) and was awarded 2 patents in his areas of interest. He is an active participant in IEEE conferences and serves as an associate editor of Springer and IEEE journals. He is also a senior member of IEEE and a founding member of the 4DIAC open-source initiative and involved in IEC SC65B/WG15, IEC TC65/WG17, and IEC SyC Smart Energy/WG6.

Acknowledgments

We would like to thank all contributing authors, particularly authors from industry and industry consortia whose job priorities usually are different from writing chapters in books.

We also wish to thank Nora Konopka, publisher of this book; Kyra Lindholm, Michele Smith, Jessica Vakili, Florence Kizza, and Karen Simon at CRC Press Taylor & Francis.

Finally, we would also like to provide our special thanks to the series editor, Dr. Richard Zurawski, for the invitation to write this book and his support.

Contributors

Sharath Chandra Akkaladevi
PROFACTOR GmbH
Steyr, Austria

Filip Andrén
AIT Austrian Institute of
  Technology GmbH
Vienna, Austria

Alfonso Blesa
Universidad de Zaragoza
Zaragoza, Spain

Robert W. Brennan
University of Calgary
Calgary, Canada

Roland Bründlinger
AIT Austrian Institute of
  Technology GmbH
Vienna, Austria

Carlos Catalán
Universidad de Zaragoza
Zaragoza, Spain

James H. Christensen
Holobloc Inc.
Cleveland Heights, Ohio

José Manuel Colom
Universidad de Zaragoza
Zaragoza, Spain

Wenbin Dai
Shanghai Jiao Tong University
Shanghai, China

Gerhard Ebenhofer
PROFACTOR GmbH
Steyr, Austria

Reinhard Hametner
Thales Austria GmbH
Vienna, Austria

George Hassapis
Aristotle University of Thessaloniki
Thessaloniki, Greece

Ingo Hegny
Vienna University of Technology
Vienna, Austria

Michael Hofmann
PROFACTOR GmbH
Steyr, Austria

Peter Idowu
Pennsylvania State University
Middletown, Pennsylvania

Srikrishnan Jagannathan
Pennsylvania State University
Middletown, Pennsylvania

Petr Kadera
Czech Technical University in Prague
Prague, Czech Republic

Gernot Kollegger
nxtControl GmbH
Leobersdorf, Austria

Arnold Kopitar
nxtControl GmbH
Leobersdorf, Austria

Matthew M. Y. Kuo
University of Auckland
Auckland, New Zealand

Georg Lauss
AIT Austrian Institute of
    Technology GmbH
Vienna, Austria

Wilfried Lepuschitz
Practical Robotics Institute Austria
Vienna, Austria

Andreas Pichler
PROFACTOR GmbH
Steyr, Austria

Matthias Plasch
PROFACTOR GmbH
Steyr, Austria

Herbert Prähofer
Johannes Kepler University Linz
Linz, Austria

Josep Maria Rams
TUROMAS Group
Teruel, Spain

Martijn Rooker
PROFACTOR GmbH
Steyr, Austria

Partha S. Roop
University of Auckland
Auckland, New Zealand

Georg Schitter
Vienna University of Technology
Vienna, Austria

Christian Seitl
AIT Austrian Institute of
    Technology GmbH
Vienna, Austria

Félix Serna
Universidad de Zaragoza
Zaragoza, Spain

Georgios Sfiris
Aristotle University of Thessaloniki
Thessaloniki, Greece

Mario de Sousa
University of Porto
Porto, Portugal

Thomas Strasser
AIT Austrian Institute of
    Technology GmbH
Vienna, Austria

Philipp Svec
AIT Austrian Institute of
    Technology GmbH
Vienna, Austria

Milan Vathoopan
fortiss GmbH
Munich, Germany

Pavel Vrba
Czech Technical University in Prague
Prague, Czech Republic

Valeriy Vyatkin
Luleå University of Technology
and
Luleå, Sweden Aalto University
Helsinki, Finland

Monika Wenger
fortiss GmbH
Munich, Germany

Yu Zhao
University of Auckland
Auckland, New Zealand

Alois Zoitl
fortiss GmbH
Munich, Germany

Part I

IEC 61499 Basics

1

Challenges and Demands for Distributed Automation in Industrial Environments

Thomas Strasser

AIT Austrian Institute of Technology GmbH

Alois Zoitl

fortiss GmbH

CONTENTS

1.1   Trends in Industrial Automation

The industrial automation domain is a key driver and supporter for other industries. With its innovations it can be seen as the backbone for many industrial sectors, like manufacturing, process technology, or the power and energy systems domain. Since production processes are more and more performed by automated machines, factories and plants, an increased level of automation is observable today. The major trend and driving factor for industrial automation in recent years is the growing need for customized and individualized products and goods. There is a need to evolve corresponding automation solutions in response to the rapidly changing demands due to new production processes and technologies. Also recent advancements in hardware and software solutions influence this trend. This requires that production lines, machines, and components have to be constructed and adapted to new products and production processes quickly [3, 10].

Such highly automated systems are mainly controlled by a vast set of embedded hardware and software components which are heterogeneous in nature. Typically, corresponding specifications and architectures lead to a tremendous increase in the design complexity. As a result of this trend, an increase of the software engineering costs from about 50% of the overall system engineering costs up to 80% in the next 15 years can be expected [11]. However, more recent studies also show that the engineering effort can be reduced up to 70% by optimizing the overall engineering process using proper methods, architectures, and corresponding tools [4]. Such an optimization result cannot really be achieved for the control software development, which is still a key problem. A further specialty of the manufacturing and also of the power and energy systems domain is the fact that each plant or system can be seen as a prototype. Hence, engineering costs are still key cost factors for the above mentioned processes and systems.

Applied design methods and approaches strongly depend on the specific application field. Well known control approaches in industrial automation are computerized numerical control (CNC), robot control, programmable logic controller (PLC), distributed control system (DCS), and supervisory control and data acquisition (SCADA). Also other approaches like field programmable gate array (FPGA)-based controllers becoming popular. These approaches are often used together to perform the automation and control tasks of a specific application [1]. The design and engineering of such heterogeneous systems is elaborate and needs the knowledge from different domains during all phases of development. For the engineering of industrial automation and control solutions, the system complexity, the domain or platform dependence, and also the design time are the most critical factors. In order to keep the complexity under control, the following key requirements and needs have to be satisfied [9]:

- The system specifications should be prepared at an easily understandable level of abstraction. Specification means should support designers of industrial automation and control systems at each level of the plant architecture to define the desired functionality and avoid to implement it on code level.

- Different domain-specific techniques and approaches should be combinable in a flexible way, overcoming potential limitations of each procedure. This allows modeling, analysis and implementation of complex, heterogeneous systems forming industrial automation and control systems.

- Methods and tools for specifying systems architectures should not depend on general languages used for embedded system design.

To cope with the challenges and demands of industrial automation and control systems a key design trend is to put components—blocks made of hardware, software and intellectual property (like algorithms and data structures)—together [15, 16] into one reusable component. This requires common, language-independent models for representing, saving and reusing the engineering artifacts of such components. Today, state-of-the art approaches, methods and corresponding tools for designing and engineering industrial automation systems are not fully capable of providing applicable solutions to the above mentioned issues [1, 15, 16]. Solutions developed for other domains like automotive or aerospace focus on the specific needs and requirements of their own sectors. They are different from the demands of industrial automation. Moreover, most of these solutions require a deep knowledge in software engineering. In the industrial automation sector, however, electrical and control engineers usally design the control systems. These domain experts may have only a basic education in software design. As software development becomes more and more important for industrial automation, methods and approaches have to be developed that allow also non-software engineering experts to effectively and efficiently develop control software.

1.2 Requirements for Future Automation Architecture

The above identified trends show the need for a more flexible and adaptable future automation system architecture and corresponding tools fulfilling requirements of different products and customers. As identified by [14] distributed architectures and dynamic reconfiguration are key drivers supporting the adaptivity of automation environments. In order to make this concept applicable, a future automation architecture has to fulfill the following key requirements:

Distributed intelligence has the potential to provide an opportunity to achieve manageability and robustness of large and high complex industrial processes and machines.

Usability is an important requirement for automation architecture. It needs to be simple to use, understandable, and maintainable as current PLC-based systems are. At the same time, design should allow a more expressive way to develop and maintain them.

Domain-specific modeling languages are useful tools allowing domain experts with little software engineering knowledge to describe automation tasks on a higher level of abstraction. This ensures reusability, reduces the engineering effort, and increases the software quality.

Easy communication configuration is needed for defining the interaction as well as the data exchange of a large number of distributed (embedded) control devices.

Predictability of distributed applications (in terms of timing behavior and its flow of execution) should be as easy to understand as the well known cyclic execution behavior typically implemented in PLC systems.

Heterogeneous execution is required to cope with the different needs of application tasks. Time-triggered execution with jitter below a defined border (especially for the execution of closed loop control algorithms), and event-based execution within a specified timing window must be supported within the same execution context or environment. Hence, the functions don't have to disturb each other and they may have to interact in a deterministic way.

Real-time execution with guarantees are needed as industrial automation and control systems interact with real-world processes, machines, and components. In case of dynamic reconfiguration, timing constraints may also have to be met deterministically.

Real-time reconfiguration with guarantees allows changing the functionality of control applications and systems during full operation. This requires that the reconfiguration process does not result in observable disturbances of the automation application. The switching or relocation of full control applications is required along with a fine grained adaption of application parts in a deterministic. The execution environment has to support reconfiguration on specific events or states of the industrial automation system as well as the switching at fixed points in time. Guarantees have to be given that the reconfiguration and switching processes do not violate the execution constraints of the running application. Typically, those processes have to be performed within the available resources (e.g., memory, processing power, network bandwidth) as well.

Durability of industrial automation and control systems needs to support system lifespans up to decades. Even if the produced products are more short-lived, the manufacturing system may also be used for a vast range of other products. This requires long-term upward and downward compatibility for replacement of components and devices.

Standards compliance is required to provide interoperability of components from different vendors, allowing configuration from any standards-compliant tools, and source-level portability for software reuse. Nonetheless, portability at binary level for the migration of software components between control devices might be necessary as wells.

Form factor, price and robustness of distributed industrial automation devices needs to be such that the devices can be mounted directly onto the electromechanical components.

1.3   Outlook

Currently the domain of industrial automation and control is subject to major technological changes. One of the key drivers behind this trend are cyber-physical systems (CPSs) [12]. CPSs are the extensions of embedded systems to fully networked embedded systems with a virtual representation of their physical entity. This leads to a key aspect of CPSs: they can be connected and interact for tasks they have not been designed for.

For example, in Germany the CPS concept has been the basis for the Industry 4.0 initiative [2]. This initiative states using CPSs in production systems will bring about the fourth industrial revolution. With Industry 4.0 the following goals and expectations are envisaged:

- Ability to manufacture individual products according to customer demands

- Increased flexibility of the production system

- Optimized decision-making

- Greater resource efficiency and productivity

- Extended business models based on new services incorporating the full value chain (i.e., from suppliers to customers),

- Improvement of work-life balance and healthier workplaces that support demographic changes in the work force and allow workers to remain in the workforce longer

- Re-industrialization of Europe

In US, China, Korea, and Japan, similar attempts are undertaken under the term *smart manufacturing*. A further, technology for evolution of industrial automation and control systems is Internet-of-Things (IoT) [7]. IoT targets connect everything from machines to products and materials to the Internet and if necessary also to the cloud.

Similar changes are also observable in the domain of power and energy systems. Due to the integration of distributed, renewable sources, improvements in energy efficiency, and market de-regulation, a change in the planning and operation of power systems is necessary to guarantee a sustainable system for the future. This trend influences also the automation and control strategies used in the power and energy domain. Similar to Industry 4.0, utilizing CPS and IoT technologies a transformation into so-called Smart Grids allowing better monitoring and control of complex power grids becomes possible [5, 6, 8].

Summarizing, these trends show that future industrial automation and control systems are composed increasingly by intelligent, interconnected devices utilizing different networking approaches. Such future automation environments are also characterized by greater amounts of software usage. As stated above, control software development is already one of the key cost factors for developing and deploying industrial automation and control systems. In order to reduce these costs, new approaches, concepts, and corresponding engineering tools for developing these systems are necessary. IEC 61499 as a reference model for highly distributed, networked industrial automation and control systems provides the basis for the realization of a future distributed architecture [13]. It has the potential of fulfilling most of the above needs.

In the remaining part of this book, the usage of the IEC 61499 reference architecture is discussed by introducing corresponding design patterns, application guidelines and rules, and implemented examples from various industrial domains and laboratory environments.

Bibliography

[1] Roadmap on real-time techniques in control system implementation – control for embedded systems cluster, EU/IST FP6 Artist2 NoE, www.artist-embedded.org, 2006.

[2] Arbeitskreis Industrie 4.0. *Umsetzungsempfehlungen für das Zukunftsprojekt Industrie 4.0.* Forschungsunion im Stifterverband für die Deutsche Wirtschaft e.V., Berlin, Oct. 2012.

[3] Artemis Strategic Research Agenda, 2005.

[4] M. Buchwitz. Neue Wege in der Software Entwicklung. *SPS Magazin*, 2012.

[5] EC. European SmartGrids technology platform: vision and strategy for Europe's electricity networks of the future. Technical report, European Commission (EC): Directorate-General for Research-Sustainable Energy Systems, 2006.

[6] H. Farhangi. The path of the smart grid. *IEEE Power and Energy Magazine*, Vol. 8, No. 1, 18–28, 2010.

[7] J. Gubbi, R. Buyya, S. Marusic, and M. Palaniswami. Internet of Things (IoT): A vision, architectural elements, and future directions. *Future Generation Computer Systems*, Vol. 29, No. 7, 1645–1660, 2013.

[8] IEA. Smart grid insights. Technical report, International Energy Agency, 2011.

[9] M. Jersak, K. Richter, and R. Ernst. Combining complex event models and timing constraints. In *Proceedings of Sixth IEEE International High-Level Design Validation and Test Workshop*, pages 89–94, 2001.

[10] Y. Koren. *The Global Manufacturing Revolution.* 2010.

[11] S. Kuppinger. Die Schlüssel zur Effizienz - Integration von Motion und Logic. *IEE (Industrie elektrik+elektronik)*, Vol. 51, No. 1, 2006.

[12] E. A. Lee and S. A. Seshia. *Introduction to Embedded Systems, A Cyber-Physical Systems Approach.* http://LeeSeshia.org, second edition edition, 2015.

[13] TC 65/SC 65B. IEC 61499: Function blocks. International Electrotechnical Commission, Geneva, 2nd ed., 2012.

[14] I. Terzic, A. Zoitl, B. Favre-Bulle, and T. Strasser. A survey of distributed intelligence in automation in european industry, research and market. In *Proceedings of IEEE International Conference on Emerging Technologies and Factory Automation*, pages 221–228, Sept. 2008.

[15] T. Tommila, J. Hirvonen, L. Jaakkola, J. Peltoniemi, J. Peltola, S. Sierla, and K. Koskinen. *Next generation of industrial automation – Concepts and architecture of a component-based control system.* VTT Research Notes 2303, 2005.

[16] D. Woll. *Setting the Stage for the Next Generation of Automation Control System Software: IEC 61499.* ARC Advisory Group on Industry Trends, 2007.

2

Basic Principles of IEC 61499 Reference Model

Thomas Strasser

AIT Austrian Institute of Technology GmbH

Alois Zoitl

fortiss GmbH

CONTENTS

2.1 Introduction

Automation technology gone through several evolutionary steps during the last decades. Centralized control based on IEC[1] 61131 has been designed for programmable logical controllers (PLCs) with one ore more tightly coupled processors [5]. Part 3 of IEC 61131 provides five well-known programming languages (textual and graphical) for PLCs [3]. This standard, however, was not designed for distributed automation and control systems, although Part 5 of IEC 61131 provides definitions of communication function blocks (FBs) for interoperability between different PLCs. Part 5 specifications are not sufficient to ensure interoperability of different vendors' PLCs. Furthermore, IEC 61131 does not sufficiently support distribution of control code to distributed devices

---

[1]International Electrotechnical Commission, Geneva, Switzerland.

and engineering is more device oriented than application oriented. However, the IEC 61131 approach, especially the programming languages as defined in Part 3, is broadly accepted and widely used in industry for the definition of control code and algorithms.

Through the introduction of the fieldbus technology in the early 1980s the trend of distributed intelligence evolved from the decentralized automation approach. Decentralized control greatly reduced the wiring costs of automation systems but no significant changes in the general architecture of control programs and PLCs have been developed. Distributed intelligence in automation goes one step further by equipping "dumb" networked peripheral units with control software [6]. This helps define new architectures with no central units since all control tasks are executed in the intelligent units.

The programming constructs of today's PLC approaches based on IEC 61131-3 are too monolithic and closed for distributed applications. The main restriction of current systems is the missing support for building control software from independent components that can be added and removed during execution. New ways for programming and modeling of control systems have to be devised. Hence, the automation standard IEC 61499 "Function Blocks" [4] introduced by the IEC Technical Committee (TC) 65 has been developed. It provides a reference model for distributed and reconfigurable automation systems. The modeling of open and distributed industrial process, measurement and control systems (IPMCSs) is also in the focus of this standard. The goal is to obtain a vendor-independent automation application and hardware configuration description to manage the increased complexity of next generation automation and control systems. The high-level goals of this standard can be summarized as the portability of automation projects as well as the configurability and interoperability of intelligent field devices in IPMCS applications. It serves as a reference architecture for distributed, modular, and flexible control systems meeting fundamental requirements of open distributed systems [7, 8]. The following main features briefly characterize this distributed automation standard [8]:

- Component or object-oriented basic building blocks called FBs

- Graphical intuitive way of modeling control algorithms through connecting inputs and outputs of FBs

- Event-based execution model

- Direct support for distribution

- Definitions for the interactions between devices of different vendors

- Basic support for reconfiguration

- Rules for the definition and portability of automation applications between software tools

- Based on existing domain standards (IEC 61131, etc.)

The main aim of this chapter is to provide a brief overview of the main elements of the IEC 61499 standard which are described in Section 2.2. In Section 2.3, the main differences between the first and second editions of the standard are discussed and must be considered in the implementation of automation and control applications.

2.2 IEC 61499 Reference Model

2.2.1 History

IEC 61499 was first published in 2000 through 2002 by IEC as a series of publicly available specifications (PASs) for trial use together with a technical report (TR) containing tutorial information. Nearly the same experts from industry, consulting, and academia who specified IEC 61131-3 have also developed the IEC 61499 reference model. Since then, it has undergone continuous improvement and development as a result of extensive testing in academic and industrial laboratories and applications. As a result of these developments, IEC 61499 was published in 2005 as an international standard in three main parts. Subsequently, work has continued on updating and improvement the three parts of the standard as results of various implementations and realized applications in the field. In 2012 the second edition of IEC 61499 was published.

2.2.2 Overview of Standard

The IEC 61499 automation standard covers a broad range of activities including the specification of different modeling elements for IPMCS, their analysis and validation. The configuration of devices and their maintenance are covered too. IEC 61499 consists of the following three parts [4, 8]:

Part 1 – Architecture defines the main elements of the reference model addressing the following, important points:

- General requirements, scope, definitions, modeling elements
- Rules for the declaration of FB types and rules for the execution behavior of FB instances
- Rules for the usage of FBs for modeling control applications of distributed IPMCS
- Rules for modeling system configurations of IPMCS (i.e., control devices and communication systems) and distributing applications to the modeled control devices

- Rules for the usage of FBs fulfilling communication requirements of distributed IPMCS

- Rules for the usage of FBs in the management of (distributed) applications, resources, and devices used in distributed IPMCS

Part 2 – Requirements for Software Tools defines requirements and needs for software tools supporting the following system engineering tasks of IPMCS:

- Specification of FB types

- Functional specification of resource and device types

- The specification, analysis, and validation of distributed IPMCS

- The configuration, implementation, operation, and the maintenance of distributed IPMCS (i.e., life-cycle management)

- The exchange of information and elements between different software engineering tools utilizing XML

Part 4 – Rules for Compliance Profiles defines rules and guidelines for the specification as well as the development of compliance profiles used in different domains and applications addressing the following attributes of IEC 61499 systems, devices and software tools:

- Portability of software tools to accept and correctly interpret library elements produced by other software tools

- Configurability of devices and their software components (selected, assigned locations, interconnected, and parametrized) by multiple software tools

- Interoperability of devices to operate together to perform the functions specified by one or more distributed applications

IEC/TR 61499-3 containing tutorial information was withdrawn as obsolete in 2007 due to the widespread availability of education and tutorial materials and books. Currently, activities are ongoing to reactivate and update Part 3 and to define a new Part 5 introducing new elements (timing behavior in service sequences, binary XML deployment, composite adapter types, namespaces, hybrid IEC 61499 and IEC 61131 platforms, etc.) into the IEC 61499 reference model.

In the following sections the essential concepts and modeling elements which are mainly covered by Part 1 of the standard are described.

2.2.3   Event-Based Execution Model

The execution model of IEC 61499 defines FBs characterized by two different input/output (I/O) types: events and data. The execution is triggered by events; needed data have to be valid at the inputs of the FB before an event arrives (denoted as the WITH construct in IEC 61499; i.e., the vertical lines between event and data inputs or outputs as shown in Figure 2.1). Only if an event occurs at an FB input, the execution environment has to process the execution of this FB. Hence, it is obvious whether an application is active and exactly which part is currently processed. Compared to PLC-based approaches like IEC 61131-3, the IEC 61499 approach doesn't define any time base or cyclic-based implicit defined execution behavior. It supports the asynchronous execution of FBs and FB networks through the usage of special event FBs and the synchronous execution of FBs is possible.

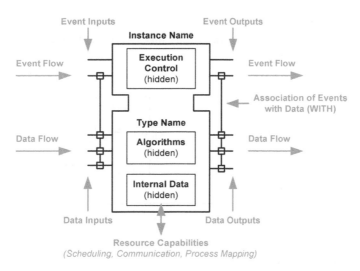

FIGURE 2.1
Characteristics of IEC 61499 FBs [4].

The standard defines a relatively simple execution model for FBs which is shown in Figure 2.2a. The execution of FBs is triggered by an event after the input data are available. As a next step, the execution control function is evaluated and the underlying scheduling function (provided by the resource) is responsible for the scheduling of the corresponding algorithm. When the execution is finished, the output data are updated and an event output is generated. Usually this triggers the execution of another connected FB. The resulting timing behavior of IEC 61499 FBs is given in Figure 2.2b.

2.2.4 Function Blocks as Main Modeling Elements

As expressed above, the main modeling elements defined in IEC 61499 are FBs. They are established for industrial applications to define robust and reusable software components. FBs principally store software solutions and algorithms for various problems. The FB definition of IEC 61499 is based on

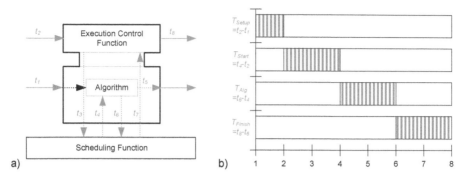

a)   b)   1   2   3   4   5   6   7   8

FIGURE 2.2
IEC 61499 execution characteristics: a) model and b) timing behavior [4].

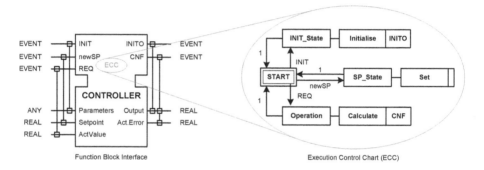

FIGURE 2.3
Example of a basic FB [4].

the specifications of IEC 61131-3 but it is extended with an event interface for execution control as already above.

The IEC 61499 standard defines three main FB types: *(i)* the basic FB for algorithm encapsulation, *(ii)* the service interface function block (SIFB) as interface to non-IEC 61499-compliant elements (communication network, process/machine, etc.), and *(iii)* the composite FB for functional aggregation. In addition to these FB types, a so-called adapter interface type allows the definition of component interaction interfaces. For the structuring of applications the so-called subapplication type is introduced.

As its name implies, the basic FB type is the "atom" from which higher-level "molecules" can be constructed. Engineers can encapsulate algorithms written in one of the IEC 61131-3 programming languages or other higher-level programming languages (e.g., Java, C++, C#). The execution of these algorithms is triggered by so-called execution control charts (ECCs) which are event-driven state machines as depicted in Figure 2.3.

Other "atomic" FBss are the SIFB types. They are used to model the interface to low-level services provided by the operating system or hardware of a controller device. SIFBs are for example used in the communication networks since specific protocols are not directly covered by the reference model. Only specific interfaces for data and information exchange between devices and resources are represented as SIFBs to the FB world. Two different high-level communication patterns are proposed and suggested in the standard: *(i)* the publish/subscribe and *(ii)* the client/server models. The first one is mainly used for unidirectional communication according to the producer and consumer concept whereas the second one is dedicated for bidirectional communication as indicated in Figure 2.4.

Engineers can also build higher-level FB "molecules" called composite FB types from existing FBs (i.e., the component FBs). This is carried out by specifying the event and data interfaces of the composite type followed by the modeling of the encapsulated FB network from internal component FBs. In this FB type, execution of the algorithms in the component FBs is controlled by the flow of events from one component to another.

2.2.5 Application, System, Device, and Resource Model

In order to define and model a distributed IPMCS with several controller devices using the above described FBs, the following additional elements are defined by IEC 61499 (see Figure 2.5) [5]:

- *Application Model:* In the architectural model, applications are built by interconnecting instances of reusable FB types with appropriate event and data connections. This is comparable to the design of circuit boards with integrated circuits. A key aspect of IEC 61499 is that applications are developed independently from the physical control infrastructure (i.e., system model).

- *System Model:* An IPMCS is modeled as a collection of devices interconnected and communicating with each other by means of a communication network consisting of segments and links. Devices are connected to network segments via links.

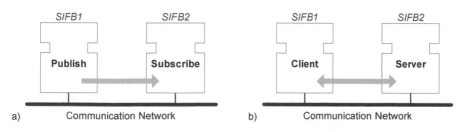

FIGURE 2.4
Communication SIFB patterns [4].

- *Device Model:* Describes the structure of a control equipment capable of executing IEC 61499 FB networks. It contains at least one interface, either the process and/or the communication interface. As execution entities devices have resources (see resource model below). The process interface provides the mapping between the physical process or machine (analog measurements, discrete I/O, etc.) and the resources. Information exchanged with the physical process is presented to the resource as data or events, or both. The communication interface provides a mapping between resources and the information exchanged with other devices via the communication network.

- *Resource Model:* A resource can be considered as a functional unit. It is encapsulated in a device and has independent control of its operation. It can be managed (created, configured, parameterized, started, deleted, etc.) without affecting other resources within the same device.

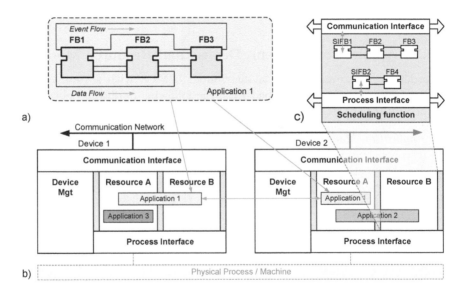

FIGURE 2.5
IEC 61499 concepts and models: a) application, b) system configuration and devices, and c) resource [4].

2.2.6   Distribution Model

One of the main intentions of IEC 61499 is enabling distributed applications used in IPMCS. The system model consists of devices connected by a communication network and applications. These applications may be distributed across and also within devices (see Figure 2.5). The IEC 61499 standard allows the separation of devices into independent resources which are execution containers for FB networks as already pointed out above. This concept allows a local distribution of an application part of a specific device into several resources. The engineering process starts with the top-level functionality that has to be realized without reference to the concrete hardware structure. By use of libraries of software components, the user models the needed applications. The last step within the engineering cycle is the mapping of the applications to concrete hardware components, independent of whether the application is executed by one device or distributed to several devices.

2.2.7 Management Model

The configuration of a distributed IPMCS applying the IEC 61499 reference model can be carried out using special management functions. They are usually provided by the control devices. For this purpose, the standard defines device management (see Figure 2.5b) together with a corresponding management application represented by a management SIFB. The generic interface is depicted in Figure 2.6.

By using this FB combined with a remote application, mutual access between different IEC 61499-compliant devices is possible. The following standardized management functions are defined within the standard and can be used to interact with and configure a device:

- Initiating the execution of elements with START, RESET

- Stopping the execution of elements with STOP, KILL

- Creating FB instances, as well as event and data connections using CREATE

- Deleting FB instances and connections using DELETE

- Parametrization of elements applying WRITE, READ

- Provision of status data about devices, resources, and FBs using QUERY

In order to access these commands, the standard provides the above-mentioned device management SIFB (see Figure 2.6) but leaves the protocol

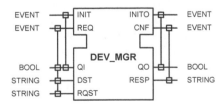

FIGURE 2.6
Device mangement FB [4].

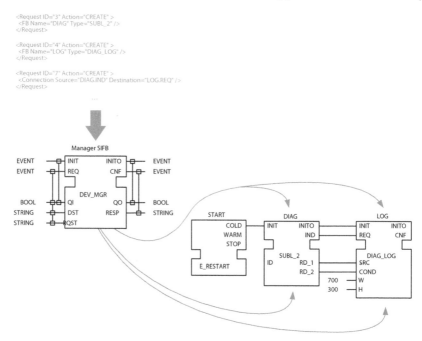

```
<Request ID="3" Action="CREATE" >
  <FB Name="DIAG" Type="SUBL_2" />
</Request>

<Request ID="4" Action="CREATE" >
  <FB Name="LOG" Type="DIAG_LOG" />
</Request>

<Request ID="7" Action="CREATE" >
  <Connection Source="DIAG.IND" Destination="LOG.REQ" />
</Request>
```

FIGURE 2.7
Configuration example using the device manager concept [1].

for accessing it open. In the "IEC 61499 Compliance Profile for Feasibility Demonstrations" an XML-based specification is suggested [1]. In Figure 2.7 an example for this XML protocol is provided.

2.2.8 Operational State Model

In addition to the above models, the IEC 61499 standard defines the operational state model. It is used to model the system's life cycle, which means that any IEC 61499-compliant system configuration has to be *(i)* specified, *(ii)* designed and modeled, *(iii)* implemented, *(iv)* commissioned, *(v)* operated, and *(vi)* maintained. Typically, a system is composed of functional units (i.e., devices, resources, applications, FBs) having their own life cycles. To support these functional units during each step of the life cycle, different actions have to be performed. Therefore, operational states are suggested in the standard to characterize which action can be carried out. Examples could be `Operational`, `Configurable`, `Loaded`, `Stopped`, etc. Each operational state specifies which actions are authorized, together with expected behaviors. Furthermore, a system could be organized that certain functional units may have or acquire the right to modify the states of other functional units.

2.3 Main Differences between First and Second Editions

Since the release of the first edition of IEC 61499 several implementations and applications have been realized in academic and in industrial applications. Some improvements were necessary to further develop this reference model for distributed IPMCS. As a result, in 2012 the second edition of IEC 61499 was published including several changes made in response to approximately 120 editorial and 40 technical comments received from national committees, with additional editorial changes to conform with IEC requirements. The most important and significant technical changes compared to the first edition are listed below [2, 4]:

- *Execution Control:* The execution model of FBs—especially the ECC—was one of the most discussed issues of the IEC 61499 specification in industry and academia since its publication. To clarify this issue, a more understandable, precise specification has been provided in the second edition. Concurrency and data consistency have been solved and a clarification of the event semantic added.

- *Temporary Variables:* Similar to IEC 61131-3, temporary variables in algorithms encapsulated into basic FBs are now possible. They are only visible within the algorithms in which the are used. This greatly improves the readability of basic FBs as variables needed only within an algorithm now need not be declared as internal variables of the FB. A typical application example is a variable used as a loop counter.

- *Network and Segment Types:* The usage of different segment types in the system model was not possible in the original specification of IEC 61499. Since various fieldbus and industrial ethernet solutions exist, their use in the system model is preferable. This issue has been addressed in the new edition, providing the possibility to specify the properties of a network type and its corresponding protocol(s). This allows a clear documentation of the overall system structure of a distributed IPMCS.

- *Interaction with PLCs:* Since IEC 61131-3 is widely used in IPMCS, a harmonization with IEC 61499 makes sense [9]. Hence, the second edition of IEC 61499 defines the use of special SIFBs that can act as clients of the PLC communication services defined in IEC 61131-5. A READ FB for synchronous reading of PLC data, a UREAD FB for asynchronous reading, and a WRITE FB for synchronous writing of PLC data are now possible. The remote triggering of PLC tasks as defined in IEC 61131-3 becomes now possible using the new TASK FB.

- *Simplified READ and WRITE Commands:* In the first edition of IEC 61499, the access path concept from IEC 61131-3 was applied. With the READ and

WRITE management commands, access to internal variables of basis FBs was possible. This is in contradiction with the object-oriented IEC 61499 approach. As a consequence, this option was removed in the new edition of IEC 61499. From now on, the READ and WRITE management commands are only allowed to access the interface of FBs, devices and resources, and all internals are hidden from them.

- *Further Changes:* In addition, a number of smaller corrections and changes have also been carried out in the second edition. For example, the forgotten RESET management command has been added to the reference model. Another change was related to the service sequences: originally defined to describe the behavior of SIFBs and now allowed for all FB types. Moreover, the adapter interfaces can now be used for all types of FBs instead of only for composite FBs.

Bibliography

[1] J. H. Christensen. HOLOBLOC.com. Function Block-Based, Holonic Systems Technology, Access Date: June 2015.

[2] J. H. Christensen, T. Strasser, A. Valentini, V. Vyatkin, and A. Zoitl. The IEC 61499 Function Block Standard: Overview of the Second Edition. In *Proceedings of ISA Automation Week, September 24-27, Orlando, Florida,* 2012.

[3] R. W. Lewis. *Programming Industrial Control Systems Using IEC 1131-3 (IEE Control Engineering Series).* Institution of Electrical Engineers, Stevenage, UK, 1998.

[4] TC 65/SC 65B. IEC 61499: Function blocks. International Electrotechnical Commission, Geneva, 2nd ed., 2012.

[5] TC 65/SC 65B. IEC 61131: Programmable controllers. International Electrotechnical Commission, Geneva, 3rd ed., 2012.

[6] I. Terzic, A. Zoitl, B. Favre-Bulle, and T. Strasser. A survey of distributed intelligence in automation in European industry, research and market. In *Proceedings of IEEE International Conference on Emerging Technologies and Factory Automation,* pages 221–228, Sept. 2008.

[7] V. Vyatkin. IEC 61499 as Enabler of Distributed and Intelligent Automation: State-of-the-Art Review. *IEEE Transactions on Industrial Informatics,* Vol. 7, No. 4, 768–781, 2011.

[8] A. Zoitl and R. Lewis. *Modelling Control Systems Using IEC 61499.* Control, Robotics & Sensors. Institution of Engineering and Technology, 2nd edition, 2014.

[9] A. Zoitl, T. Strasser, C. Sünder, and T. Baier. Is IEC 61499 in harmony with IEC 61131-3? *IEEE Industrial Electronics Magazine*, Vol. 3, No. 4, 49–55, Dec. 2009.

Part II

Design Guidelines and Application Development

3

Design Patterns, Frameworks, and Methodologies

James H. Christensen

Holobloc Inc.

CONTENTS

3.1 Introduction, Motivation, and Overview

A *design pattern* is defined as "the formalization of an approach to a common problem within a context" [4, p. 219]. In our case, the context is given by the IEC 61499 architecture for functional encapsulation, reuse, and deployment of software in *distributed automation and control systems* [2]. The adaptation of familiar design patterns in this context can decrease the time required for learning and for implementing applications within this architecture.

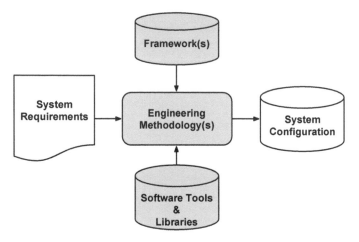

FIGURE 3.1
Application of design pattern elements (shaded).

As illustrated in Figure 3.1, for each design pattern this adaptation consists of three major steps:

1. Defining the *problem* to be solved.

2. Defining an appropriate *framework* for the solution of the problem. In object-oriented design, a framework is considered to be a skeletal structure that must be fleshed out to build a complete application. In the IEC 61499 architecture [6, 7], a framework will consist of a set of related types including, among others:

   (a) Function block types

   (b) Data types

   (c) Resource types

   (d) Device types

3. Defining an *engineering methodology* comprising procedures for using the types in the framework to:

   (a) Construct appropriate *devices* and *resources*

   (b) Populate the devices and resources with appropriate *function blocks*

   (c) Establish appropriate *event* and *data connections* among the function blocks

   (d) Appropriately *configure* the devices, resources, and function blocks

   (e) *Install, commission, and operate* the developed configurations in physical devices

It may be necessary to specify the requirements for *software tools* to support the design pattern.

This chapter provides brief descriptions of a number of typical design patterns and frameworks for the IEC 61499 architecture, including:

- Distributed application methodology

- Proxy pattern

- Layered model/view/controller/diagnostics (MVCD) pattern

- Local Multicast Pattern

- Tagged data pattern

- Matrix framework

In the following sections these important IEC 61499-related design patterns are discussed in more detail.

Detailed information and worked examples are available online at [1]. Furthermore source and executable codes for the examples given in this chapter are included with the Function Block Development Kit (FBDK) [5]. Additional and alternative design patterns with their associated frameworks, tool support and engineering methodologies are provided by other software tools and runtime environments for the IEC 61499 architecture, such as the open-source 4DIAC-IDE and its associated FORTE runtime environment, for which details and downloads are also available online [8].

3.2 Distributed Application Methodology

3.2.1 Purpose

This pattern assists engineers in the *configuration* of distributed *applications* in the IEC 61499 architecture.

3.2.2 Methodology

In order to apply this pattern the following steps have to be followed:

1. Develop and/or acquire *libraries* of *function block types* available to build an application to be distributed, along with the *device types* and *resource types* necessary for system implementation.

2. Define the *application(s)* to be used to implement the required system functions as networks of appropriately interconnected and parameterized function block *instances*.

3. Perform a *mapping* of function block instances and their associated connections from *applications* to the appropriate *resources*. Figure 3.2 illustrates this step for the mappings defined by the IEC 61499-1 textual declarations

```
MAPPING
SORT.SENSOR ON SENSOR.R1.SENSOR;
SORT.GATE ON SENSOR.R1.LOGIC;
SORT.ACTUATOR ON ACTUATOR.R1.VALVE;
END_MAPPING
```

4. *Configure device* and *resource instances* in the system.

5. Configure and interconnect communication *service interface function blocks* (SIFBs) to implement the event connections and data connections of the application across resource and device boundaries.

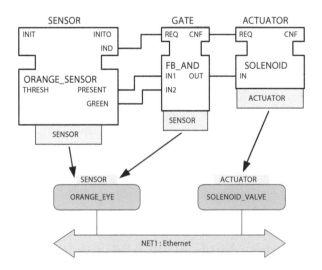

FIGURE 3.2
Mapping of an application's function blocks to devices (example: APPLICATION SORT).

3.2.3   Tool Support

Figure 3.2 provides an example of *tool support* for the graphical display of the mapping between the SORT application and the SENSOR and ACTUATOR devices via the labels attached to the bottom of the mapped function blocks. Since this tool support is not defined in the IEC 61499 standard, various tools, e.g., FBDK [5] and 4DIAC [8], may provide differing kinds of support.

3.3 Proxy Pattern

3.3.1 Purpose

This pattern "decouples clients from their servers by creating a local proxy, or stand-in, for the less accessible server. When the client needs to request a service from the server, such as retrieving a value, it asks its local proxy. The proxy can then marshal a request to the original server..." [4, p. 224].

3.3.2 Methodology

In order to apply this pattern the following steps have to be followed:

1. Document the local interface and observable behaviors of the proxy as the *interface* and *service* (i.e., set of *service sequences*) of an IEC 61499 *SIFB type*.

2. Implement the desired behaviors as for a *basic* or *composite function block type* (the latter is preferable for reuse of *communication SIFB types* for standardized communication with the remote server).

3.3.3 Example

Figure 3.3 illustrates the use of proxy SIFBs in two separate devices, each of which is communicating via its own proprietary link with the sensors and actuators of an associated machine. Local control of each machine is performed by an application in its own control device, and the operation of the machines is coordinated by communication between the the two control devices.

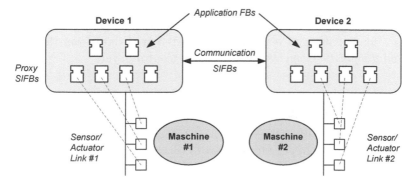

FIGURE 3.3
Proxy pattern example [3].

3.4 Layered Model/View/Controller/Diagnostics (MVCD) Pattern

3.4.1 Purpose

This pattern extends the traditional object-oriented MVC pattern in order to enable the integration of *simulation, visualization, diagnostics*, and *fault recovery* of the controlled machine or process into the control and automation systems design process using the IEC 61499 architecture [4, pp. 262ff].

3.4.2 Framework

As shown in Figure 3.4, the MVCD framework includes the *device types, resource types*, and *FB types* required to implement the following functional layers:

The Model layer implements the time-dependent logical behavior of the system or device being controlled. This layer is typically replaced by interfaces to the actual physical system once the corresponding *Control* layer has been validated.

The View layer provides the graphical display and possibly user input associated with the elements of the *Model* layer.

The Control layer implements the control functions to be performed on the elements of the *Model* layer, including appropriate event and data interfaces for integration of controller blocks.

The Diagnostic layer performs detection and diagnosis of, and possibly recovery from, equipment or device faults.

The HMI layer contains the elements necessary for human-machine interface via setting and reading of parameters and data from the *controller* and *diagnostic* elements present in the *Control* layer.

3.4.3 Methodology

The abbreviated methodology described below takes advantage of the highly visual nature of the MVCD framework to establish and maintain a visual representation of the operation of the controlled process or machine throughout the entire systems engineering process.

1. From an initial sketch and verbal description of the desired behaviors of the process or machine to be controlled, develop the contents of the *View* layer, along with a suitable *HMI* layer to receive user inputs from, and send

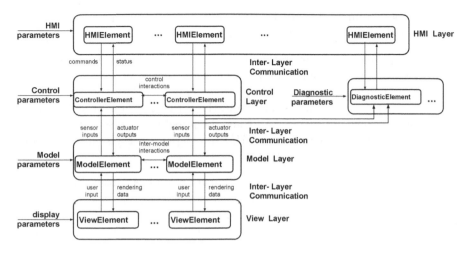

FIGURE 3.4

Model/view/controller/diagnostics framework.

data values to be rendered, to the view. Use this configuration to develop a sequence of snapshots representing the desired sequence of system states.

2. Develop a *Model* layer to simulate the dynamic behavior of the associated machine or process equipment. The *model/view* interface should replace the previous *HMI/view* interfaces, and a new *HMI* layer should be developed to receive simulated sensor inputs from and present simulated actuator outputs to the Model layer. The combined *HMI/model/view* configuration can then be used to observe the responses of the Model layer to the simulated actuator inputs and refine the contents of the Model layer as necessary.

3. Populate the *Control* layer with interconnected instances of *controller* FB types that encapsulate the necessary control logic to achieve the required system behaviors. Replace the sensor and actuator connections from the prior *HMI/model* interfaces with corresponding interconnections at the new *controller/model* interface. Restructure the *HMI* layer to contain those elements necessary to issue commands to and receive status notifications from the *Controller* layer, and use this new HMI to test that the *Controller* layer produces the required system behaviors in response to commands from the HMI.

4. Add *diagnostic elements* to the *Control* layer to monitor actuator outputs to and sensor responses from, the Model layer to detect whether proper system behaviors are exhibited. If invalid behaviors are observed, the *diagnostic elements* should report this to the *HMI* layer via error status messages, and implement fault recovery procedures either automatically

or upon command from the HMI (alternatively, exception recovery may be built into controller elements).

5. Once proper system operation, fault detection and recovery have been validated with the simulated system at the *Model* layer, *physical design* can be implemented by replacing the *Model* and *View* layers with an *Interface* layer between the *Control* layer and the actual physical process or mechanism, as shown in Figure 3.5 (it may be possible to reuse some elements of the *View* layer as part of the *HMI*, e.g., to provide a visual representation of the current process status).

6. Additional steps may include the allocation of blocks from their respective layers as appropriate to distributed devices using the *distributed application* pattern, and the allocation of *controller* and *diagnostic* elements to embedded controllers tightly coupled to local mechanisms, i.e., mechatronic design.

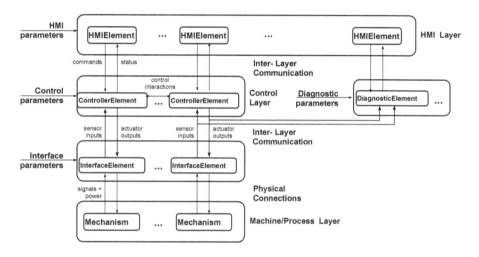

FIGURE 3.5
Physical design.

3.5   Local Multicast Pattern

A *local multicast* pattern, framework and methodology are available to provide efficient intraprocess multicasting of data without resorting to more expensive interprocess communication methods such as UDP multicast. As shown in

Figure 3.6, this efficiency is achieved by replacing the encoding and decoding of data for serial multicast with simple copying of data in local multicast.

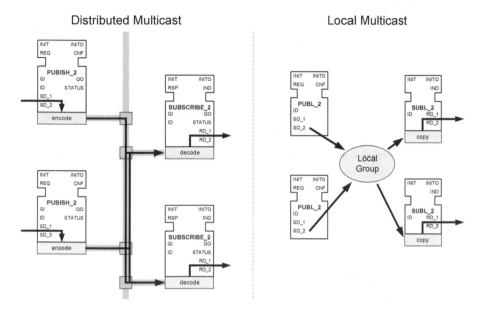

FIGURE 3.6
Local versus distributed multicast [3].

3.6 Tagged Data Pattern

A *tagged data* design pattern, framework, methodology and tool support may be employed to help ensure that:

- Data used in a local multicast channel is consistent with that used in the corresponding distributed multicast channel.

- Data subscribed from a local or distributed multicast channel is consistent with that published on the channel.

3.7 Matrix Framework

A matrix framework is available to support operations on two-dimensional matrices of floating-point values. This framework comprises:

- A MATRIX data type, which can be implemented internally using a high-level language such as Java or C

- Function block types to perform common matrix operations such as addition, subtraction, multiplication, and inversion

- A SIFB type for initializing and performing enumerable operations on an encapsulated matrix

- SIFB types for table-oriented matrix input and output

3.8 Conclusions

Design patterns and their associated frameworks, engineering methodologies, and software tool support are required to make effective use of the architecture defined in the IEC 61499-1 standard for encapsulation, reuse, and deployment of intellectual property (IP) in distributed control and automation systems. Fortunately, the software tool requirements defined in IEC 61499-2 and the compliance profile requirements defined in IEC 61499-4 make the development of such frameworks and software tool support a straightforward matter of systematic software engineering practice, as illustrated by the examples in this chapter.

Bibliography

[1] J. Christensen. IEC 61499 Design Patterns.
    http://www.holobloc.com/doc/despats, access date July 2015.

[2] J. Christensen and A. Zoitl. IEC 61499: A Standard for Software Reuse in Embedded, Distributed Control Systems.
    http://www.holobloc.com/papers/iec61499/overview.htm, access date July 2015.

[3] J. H. Christensen. Automation modeling, design and programming with the iec 61499 function block standard. Helsinki Technical University, June 2004.

[4] B. Douglass. *Real-Time UML*. Addison Wesley Longman, 1998.

[5] Holobloc Inc. FBDK - Function Block Development Kit. http://www.holobloc.com/doc/fbdk/, access date July 2015.

[6] TC 65/SC 65B. IEC 61499: Function blocks. International Electrotechnical Commission, Geneva, 2nd ed., 2012.

[7] A. Zoitl and R. Lewis. *Modelling Control Systems Using IEC 61499*. Control, Robotics & Sensors. Institution of Engineering and Technology, 2014.

[8] A. Zoitl, T. Strasser, and A. Valentini. Open source initiatives as basis for the establishment of new technologies in industrial automation: 4DIAC a case study. In *Proceedings of IEEE International Symposium on Industrial Electronics*, pages 3817–3819, 2010.

4

Applying IEC 61499 Design Paradigms: Object-Oriented Programming, Component-Based Design, and Service-Oriented Architecture

Wenbin Dai

Shanghai Jiao Tong University

Valeriy Vyatkin

Luleå University of Technology and Aalto University

James H. Christensen

Holobloc Inc.

CONTENTS

4.1   Introduction

With the growth of requirements for integration and interaction with machines, robots and other systems, the complexity of industrial automation systems has continuously increased over the past decade. Due to performance limitations and redundancy concerns, modern industrial automation systems are often controlled by several programmable logic controllers (PLCs) interconnected via Ethernet-based industrial networks. Traditional industrial automation systems are controlled by IEC 61131-3 PLCs [10]. Due to the limitation of the IEC 61131-3 software model, legacy PLC control programs are designed based on a per-controller basis. Interactions between multiple PLCs must be handled implicitly. As a result, reusability and flexibility of existing IEC 61131-3 PLC programs are largely reduced.

On the other hand, another international standard, the IEC 61499 standard [9], is intended to provide system level design methodology for distributed automation systems. Systems could be configured even when hardware deployment information is missing with the IEC 61499 standard. Block-based design provides great reusability by encapsulating control logics, hides complexities by providing hierarchical software structure, and reflects visual image of control programs. However, without applying certain design methodologies, middle to large scale function block network diagrams could be messy and reduce efficiencies of software design significantly. To avoid wasting massive development time, mature design paradigms are necessary for IEC 61499 function blocks. In this chapter, several software design paradigms will be investigated for the IEC 61499 standard and questions such as how and where to apply these paradigms will also be discussed.

The remainder of this chapter is organized as follows: In Section 4.2, essential elements of function block design are introduced. Following that, three major design paradigms, the object-oriented programming paradigm, the component-based design paradigm and the service-oriented architecture paradigm are introduced in Sections 4.3, 4.4, and 4.5 respectively. These paradigms are illustrated using an airport baggage handling system example and finally comparison and discussion for these paradigms are provided in Section 4.6.

4.2 Essential Elements in IEC 61499 Function Block Designs

From a software design perspective, paradigms are models for implementing a group of applications sharing common properties [20]. A major purpose of software design paradigms is the improvement of the efficiency of software design, implementation, and deployment by increasing the reusability and portability of existing control programs. Specifically developed for distributed automation systems, the IEC 61499 standard [9] aims to resolve the well-known deficiencies with respect to distributed control in the IEC 61131-3 standard [10]. The core feature of the IEC 61499 standard is the event-triggered function block in which control algorithms must be encapsulated in function blocks and can only be executed by specific, event-driven request mechanisms [18]. In order to apply design paradigms, some common elements used in the IEC 61499 software model must be summarized.

In an IEC 61499 design, elements are grouped into two essential components, namely modularity and communication as shown in Figure 4.1. Modularity is the fundamental element for design of distributed automation programs. A program based on modular units provides the ability for those components to be utilized repeatedly with minimal configuration effort [7]. The modular-based design methodology aims at better efficiency of software development and higher quality of the resulting software, achieved through the reuse of proven modules. Communication provides mechanisms of exchanging data between modules such as function calls, events, and messaging. Communication aims for better compatibility and interoperability between modules. Modularity combined with communication forms a distributed software design paradigm. A software design paradigm is a model for handling a class of problems that share a set of common characteristics [8]. A software de-

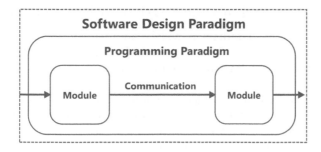

FIGURE 4.1
Essential elements of distributed automation programs.

sign paradigm must indicate how to construct individual modules and how to combine these modules into complete programs.

There are many programming paradigms already developed in the computer science domain, such as structured, imperative, declarative, functional, logical, and object-oriented programming paradigms [16]. These programming paradigms usually fit well for one or more specified domains. Programs may be developed in multi-paradigm styles and a programming language may support many paradigms. Some paradigms have already made significant impacts in the industrial automation domain. For example, the object-oriented paradigm is officially supported by the IEC 61131-3 standard third edition [21]. The object-oriented paradigm is also applied to the IEC 61499 standard in several applications such as airport baggage handling systems [2], process control [1], and smart grid [22] and building automation systems [19]. More importantly, a generic design guideline is expected to be summarized from existing approaches in order to provide comprehensive guidelines for applying various design paradigms in distributed automation programming.

In the next several sections, three major paradigms will be illustrated. First of all, the most commonly adopted design paradigm, object-oriented programming (OOP), is discussed. Secondly, the component-based design (CBD) paradigm is covered as IEC 61499 function blocks are natural components. Finally, the service-oriented architecture (SOA) paradigm is introduced for industrial automation. There is not yet any existing work of adopting SOA on design level rather than execution level. All paradigms are illustrated along with a case study of an airport baggage handling system example.

4.3 Applying Object-Oriented Programming Paradigm in IEC 61499 Function Blocks

4.3.1 Background of Object-Oriented Programming

The object-oriented paradigm is widely adopted in computer programming languages such as C++, Java, C#, and Python. The OOP paradigm was initially proposed for improving reusability and efficiency of programming languages.

The key concept of the object-oriented paradigm is the object [11]. The object is a reusable module that consists of two parts: code and data. The code part involves functional behavior of the object and the data part contains current state of the object. Typically, for each kind of object, a class is defined, which contains a single copy of the code part and a definition of the data part. A new instance, i.e., copy of the data part and reference to the common code part, is initialized for every specific object. The key feature of the OOP is the encapsulation by which systems are divided into modular objects whose data are physically isolated. Methods, as interfaces between objects, are functions that are used to access the data encapsulated in the object instance. Inheritance is the ability to extend object classes into subclasses with new methods and data.

Those features ensure certain part of existing code could be reused, but there are some exceptions. An object could have numerical definitions as requirements vary for each project. It becomes extremely difficult to manage all variations of objects in long term development using the OOP paradigm. Also any interface change in an object method will require massive redevelopment time. It is hard to determine whether any time is really saved from applying the OOP paradigm.

4.3.2 Applying Object-Oriented Programming Concepts in IEC 61499 Standard

Applying the object-oriented paradigm in the IEC 61499 standard is based on considering IEC 61499 function blocks as objects (as demonstrated experimentally by [2, 19] etc.). First of all, each IEC 61499 function block type is mapped to a class. There are three types of function blocks defined in the IEC 61499: basic function block (BFB), service interface function block (SIFB) and composite function block (CFB). Both BFBs and SIFBs are atomic function blocks but aim for different purposes. BFBs are commonly used for presenting state machine-based control algorithm whose behaviors are driven by execution control charts (ECC). SIFBs are utilized as "black boxes" that are driven by service sequences. Both BFBs and SIFBs can be used for defining classes as shown in Figure 4.2.

As all function block types have identical interfaces, each event input of any function block type is mapped to a method. Data inputs associated with this event input are used as method input parameters. Methods are invoked by event connections in the IEC 61499 approach. In BFBs, methods are implemented by EC algorithms which are programmed in IEC 61131-3 programming languages. In SIFBs, methods could be implemented in any programming language driven by service sequences. To return parameters from methods, an event output must be created and data outputs associated with this event output are considered as return parameters. By creating a new FB instance, data are stored as internal variables in BFB and SIFB.

CFBs provide a means for achieving inheritance by parts in IEC 61499 applications. As indicated in Figure 4.3, the original BFB or SIFB is encapsulated by a new CFB and the original interface is directly mapped to a new CFB interface. New methods could be implemented in a separated BFB or SIFB as part of the CFB internal FB network. However, there is one minor issue: how to handle data between base and extended FBs. As there is no internal variable available in CFBs, data variables must remain in a base class (BFB or SIFB). To access those internal data variables, extra PUBLISH and SUBSCRIBE channels between original and extended FB are introduced as shown

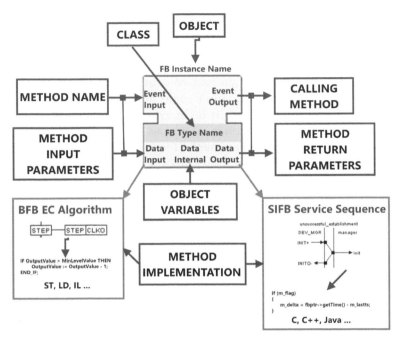

FIGURE 4.2
Mapping IEC 61499 FBs for OOP features.

in Figure 4.3. When some internal variables are required by extended FBs, values are fetched from the base FB and updated variables are returned to the base FB by using a dedicated pair of PUBLISH and SUBSCRIBE SIFBs. Internal variables introduced by new methods will be placed in extended FBs. If further extension is needed, extra extended function blocks are inserted as well as a pair of PUBLISH/SUBSCRIBE communication channel.

Alternatively all internal variables can be placed in a separated data FB. As shown in Figure 4.4, no matter variables are requested from the base FB or extended FB, and values will be passed to that particular FB via PUBLISH/SUBSCRIBE communication channels. Data variables will be written back into the data FB when their values are modified during processes. This design pattern provides a generic solution for all base and inherited objects, however there are some downsides. First, more time will be spent on design and development as all data variables must be manually configured with PUBLISH/SUBSCRIBE SIFBs. Second, separated data and code will create more communication overhead that affects system performance.

For both approaches, a base function block is compulsory for adapting extended functions to enable generic inheritance. The first approach creates less communication overhead but more manual efforts are expected to cover customized communication between base and extended function block. The

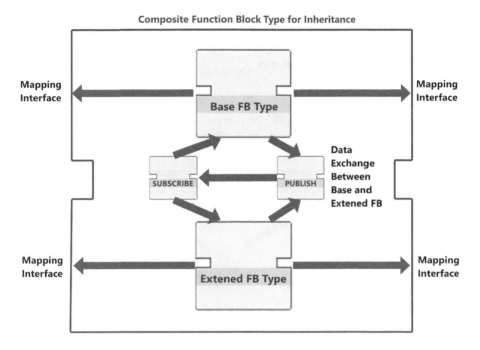

FIGURE 4.3
Mapping CFBs for OOP inheritance.

second approach is suitable for all scenarios, but the execution speed will be affected due to larger volume of data exchange in inherited function block type.

One common approach is to consider an object in IEC 61499 function blocks as a physical device in a plant. As objects are defined as passive functions, control units are needed for organizing those objects in a logical order. Extra function blocks are introduced to perform scheduling functions for plant processes. Those scheduling function blocks are responsible for bridging device models with physical operations by invoking methods from objects. Scheduling function blocks could be deployed to multiple PLCs for providing distributed control.

4.3.3 Case Study of Object-Oriented Programming Paradigm

The object-oriented paradigm will be illustrated on a case study of an airport baggage handling system (BHS). The layout of the inbound airport BHS is given in Figure 4.5. Bags offloaded from aircraft will be transferred to the inbound BHS from the IB101 conveyor. Bags travel from conveyor IB101 to IB105 via security door IB1D1 and finally reach the baggage reclaim loop

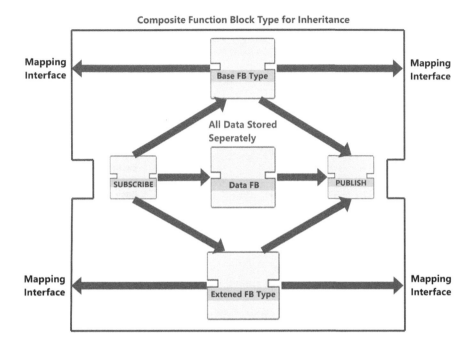

FIGURE 4.4
Alternative data handling solution for OOP inheritance.

IB1. There are two zones defined in this inbound BHS: air side from IB101 to IB103 conveyor with security door IB1D1 and public side from IB104 to IB1. There are three emergency stop buttons installed for handling emergency situations. By pressing an emergency stop button all devices in this zone will halt immediately. There are two emergency stop buttons installed on the public side (ESTOP1 and ESTOP2) and one on the air side (ESTOP3).

The inbound BHS OOP design is given in Figure 4.6. At the beginning of the event chain, A SIFB FB_BHS_INPUTS polls input values from fieldbuses in a fixed period. Such inputs include photo eye sensors and motor running feedback signals from conveyors, emergency stop button status, proximity sensors and motor running feedback signals from doors and system control buttons (start, stop and reset). On the end of the event chain, another SIFB FB_BHS_OUTPUTS writes output values back to fieldbuses which contain motor run signals for conveyors and motor run forward and reverse signals for doors. Three classes (function block types) are defined in the OOP approach: emergency stop, conveyor, and door. The emergency stop function block type (namely FB_EStop_OOP as shown in Figure 4.6) takes E-Stop button inputs and update emergency stop button status which are stored internally.

The design of the conveyor function block type FB_Conveyor_OOP is

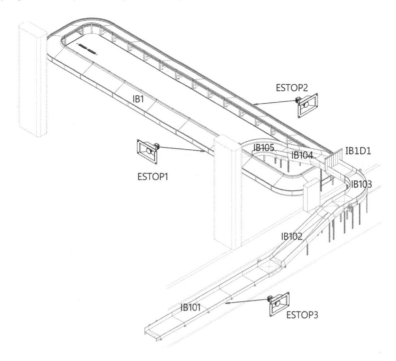

FIGURE 4.5
Airport baggage handling systems example layout.

illustrated in Figure 4.7. It uses a photo eye sensor and motor running feedback signal as data inputs and generates motor run signal output as well as conveyor current status. According to the IEC 61499 OOP paradigm definition, there are four event inputs (methods) listed in the interface: INIT, RUN, STOP and UPDATE. Each event input will switch to an individual EC state with dedicated EC algorithm and return to the START state immediately after this event is handled. The RUN, STOP and UPDATE states share one event output CNF to update motor run signal output and conveyor status. The RUN and STOP states are responsible for starting and stopping the running conveyor. When input values are polled from fieldbuses, the UPDATE event will be triggered to refresh input values stored in the function block.

The door class FB_Door_OOP has similar design compared to the conveyor FB design. One extra feature required by security doors is that the door must run in both forward (open) and reverse (close) directions. As illustrated in Figure 4.8, instead of using one RUN state in the ECC design, there are two EC states, RUNF and RUNR, for controlling motor running directions of doors.

Now all objects are created, but still needs to be a scheduler to provide system level functionalities by linking objects. The scheduler FB FB_SEQ_OOP

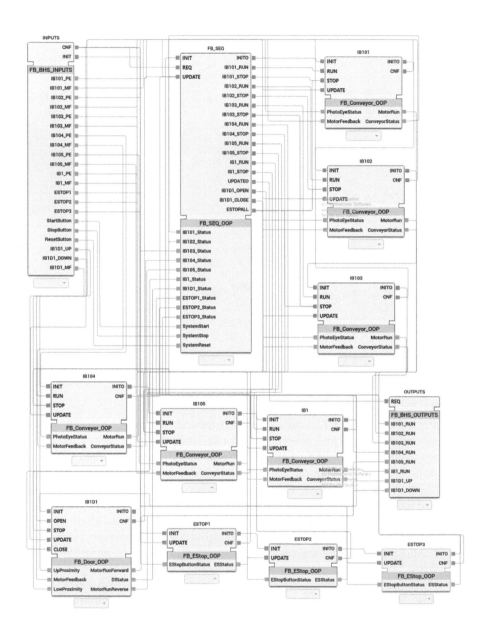

FIGURE 4.6
Inbound BHS OOP design example.

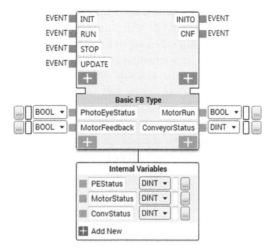

(a) Interface Design

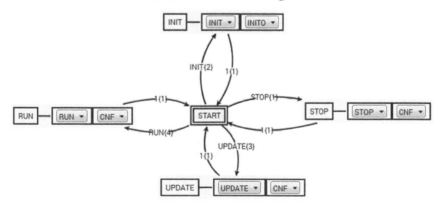

(b) ECC Design

FIGURE 4.7
Inbound BHS OOP conveyor FB design.

is introduced as demonstrated in Figure 4.9. It uses current status of all conveyors deployed and system control signals as inputs and generates control signals for all devices including conveyors and doors. On every I/O scan cycle, the UPDATE event is activated by the FB_BHS_INPUTS SIFB and current status of field devices is refreshed. If any emergency stop button is pressed, the ECC will jump to the ESTOP state and the event output ESTOPALL will be emitted to direct all devices to stop operation immediately. Alternatively, if no emergency occurs, the ECC will jump back to the START state.

There are two process sequences defined in the ECC: start sequence and stop sequence. For the start sequence, conveyors start in reverse order and the most downstream conveyor or cascade start operates first. Cascade start is designed to avoid bag jams caused by bags crashing into stopped downstream conveyors. Once the system start button is pressed, the ECC will open the security door IB1D1 and start all conveyors from IB1 to IB101. For the stop sequence, conveyors stop in forward order so that the most upstream conveyor stops first. This feature is to clear all bags on a conveyor before it stops. Once the system stop signal is received, conveyors stop one-by-one from IB101 to

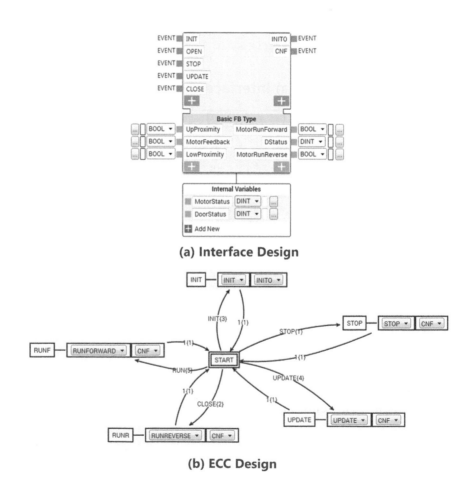

(a) Interface Design

(b) ECC Design

FIGURE 4.8
Inbound BHS OOP door FB design.

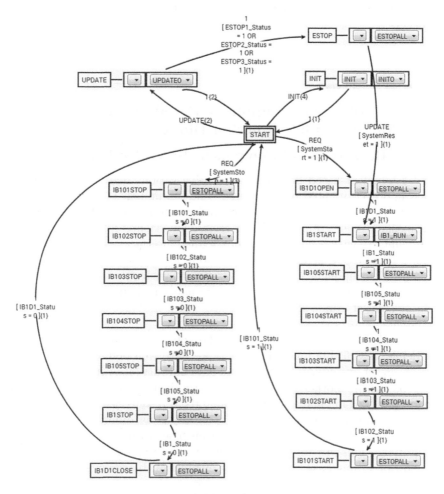

ECC Design

FIGURE 4.9
Inbound BHS OOP sequence control FB design.

IB1. Finally the security door IB1D1 is shut after all conveyors are stopped. Also after emergency stop status is cleared, the inbound BHS will restart in reverse order to resume normal operation.

Overall, the OOP paradigm starts from individual object class design. For each object function block type, all input values must be cached inside this function block and status must also be kept as local variables. As function blocks are passive, a scheduler is introduced to generate events for triggering objects.

4.4   Adoption of Component-Based Design Paradigm for IEC 61499 Function Blocks

4.4.1   Background of Component-Based Design

The component-based design aims to solve various definitions for the same class in object-oriented programming. The component-based paradigm is designed for separating functionalities and an individual software component encapsulates a set of semantically related functions and data. Components communicate with each other via pre-defined interfaces [5]. Interfaces provide "black-box" usage of components without identifying details of implementations. A component can be easily replaced with other components during design time and runtime as its interface remains unchanged. A software library can be created from a set of components which provides great reusability and reliability.

The IEC 61499 standard follows the component-based design paradigm principle. IEC 61499 function block types have well defined interfaces: event inputs, event outputs, data inputs and data outputs [5]. A function block could be replaced by another function block with identical interfaces seamlessly by sending IEC 61499 management commands to resources, even alogorthms encapsulated in those function blocks are completely different [3, 14, 17]. Any modification of algorithms inside function blocks will not affect system configurations. This provides better flexibility as software architects could design control programs based on interfaces of function blocks even if implementations of these software components are not ready.

4.4.2   Component-Based Design Principles for IEC 61499 Function Blocks

The design principles of the component-based paradigm for IEC 61499 applications can be summarized as two points. Firstly, the component-based paradigm aims for grouping semantically related data and functions to minimize data exchange between modules. Differing from object-oriented paradigm, the component-based paradigm is focused on splitting systems by functionalities rather than physical objects. As most data required for an IEC 61499 function block in CBD paradigm are stored internally, overall quantity of data between function blocks is reduced significantly. Hierarchical software architecture is commonly adopted to minimize connections between function blocks.

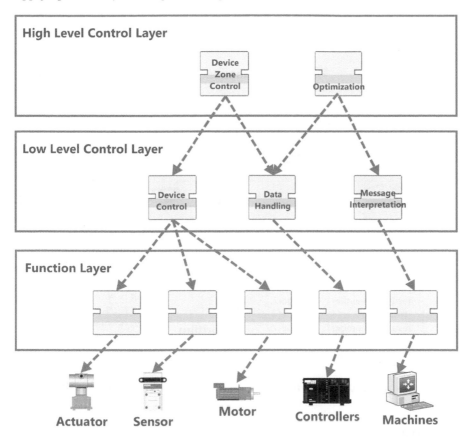

FIGURE 4.10
Multi-layered CBD paradigm for IEC 61499.

As illustrated in Figure 4.10, a multi-layered software architecture is proposed for the IEC 61499 component-based design paradigm. Three layers are the minimum requirements for applying IEC 61499 CBD: the function layer, the low level control (LLC) layer and the high level control (HLC) layer. The function layer contains function blocks for accessing sensors, actuators, motors and drives via fieldbuses handling external communication gateways, and reusable function library for string processing and complex math calculation. All FBs used to provide direct control to the physical plant are placed on the function layer.

The middle layer is the low level control layer. The LLC layer is responsible for providing basic control functionalities for plants. Function blocks on the LLC layer usually represent a physical device on the shop floor. A process step may require cooperation between several function blocks from both the

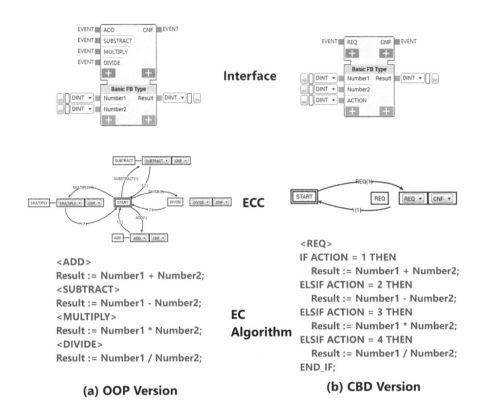

FIGURE 4.11
OOP versus CBD function block design example.

function layer and the LLC layer, for example, providing conveyor control for an airport BHS. Those steps usually focus on individual devices so that cooperation between various devices is not necessary.

The top layer is the high level control layer responsible for achieving complex functionalities by coordinating multiple devices, for instance, load sharing in an airport BHS. The HLC layer coordinates components to accomplish system level functionalities. By applying this multi-layered architecture for CBD in IEC 61499 function blocks, connections between function blocks are largely reduced.

Secondly, interfaces between function blocks should be abstract and generic. Instead of using a dedicated event and algorithm for every method call in OOP, multiple functions share a single algorithm with one generic event input in CBD. For instance, a function block for performing simple calculation between two input numbers and return a result number is shown in Figure 4.11a. In the OOP approach, a separated event input is created for

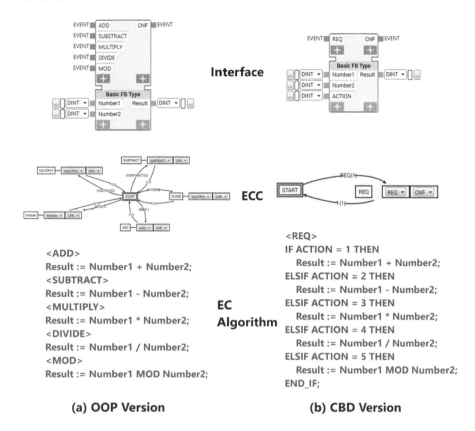

(a) OOP Version   **(b) CBD Version**

FIGURE 4.12
OOP versus CBD function block design insert new functionality example.

adding, subtracting, multiplying and dividing functions. In the CBD version (Figure 4.11b), those event inputs are combined into a single event REQ associated with a data input ACTION to specify selection of actions (1-Add, 2-Subtract, 3-Multiply, 4-Divide). All algorithms are placed in one EC algorithm and distinguished by interpreting the ACTION variable value.

Although complexities of individual algorithm increase, IEC 61499 function blocks still could be easily extended without modifying any connections in system configurations. For example, if modulus calculation needs to be inserted for the FB provided in Figure 4.12, in the OOP approach, a new event input MOD is inserted and the ECC is modified with a new introduced EC state MODULO. In the CBD system, there is no change required for the FB interface and a new pre-defined value (for instance, 5) for the data input ACTION is introduced. In the EC algorithm, the modulo operation is inserted by checking ACTION value (equals 5 in this case). As no change is needed

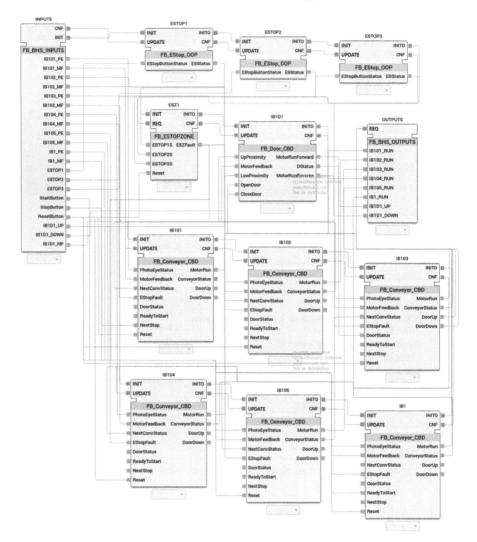

FIGURE 4.13
Inbound BHS CBD design example.

for the interface, new functionality is added without any modification to the function block network design.

4.4.3   Case Study of Component-Based Design Paradigm

In order to explain how to apply the proposed CBD architecture, the airport BHS example used as the reference in the previous section is modified accord-

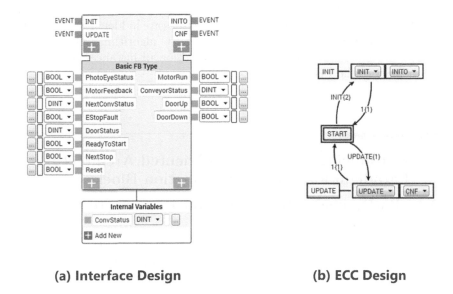

(a) Interface Design   **(b) ECC Design**

FIGURE 4.14
Inbound BHS CBD conveyor FB design.

ing to the CBD paradigm (as shown in Figure 4.13). In the CBD approach, there is no scheduling function block to coordinate control systems. Instead, coordination is spread over several distributed software components. A new FB component FB_ESTOPZONE is added to provide emergency stop zone control for informing the BHS when any ESTOP button is pressed. The system will resume operation once the ESTOP button is released and the system reset button is pressed.

The CBD version of the conveyor control function block type is given in Figure 4.14. One obvious change from the OOP version to the CBD version is that all control signals use data variables instead of events. Event inputs and outputs and ECC are simplified by applying unified interfaces. There are only two states listed in the ECC: INIT which is responsible for assigning initial values and UPDATE in which all control logic is implemented. All information required by conveyors is listed as data inputs in the FB interface, for example, downstream conveyor status, emergency stop status, associated security door status, and system start, stop, and reset signals. As seen in Figure 4.13, some data inputs and outputs may not be required by particular devices. For instance, the IB1D1 security door is attached to the IB103 conveyor, the other conveyors are not configured with any security door. Also the system start signal is only attached to the IB1 conveyor for cascade starting and the system stop signal is only attached to the IB101 conveyor for clearing all remaining bags in the system.

To summarize the CBD paradigm, the reusability is improved by applying unified interfaces for function blocks. Complexities are hidden from application level as all control logics are encapsulated in EC algorithms. One downside is that not every input and output is required for all instances but they still must be part of the FB interface.

4.5 Introducing New Service-Oriented Architecture Paradigm for IEC 61499 Function Blocks

4.5.1 Background of Service-Oriented Architecture with IEC 61499

With adoption of Ethernet-based communication technologies at shop floor level in recent years, the service-oriented architecture (SOA) is becoming appealing as a framework for distributed automation software systems [12]. The SOA paradigm consists of a set of software components whose interface descriptions can be published and discovered [6]. Service providers and consumers are loosely coupled to ensure minimum dependencies between services. The interaction between services is defined in the service contract, which can be published to the service repository. Services are combinable with other services (called service orchestration and choreography) and remain stateless when there is no message present.

Such flexible paradigms can also be represented using the artifacts of IEC 61499. We consider each IEC 61499 function block as a service and its interface is defined in a service contract. As discoverability and service repository must be supported by execution environment, how to publish and discover service contracts will not be discussed in this chapter. Due to services that use only messaging mechanisms to exchange data, similar to the OOP approach, each event input is defined as a message type and all data inputs associated with this event are considered message contents. Because connections in the IEC 61499 standard are one-directional, an IEC 61499 adapter connection requires separated pair of event and data connections for returning response messages.

4.5.2 Apply Service-Oriented Architecture Concepts in IEC 61499 System Design

As shown in Figure 4.15, the SOA paradigm for IEC 61499 is also proposed as a multi-layer architecture. Although services shall be accessed from anywhere, total flexible system design would cause performance degradation. Four layers are the minimum required for applying the IEC 61499 SOA paradigm. The base layer is where data processing is performed required. As services are stateless when they are not activated, the data layer is responsible for storage of services. Only BFBs and SIFBs are allowed on the data layer as there is no internal variable available in CFBs.

On top of the base layer is the component layer where frequently used functions are implemented. Such components include message interpretation functions, I/O and data handling functions, and other reusable logic for services. The next layer up is the service layer where control logics for automation systems are located. Function blocks on the service layer may invoke other function blocks on the data layer and the component layer.

Finally, the top layer is the process layer where plant physical processes are described. On the process layer, CFBs are used for encapsulating function

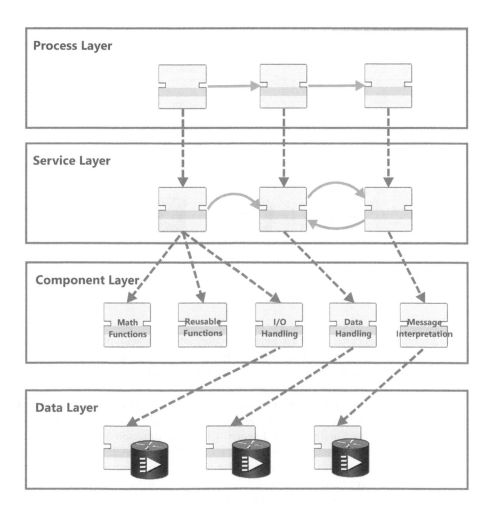

FIGURE 4.15
SOA paradigm architecture for IEC 61499.

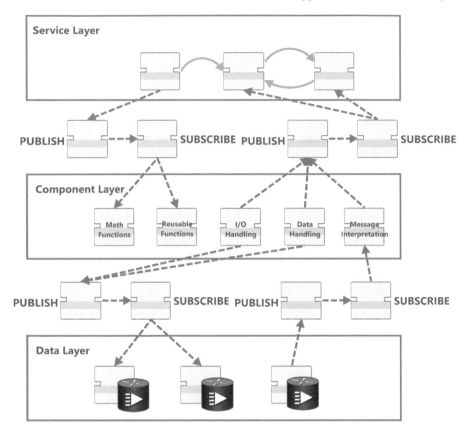

FIGURE 4.16
Data handling in IEC 61499 SOA paradigm.

blocks on the service layer to perform complete physical processes, known as service orchestration and choreography. In the SOA paradigm, only plant physical processes are illustrated in IEC 61499 system configurations and complexities of control logic are hidden.

As all data are stored apart from control logic, data handling between function blocks on the data layer, the component layer and the service layer are challenges. As illustrated in Figure 4.16, if a function block is invoked by more than one function block, the PUBLISH/SUBSCRIBE mechanism introduced earlier in this chapter must be applied. Data outputs from source function blocks are transferred by PUBLISH SIFBs with identical IDs to the same SUBSCRIBE SIFB with unique IDs for data inputs on the other side. This is due to the limitation that only one connection is allowed for each data input of any IEC 61499 function block for data consistency and semantic robustness.

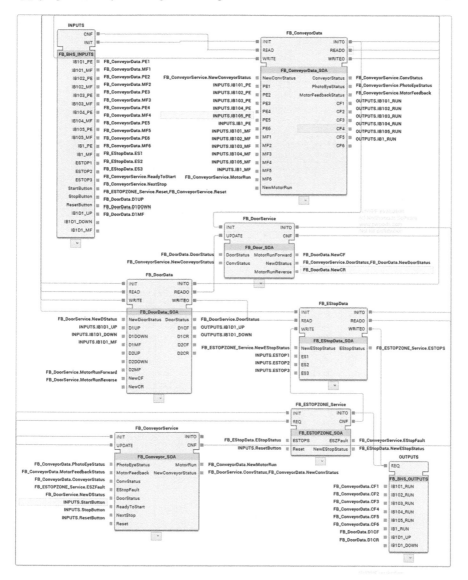

FIGURE 4.17
Inbound BHS SOA design example.

4.5.3 Case Study of Service-Oriented Architecture Paradigm

After the SOA paradigm for IEC 61499 is described, the airport BHS case study example is modified again as Figure 4.17 to suit the SOA paradigm. According to the SOA paradigm, each device type is divided into function

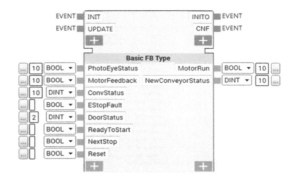

(a) Logic Part Interface Design

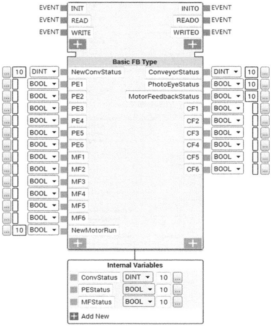

(b) Data Part Interface

FIGURE 4.18
Inbound BHS SOA conveyor FB design.

blocks for logic and data. The data function block stores both input and output values and current status of devices as local variables. Unlike the CBD paradigm, arrays are utilized to group data with identical device type. The number of function block instances in IEC 61499 applications is reduced by applying the SOA paradigm. The SOA version logic FB interface design has large similarities compared to the CBD version. There are only two minor changes: first, data variables such as device status are declared as arrays; second, all variables processed externally must be return to data function blocks.

The conveyor control service is illustrated in Figure 4.18. The ECC design of a logic handling function block for conveyor control is identical to its CBD version and the interface remains identical except that arrays are introduced for photo eye sensor status, motor running feedback status, conveyor status and security door status. Also a new data output is added for returning updated conveyor status back to the data handling function block. The data handling function block design is given in Figure 4.18b. Three EC states are defined in the ECC and two tasks are implemented for read and write processes. The read process takes latest input values of photo eye sensors, motor running feedbacks and refresh photo eye status and motor status. The write process overwrites existing conveyor status and motor run signals with new values from logic handling function blocks.

From the CBD paradigm to the SOA paradigm, only one function block instance is obligatory for each type of device. Status variables of the same device type are grouped as an array and updated by logic handling function blocks. Status variables must be returned to data handling function blocks as logic handling function blocks do not store any data. By splitting logic and data, modification of one part will not affect the other part. Also the number of function block instances and connections are largely reduced by applying the SOA paradigm.

4.6 Summaries of IEC 61499 Design Paradigms

Three common design paradigms for IEC 61499 and their applications in airport baggage handling systems are illustrated in previous sections. In this section, some experiences are summarized and design guidelines are provided.

First, the object-oriented programming paradigm is ideal for representing physical layouts of automation systems. Devices are one-to-one mapped to function blocks in IEC 61499 applications and their physical locations may be reflected by function block networks. The OOP paradigm provides intuitive feelings of system design and is applicable for many industrial domains where locations of devices must be quickly identified, for example, material handling systems, building automation systems and process control. Reusability is largely improved by considering objects as function blocks compared to monolithic code.

However a good object-oriented design does not necessarily make a good component-based design in the computer science and industrial automation domains [15]. Class-to-component and method-to-event mapping techniques are commonly adopted when shifting from the OOP paradigm to the CBD paradigm [13]. Major issues for the OOP paradigm are that objects are too specific that contains lots of unnecessary details, for every new project with slightly different requirements, classes must be adjusted before they can be reused. It is challenging to manage variations of the same object. By applying the CBD paradigm, dependencies between components are largely reduced so that components may be reused without interfering with each other. Development costs are reduced by building systems from standard component libraries. As components are well-proven independent modules and systems built from such components are more reliable rather than customization each time. Another issue for the OOP paradigm is that some functions may not have to be implemented as objects or cannot be reflected on physical layout. For instance, straightforward mathematic calculations are just monolithic codes and communication handlers are virtual objects that do not exist physically.

The CBD paradigm could be applied to any industrial automation systems theoretically but a couple of guidelines must be followed. First, application-specific methods must be removed in order to make components generic. For example, extra requirements could be encapsulated in separated function

TABLE 4.1

Paradigm Comparison

Features	OOP	CBD	SOA
Software Unit	Classes (Objects)	Software Component	Software Services
Features	Inheritance, Polymorphism	Encapsulation	Loose Coupling, Discoverability
Interface	Methods Call	Predefined Interface	Service Contract
Reusable Source	Object Library	Component Library	Service Repository
System Hierarchy	Nested Classes	Nested Components	Service Orchestration and Composition
Meta Model	Class Diagram (UML)	Network of Components	Business Process Execution Language (BPEL), Service Sequence Diagram (SSD)
Flexibility	Low	Medium	High

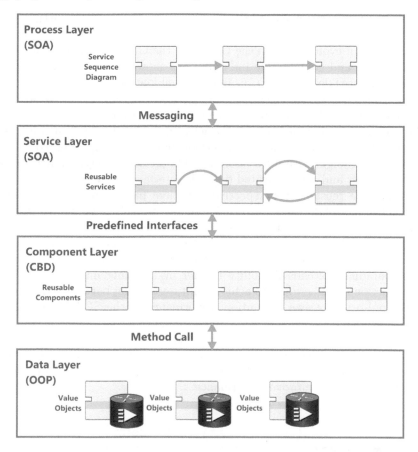

FIGURE 4.19
Generic service-oriented design for IEC 61499.

blocks even if their functionalities belong to another function block. Those components might be integrated by encapsulation into a CFB to reduce dependencies. Second, names of events should be generic to avoid modifying interfaces frequently. In the OOP paradigm, events are mapped as methods. But in the CBD paradigm, to maintain interface, events are replaced by data inputs for selecting appropriate methods. In transition from the OOP paradigm to the CBD paradigm, reusability and flexibility are improved as indicated in Table 4.1.

The CBD and the SOA paradigm are intended to provide loosely coupled and interoperable software modules. Although both paradigms have similarities, the SOA paradigm provides enhanced flexibility and interoperability [4]. A single instance of a component in CBD manages only one resource and a service manages a set of resources which is mostly stateless. A service runs as a manager that is responsible for creating, modifying and deleting a set of instances of the same type or even multiple types. The SOA paradigm can be considered an evolution from the OOP and the CBD paradigms with flexible architecture and configurable interfaces. As shown in Figure 4.19, a generic service-oriented design for IEC 61499 is a combination of OOP, CBD and SOA paradigms. On the data layer of the SOA paradigm, function blocks are implemented using the OOP paradigm where data are grouped by object types. On the component layer, function blocks act as gateways for bridging objects and services by providing interfaces. Function blocks are implemented following the CBD paradigm with fixed interfaces. On the service layer, function blocks fetch, process and return related data via interface gateways.

From the reusability perspective, all three paradigms are based on the concept of modular design but provide different levels of flexibility. The CBD paradigm improves design efficiency by specifying interfaces between objects proposed by the OOP paradigm. Provided that interfaces remain unchanged, modifications of individual function blocks will not affect system design. The SOA paradigm extends flexibility further by handling multiple instances in one function block service. To achieve that, separation between logic and data is compulsory. However, execution performance of the SOA paradigm is reduced compared to the OOP and CBD paradigms, as data must be transferred around function block networks. The balance between design efficiency and execution performance must be considered: the aim is to divide functionalities into modules to improve efficiency while minimizing connections between function blocks to improve performance.

4.7   Conclusions

In this chapter, three commonly adopted design paradigms in software engineering, object-oriented programming, component-based design, and service-oriented architecture, are introduced for the IEC 61499 standard. For each design paradigm, definitions and examples are provided. The OOP paradigm connects physical layout with its cyber view by mapping devices to software objects. The reusability and flexibility are improved by introducing pre-defined interfaces between objects in the CBD paradigm. The SOA paradigm provides best flexibility by using multi-layered architecture although performance is affected due to heavy data handling in function block networks. There is not a conclusion that one paradigm is significantly better than others. Based on the specific requirements of various industrial automation systems, a proper software design paradigm should be selected by evaluating reusability, flexibility, and performance of the paradigms.

Bibliography

[1] F. Basile, P. Chiacchio, and D. Gerbasio. On the Implementation of Industrial Automation Systems Based on PLC. *IEEE Transactions on Automation Science and Engineering*, Vol. 10, No. 4, 990–1003, 2013.

[2] G. Black and V. Vyatkin. Intelligent Component-Based Automation of Baggage Handling Systems with IEC 61499. *IEEE Transactions on Automation Science and Engineering*, Vol. 7, No. 2, 337–351, 2010.

[3] R. Brennan, M. Fletcher, and D. Norrie. An Agent-Based Approach to Reconfiguration of Real-Time Distributed Control Systems. *IEEE Transactions on Robotics and Automation*, Vol. 18, No. 4, 444–451, 2002.

[4] A. Brown, S. Johnston, and K. Kelly. Using Service-Oriented Architecture and Component-Based Development to Build Web Service Applications, Rational Software Corporation. 2012.

[5] L. De Alfraro and T. A. Henzinger. *Interface theories for component-based design.* Springer, Heideberg, 1992.

[6] T. Erl. *Service-oriented architecture: concepts, technology and design.* Prentice Hall, 2005.

[7] R. Hametner, A. Zoitl, and M. Semo. Automation Component Architecture for the Efficient Development of Industrial Automation Systems. In *IEEE Conference on Automation Science and Engineering*, pages 156–161, 2010.

[8] D. Helic and N. Scherbakov. Software Paradigms: Introduction and Procedural Programming Paradigm, 2013. 707.023 Lecture Notes, Graz, Austria.

[9] International Electrotechnical Commission. IEC 61499-1: Function blocks, part 1: Architecture, 2012. Geneva, Switzerland.

[10] International Electrotechnical Commission. IEC 61131-3 programmable controllers, part 3: Programming languages, 2013. International Electrotechnical Commission, Third Edition.

[11] I. Jacobsen, C. Magnus, J. Patrik, and O. Gunnar. *Object Oriented Software Engineering.* Addison-Wesley, 1992.

[12] F. Jammes and H. Smit. Service-Oriented Paradigms in Industrial Automation. *IEEE Transactions on Industrial Informatics*, Vol. 1, No. 1, 62–70, 2005.

[13] M. Koskela, M. Rahikainen, and T. Wan. Software development methods: SOA vs CBD, OO and AOP, 2007. Technical Report, Aalto University, Finland.

[14] W. Lepuschitz, A. Zoitl, M. Vallee, and M. Merdan. Toward Self-Reconfiguration of Manufacturing Systems Using Automaton Agents. *IEEE Transactions on Systems, Man and Cybernetics, Part C: Applications and Reviews*, Vol. 41, No. 1, 52–69, 2011.

[15] D. Lorenz and J. Vlissides. Designing Components versus Objects: A Transformational Approach. In *Proceedings of the 23rd International Conference on Software Engineering*, pages 253–263, 2001.

[16] P. V. Roy. *Programming Paradigms for Dummies: What Every Programmer Should Know*. Number 104. New Computational Paradigms for Computer Music, 2009.

[17] T. Strasser and R. Froschauer. Autonomous Application Recovery in Distributed Intelligent Automaton and Control Systems. *IEEE Transactions on Systems, Man and Cybernetics, Part C: Applications and Reviews*, Vol. 42, No. 6, 1054–1070, 2012.

[18] C. Sünder, A. Zoitl, J. Christensen, H. Steininger, and J. Rritsche. Considering IEC 61131-3 and IEC 61499 in the Context of Component Frameworks. In *6th IEEE International Conference on Industrial Informatics*, pages 277–282, 2008.

[19] V. Vyatkin, C. Pang, Y. Deng, M. Sorouri, and H. Mayer. System-Level Architecture for Building Automation Systems: Object-Oriented Design and Simulation. In *39th Annual Conference of the IEEE Industrial Electronics Society (IECON)*, pages 5334–5339, 2013.

[20] B. Warboys. The software paradigm. *ICL Technical Journal*, Vol. 10, , 71–79, 1995.

[21] B. Werner. Object-Oriented Extensions for IEC 61131-3. *IEEE Industrial Electronics Magazine*, Vol. 3, No. 4, 36–39, 2009.

[22] G. Zhabelova and V. Vyatkin. Multiagent Smart Grid Automation Architecture Based on IEC 61850/61499 Intelligent Logical Nodes. *IEEE Transactions on Industrial Electronics*, Vol. 59, No. 5, 2351–2362, 2012.

5

New Design Patterns for Time-Predictable Execution of Function Blocks

Matthew M. Y. Kuo

University of Auckland

Partha S. Roop

University of Auckland

CONTENTS

5.1 Introduction

IEC 61499 function blocks are used to design many safety-critical and time-predictable systems in the fields of smart grids [31], robotics [9] and medical devices [32]. The correctness of real-time systems relates to their functionality and also their timing behavior [7, 29]. As an example, let us consider the robotic arm presented in Figure 5.1. The robotic arm is responsible for moving objects from a pickup conveyor to a drop-off conveyor. For safety, it employs a light curtain safety sensor to prevent harm to the operators. Table 5.1 presents

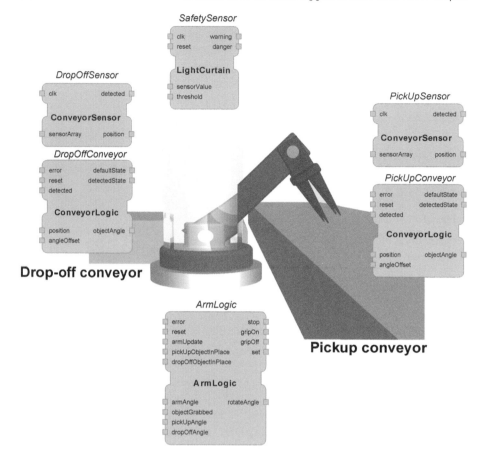

FIGURE 5.1
Robotic arm example.

a brief description of each function block in the robotic arm. The correct operation of the robotic arm system is not limited to its functionality, that is, the robotic arm must be moved to its correct position in the correct time frame. Real-time requirements are classified into two main categories: *hard* real-time and *soft* real-time [17]. Hard real-time systems cannot tolerate any missed deadlines, as these can result in catastrophic consequences, such as the loss of life, financial liability, or significant damage to assets. For example, if the safety sensor of the light curtain takes too long to respond to an over-reaching operator, the robotic arm would not be able to stop in time and could harm or seriously injure the operator. Soft real-time systems, on the other hand, may tolerate some missed deadlines. Missing deadlines in such systems typically results in lack of responsiveness or a lower quality of service without serious consequences. For example, consider the conveyor side of the

TABLE 5.1

Description of Function Blocks in the Robotic Arm System

Function Block Instance	Function Block Type	Functionality
SafetySensor	LightCurtain	Detects intrusion of the light curtain
PickUpSensor	ConveyorSensor	Detects whether object is ready for pickup
DropOffSensor	ConveyorSensor	Detects whether object is dropped off
PickUpConveyor	ConveyorLogic	Controls the movement of the pickup conveyor
DropOffConveyor	ConveyorLogic	Controls the movement of the dropoff conveyor

robotic arm system. Slow processing may result in minor spillage and material or cargo damage due to collisions caused by missed events but will not cause catastrophic events.

There are limited approaches [9, 33] that act as guidelines for the systematic design of safety-critical and time-predictable systems using IEC 61499. To design a time-predictable system, all hardware and software components must be time-predictable (illustrated in Figure 5.2). This chapter will focus on the time-predictable design of IEC 61499 applications and implementations. The design and philosophy of precision timed hardware can be found in [11].

The expressiveness of the IEC 61499 standard allows end users in their IEC 61499 applications to create and implement their own algorithms (within ECC) and specialized custom function block called service interface function blocks. This means the user can write and execute any arbitrary code within the function block model. To achieve time predictability, it is essential to take extra care when designing such algorithms. Therefore, this chapter will begin by presenting coding guidelines to aid the end user in designing time-predictable applications in Section 5.2.

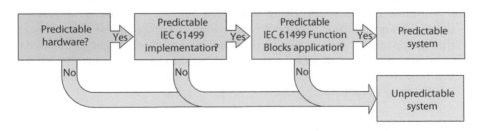

FIGURE 5.2
Time-predictable system.

However, the behavior of applications is dependent on code and also on the underlying IEC 61499 implementation. To address this, Section 5.3 presents new design patterns for designing towards time-predictable IEC 61499 implementation.

5.2   Coding Guidelines for Time Predictability

This section will present coding guidelines for common pitfalls encountered when designing time-predictable IEC 61499 applications. Our guidelines are inspired by well known principles articulated from the MISRA-C coding standard [2], the Jet Propulsion Laboratory (JPL) coding standard [1], the Power of 10 [15], and a paper by Gebhard et al. on software structure and WCET predictability [13].

5.2.1   Loops

Loops are used for iterative computing and also for busy-waiting on input events or communication packets. It is essential to statically determine the maximum number of iterations of a loop, known as the *loop bound*, to achieve time predictability. No guarantees regarding its longest execution time can be provided for an application with loops with unknown bounds. The programmer should keep the loop structure as simple as possible to ensure that a loop bound can be determined. The variables used for a loop iteration counter should not be modified within the loop body. *Continue* statements should be replaced with *if* and *else* statements [13].

Some programming languages, such as PRET-C [3], allow programmers to specify loop bounds directly within the source code. Code is then generated to force the loops to terminate within the specified bounds.

Figure 5.3 presents two functions that compute the greatest common divisor (GCD) using the Euclidean algorithm [16]. The function on the left is the original Euclidean algorithm where the number of iterations depends on the input and hence is difficult to determine at compile time. Therefore, no timing guarantees can be provided. The function on the right implements a clear bound. If a result is not determined within the set bound, the function will return −1, indicating an error. The error will need to be handled suitably by the programmer.

5.2.2   Real Numbers

Real numbers can be represented in floating point or fixed-point representation. Floating point numbers are, in general, more popular because they

```
                            int gcd(int a, int b) {
                              int i = 0; int ret = -1;
                              for (i = 0; i < [bound]; i++) {
int gcd(int a, int b) {           if (a != b) {
    while (a != b) {                  if (a > b) {
        if (a > b) {                      a = a - b;
            a = a - b;                }else{
        }else{                            b = b - a;
            b = b - a;                }
        }                         }else{
    }                                 ret = a;
    return a;                     }
}                               }
                              return ret;
                            }
```

FIGURE 5.3
Translating an unbounded loop (left) to a bounded loop (right).

are primitive data types in programming languages. However, extra care is required when using floating points in time-predictable systems. The programmer should first determine the precision of the floating point arithmetic unit (single or double) of the target platform. If the target platform does not include a floating point arithmetic unit or the programmer is required to use a precision beyond the hardware support, compilers will utilize software libraries to perform floating point computations. Generally, these software libraries are designed with average case execution in mind [13] and contain loops that are hard to bound statically. Therefore, such libraries are unsuitable for time-predictable systems. Instead, fixed-point arithmetic should be used when hardware support for floating point computation is limited or not available. Fixed-point arithmetic is essentially the manipulation of integer values. These are natively supported by most processor architectures.

Exercise

If both values of *a* and *b* in Figure 5.3 are 8-bit numbers, what is the smallest value for *bound* such that a valid result will always be returned?

Hint: the value is between 1 and 256.

5.2.3   Memory Allocation

Memory allocation can be achieved statically or dynamically. Dynamic memory allocation allows applications to assign the required amount of memory during runtime. In contrast, static memory allocation preassigns regions of memory during compile time for specific tasks. In general, dynamic memory allocation is more resource efficient in terms of memory usage than static allocation. However, for time-predictable systems, dynamic memory allocation should be avoided unless specialized time-predictable allocation algorithms are used [14, 22]. There are three main reasons for this. First, dynamic memory allocation requires the system to search for an available space during runtime. The time it takes to allocate the memory varies greatly. Therefore, to provide a guarantee on execution time, the worst-case allocation time will have to be assumed for every memory allocation. This assumption will severely over-estimate the worst-case execute time of the system, resulting in impractical timing values. Second, for systems with a memory hierarchy, it is hard to derive a precise bound for memory accesses. Therefore, result may be an overestimation of execution time. Third, dynamic memory allocation can lead to runtime errors when not enough available memory can be allocated (i.e., due to fragmentation or lack of memory).

5.2.4   Branches

There are two categories of branch statements: static branches and computed branches. Static branches are statements such as function calls or *if* and *else* statements. These branch statements are suitable for time-predictable systems. In contrast, computed branches are statements such as function pointers and computed goto statements. These branch statements should be avoided for time-predictable systems because they pose a difficulty for static analysis [13].

5.2.5   Interrupts

Interrupts allow applications to switch context to service a request from the environment (context-switching). In general, interrupts can occur any time during the lifetime of an application and the number of occurrences is usually unrestricted. Unless there is a bound on the frequency at which interrupts can occur, they are unsuitable for time-predictable systems. Alternatives to interrupts are polling and reactive functional units. Polling uses software to periodically check for requests from the environment. The rate at which an application should poll (sample) the environment should be faster than the rate at which the environment can make requests. Polling is most suited for continuous signals such as monitoring the temperature of the environment. Discrete signals, such as button presses, can be missed if they occur when the software is not checking for inputs. Reactive functional units are found in special processors, called reactive processors [20, 24, 30], in which the inputs are latched when the environment makes a request. Therefore, by using reactive processors, discrete inputs will not be missed as long as the inputs are serviced faster than their inter-arrival times. Each time the inputs are serviced, the input latches are cleared at predetermined points in an application, making them ideal for the design of time-predictable systems.

5.3   Design Patterns

Design patterns are reusable solutions that solve recurring design problems by using a well defined structure [6, 12]. When presented with a new task, it is typical practice for designers to reuse and adapt proven solutions. There are many categories of patterns ranging from access control to communication and interactive systems [6]. Design patterns can be programming language dependent or programming language independent. One of the most popular programming language independent design pattern is the *model view control* (MVC) pattern [18]. Traditionally, the MVC pattern was designed for object-orientated programming for implementing user interfaces, by separating the software into three components. The *model* implements the behavior of the application and is independent of the visual representation. The *view* implements the visual representation of some or all of the data of the model. The *controller* manages inputs from the user or the environment and sends them as commands to the model. The MVC pattern has been adapted for IEC 61499 with some modifications [8]. The *model* is a function block that implements the behavior of the system being controlled (i.e., the plant). The *view* is a function block that implements the graphical interface associated with the model (i.e., the visualizer). The *controller* is a function block that implements the controller of the system (i.e., the controller).

There are also design patterns tailored specifically for IEC 61499. For example, Serna et al. proposed a design pattern for the failure management of function blocks by introducing failure zone function blocks, failure managed function blocks, and failure supervisor function blocks [25]. Failure zone function blocks are responsible for reporting the failures and for recovering from failures through failure managed function blocks. Failure supervisor function blocks are responsible for determining if a failure has occurred due to the collective effects of multiple function blocks. When the failure supervisor detects a failure, it sends a failure event to the appropriate failure zones. Failure zone function blocks can be connected in a hierarchy to correspond to failure dependences within the network. Failure events are propagated though the hierarchy so that appropriate actions can be taken by the function block network.

Dubinin and Vyatkin proposed a semantic robust design pattern for IEC

61499 by introducing special service blocks to be placed between function block connections [10]. The service blocks control event propagations. Since function blocks are event triggered, the service blocks effectively control the network execution.

While a range of execution semantics exists for IEC 61499 [26], there are no guidelines on choosing a semantic based on application requirements. As design patterns offer effective guidelines to aid design, we need to develop design patterns to address this gap. Specifically, we intend to develop design patterns that aid the design of time-predictable systems. Our approach can be extended in the future to match other application requirements to appropriate execution semantics through design patterns. In this chapter, we will first qualify the application requirements between the centralized and the distributed function block models. Following that, two design patterns are proposed to aid the designer in choosing the appropriate execution semantics when designing time-predictable systems. One of the design patterns is tailored to centralized systems and the other is tailored to distributed systems.

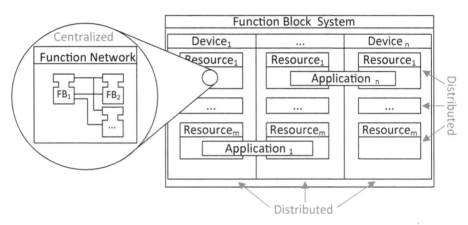

FIGURE 5.4
Overview of IEC 61499 models.

5.4   Requirement Qualification

Figure 5.4 illustrates an overview of the function block models. A succinct description of these models can be found in the introductory chapters and [29]. In this chapter, we will focus on the execution of function blocks within resources, called *intra-resource function blocks*. We will also present the execution of function blocks across multiple resources, called *inter-resource function blocks*. Table 5.2 summarizes the key distinctions between intra- and

TABLE 5.2

Centralized versus Distributed Systems

Centralized	Distributed
Fewer execution overheads	More execution overheads
Scheduled concurrently	Scheduled in parallel
Single device	Multiple devices
Fast on-chip buses or memory	Slow network communication

inter-resource function blocks. Intra-resource function blocks are centralized systems, in which the function blocks share the same hardware resources, such as the processor and the associated memory. Centralized systems aim to execute efficiently with minimal overheads. Communication between intra-resource function blocks usually takes a few processor instructions as they share the same resource. Inter-resource function blocks are distributed systems in which function blocks are deployed on separate virtual/actual devices. For distributed systems, modularity is important to avoid strong dependencies between resources and devices. There is often extra overhead associated with modular design. Communication between resources and devices is distributed using an appropriate network or local sockets. The communication delay is expected to be significantly more compared with intra-resource function blocks.

To consider the different needs of intra- and inter-resource function blocks, we propose two separate design patterns to achieve a time-predictable execution of IEC 61499 systems. Both design patterns are inspired by the synchronous programming paradigm [4, 29]. The advantages of the synchronous approach are determinism and reactivity [21]. Determinism, in this context, refers to systems that produce the same output sequence when stimulated by the same input sequence. Reactivity, on the other hand, refers to systems that always remain responsive to input stimuli from the environment.

Synchronous programming is inspired by synchronous circuits. In synchronous circuits, the components are driven using an actual *clock*. The inputs and outputs of the circuits change instantaneously relative to the clock edges. Likewise, synchronous programs execute in terms of *logical clock ticks*. During each *tick*, the application samples its inputs, computes the results and emits the corresponding outputs (see Figure 5.5). The logical ticks are assumed to be instantaneous, taking zero time (similar to the clock edges). However, in practice, all computations take time. Therefore, for practical applications, a sufficient condition to achieve a real-time response is that the longest duration among all possible ticks must be equal to or faster than the time it takes for the environment of the system to change the inputs. The longest duration among all possible ticks is commonly referred to as the *worst-case reaction time* (WCRT) of a synchronous program [3, 5]. A similar condition can also be observed in synchronous circuits in which all electrical signals must stabilize before the next clock edge.

The design pattern we propose for time-predictable execution of intra-

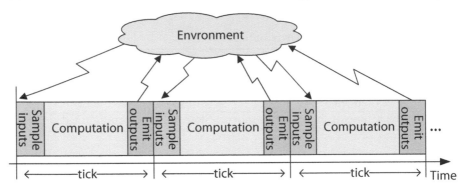

FIGURE 5.5
Synchronous execution.

resource function blocks is called the *ordered synchronous design pattern*. The design pattern we propose for time-predictable execution of inter-resource function blocks is called the *delayed synchronous design pattern*. These two design patterns describe how function block implementations should be scheduled and executed for time predictability. The ordered synchronous design pattern is inspired by a synchronous programming language called PRET-C [3]. As the name implies, this pattern enforces a sequential scheduling order of function blocks during every tick. Communication between each function block is instantaneous (i.e., occurs within a single tick). The delayed synchronous design pattern is inspired by [28]. As the name implies, that the communication between resources and devices is not instantaneous (i.e., does not occur within the same tick) but requires a unit delay. This delay allows the inter-resource function block to execute in parallel and independently from other resources or devices during each tick.

Figure 5.6 presents the mapping of the proposed design patterns to the function block models. The *ordered synchronous design pattern* applies to the models shaded in dark gray (intra-resource function blocks). The *delayed synchronous design pattern* applies to the models shaded in light gray (inter-resource function blocks).

5.5 Ordered Synchronous Design Pattern

For intra-resource function blocks, we propose the ordered synchronous pattern. The word *ordered* refers to the fact that a fixed scheduling order is required. To draw an analogy for the ordered synchronous design pattern, consider the ripple counter presented in Figure 5.7. The ripple counter is constructed using flip-flop connected in series. The system is driven by a single

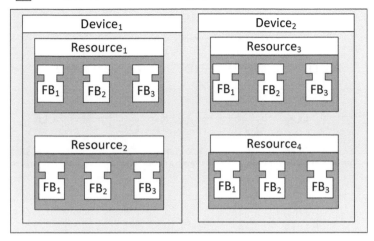

FIGURE 5.6
Mapping of proposed design patterns to function block models.

clock source. The ripple counter "executes" in a strict order from the first to the last. The communication, sampling of inputs and emission of outputs are instantaneous (i.e., occur before the next clock edge). Any back edges are read only in the next clock edge. The connection from $\overline{Q}$ to Q inverts the value stored by each flip-flop whenever a rising edge is detected. For ease of reference, the Q port of the first flip-flop $D1$ will be referred to as Q_{D1} and so on. The ripple counter works by inverting the output Q_{D1} at each rising edge of the input clock and inverting the output Q_{D2} at every second rising edge and inverting the output Q_{D3} at every fourth rising edge.

Initial: All three D flip-flops have zero values as their inputs; therefore, the value of Q is zero and the value of $\overline{Q}$ is one for all three flip-flops. The decimal value for the counter is zero (000_b).

First rising clock edge: $D1$ reads the value one from D_{D1} (previous value of $\overline{Q_{D1}}$) and sets Q_{D1} to one and $\overline{Q_{D1}}$ to zero. Both $D2$ and $D3$ do not change value because no rising edge is detected since $\overline{Q_{D1}}$ is zero. The decimal value for the counter is now one (001_b).

Second rising clock edge: $D1$ reads the value zero from D_{D1} and set Q_{D1} to zero and $\overline{Q_{D1}}$ to one. $D2$ detects the rising edge set by $D1$ and reads the value one from D_{D2} (previous value of $\overline{Q_{D2}}$) and sets Q_{D2} to one and $\overline{Q_{D2}}$ to zero. $D3$ does not change in value because no rising edge is detected. The decimal value for the counter is now two (010_b).

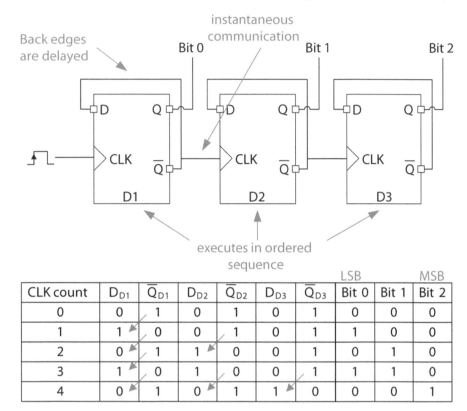

FIGURE 5.7
Analogy of ordered synchronous design pattern using a ripple counter.

Third rising clock edge: $D1$ reads the value one from D_{D1} and sets Q_{D1} to one and $\overline{Q_{D1}}$ to zero. $D2$ and $D3$ do not change in value because no rising edge is detected. The decimal value for the counter is now three (011_b).

Fourth rising clock edge: $D1$ reads the value zero from D_{D1} and sets Q_{D1} to zero and $\overline{Q_{D1}}$ to one. $D2$ detects the rising edge set by $D1$ and reads the value zero from D_{D2} and sets Q_{D2} to zero and $\overline{Q_{D2}}$ to one. $D3$ detects the rising edge set by $D2$ and reads the value one from D_{D3} and sets Q_{D3} to one and $\overline{Q_{D3}}$ to zero. The decimal value for the counter is now four (100_b).

Here, the D flip-flops are similar to intra-resource function blocks under ordered composition. The interconnection between the flip-flops represents the function block network in which communication between function blocks is instantaneous and back edge communications are delayed to the next tick. The clock pulses are analogous to the logical ticks of synchronous programming.

We will use the symbol $\otimes$ to represent the composition under the ordered synchronous pattern. For example, $D1 \otimes D2 \otimes D3$ indicates that function block $D1$ is concurrently composed with function blocks $D2$ and $D3$ and they are executed using the ordered synchronous pattern (*ordered composition*). The fixed scheduling order goes from left to right of the composition, that is, $D1$ followed by $D2$ and then $D3$.

Let us consider a concrete function block network within a given resource. Figure 5.8 presents a centralized robotic arm. The dashed connections in ①

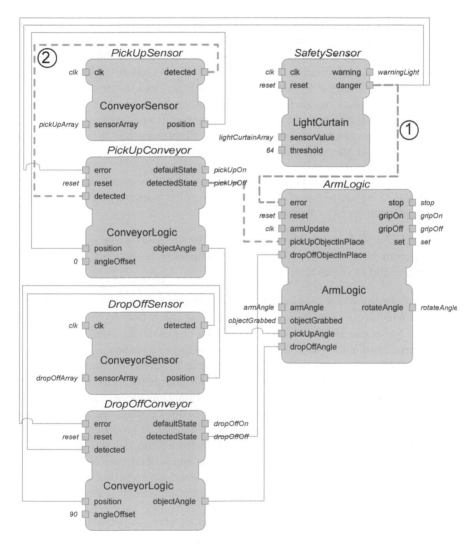

FIGURE 5.8

Robotic arm function block network.

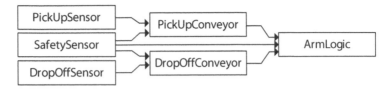

FIGURE 5.9
Input and output dependencies of robotic arm network.

and ② that will be used later to illustrate the timing properties of the ordered synchronous design pattern.

In order to apply this design pattern on the function block network, an explicit ordering of function blocks similar to the left-to-right order in the ripple counter example is needed. In a function block network, such a total order is not explicit so such an order needs to be defined. The first step is to define a partial order by analyzing the input and output dependencies in the function block network. Figure 5.9 illustrates the input and output dependencies for the robotic arm running example.

$FB_1 > FB_2$: A function block (FB_1) is ordered before another function block (FB_2) if its output port is connect the other's input port. For example, PickUpConveyor > PickUpSensor.

$FB_1 = FB_2$: A function block (FB_1) can be ordered before or after another function block (FB_2) if there are no connections between them. For example, SafetySensor = PickUpSensor.

For function block networks with cyclic dependencies, the system designer decides which function block is ordered before the others to break the dependency. Cycles are similar to the back edges of the D flip-flops in Figure 5.7. The partial order for the robotic example is SafetySensor = PickUpSensor = DropOffSensor > PickUpConveyor > DropOffConveyor > ArmLogic.

The second step is to create an execution schedule by converting the partial order (in the first step) into a total order. Function blocks that are equal in the partial order are arbitrarily assigned an order between them, for example, SafetySensor > PickUpSensor > DropOffSensor. The order between these function blocks does not matter because there are no dependencies between them. The result is a total order of function blocks that can be directly used to create the ordered composition. The ordered composition for the robotic arm is SafetySensor ⊗ PickUpSensor ⊗ DropOffSensor ⊗ PickUpConveyor ⊗ DropOffConveyor ⊗ ArmLogic.

Once the ordered composition is determined, the execution behavior within each tick is as follows. First, the system samples the environmental inputs. Next, each function block in the composition executes according to the defined order. Finally, at the end of the tick, outputs are emitted.

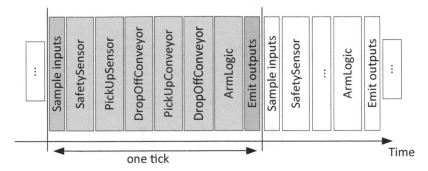

FIGURE 5.10
Ordered synchronous execution of the robotic arm.

Figure 5.10 depicts how the robotic arm network is executed during each tick. First, the inputs from the environment are sampled. Next, the function blocks are executed in the order they are composed, that is, `SafetySensor` then `PickUpSensor` and so on. Finally, the outputs are emitted to the environment at the end of the tick.

The key feature of the ordered synchronous pattern is that events and data propagate instantaneously (within a single tick) through the function blocks in a statically defined and predictable order. This means the time it takes for the system to respond to an environmental input and emit a corresponding output (the response time) is **always** a single tick for non-cyclic systems (i.e., function block networks without cycles). The response time for cyclic systems (i.e., function block networks with cycles) is at most two ticks. For example, Figure 5.11 depicts how events and data are propagated though the network in the robotic arm example. Each horizontal line represents a tick. Figure 5.11a presents a scenario where the `SafetySensor` detects an intrusion in the workspace. The intrusion is detected by `SafetySensor` and the error event is emitted to `ArmLogic`. The `ArmLogic` stops the robotic arm to prevent possible injuries by emitting a stop event. The network connection contributing to this scenario is highlighted in dashed connection ① in Figure 5.8. Figure 5.11b presents a scenario in which an object arrives at the pickup location. The `PickUpSensor` detects the object and notifies the `PickUpConveyor` by emitting the detected event. The `PickUpConveyor` pauses the conveyor and notifies the `ArmLogic` function block to pick the item up. The `ArmLogic` then emits a set event to set the robotic arm into position. The network connections contributing to this scenario are highlighted in dashed connection ② in Figure 5.8. For both scenarios, the response time is a single tick even though the input event is propagated through different numbers of function blocks, that is, two for Figure 5.11a and three for Figure 5.11b.

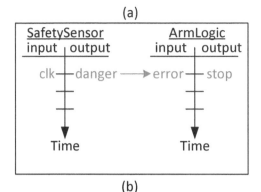

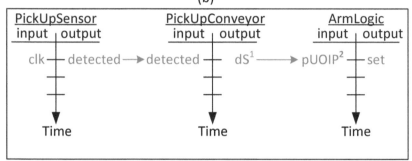

¹detectedState ²pickUpObjectInPlace

FIGURE 5.11
Response time for the robotic arm example using the ordered synchronous pattern.

5.6   Delayed Synchronous Design Pattern

Inter-resource function blocks (i.e., devices and resources) are distributed components that can execute in parallel. Therefore, the ordered synchronous design pattern is unsuitable for inter-resource function blocks because a strict execution order is restrictive for distributed execution. Thus, for inter-resource function blocks, we propose the delayed synchronous design pattern. The word *delayed* refers to the fact that a delay is implied for any communication (i.e., communication in forward and backward directions) between function blocks across multiple resources. Figure 5.13 shows an example where the robotic arm is distributed into two resources named $Resource_{conveyor}$ and $Resource_{arm}$. $Resource_{conveyor}$ contains the conveyor related function blocks and $Resource_{arm}$ contains the ArmLogic and the SafetySensor function blocks. Communication between these two resources is achieved using publisher blocks and subscriber blocks, for example, PubDropOff and

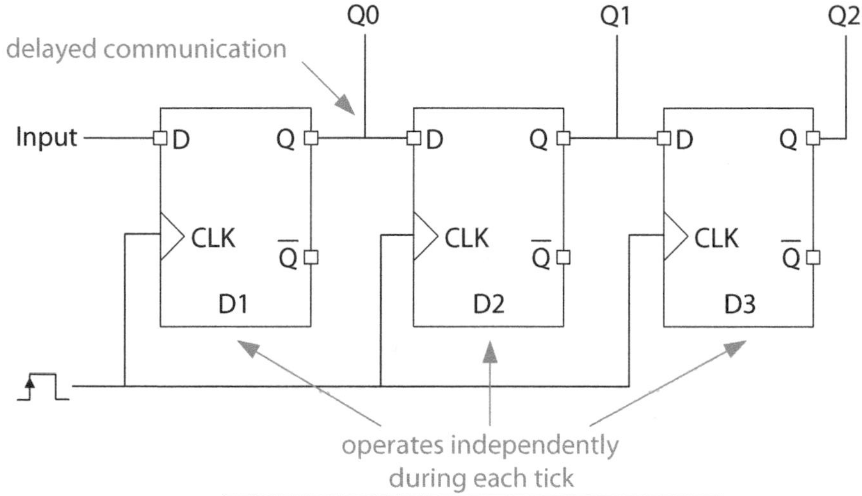

CLK count	Input	Q_{D1}	Q_{D2}	Q_{D3}
0	0	0	0	0
1	1	1	0	0
2	0	0	1	0
3	0	0	0	1

FIGURE 5.12
Analogy of delayed synchronous design pattern using a shift register.

`SubDropOff`. For a detailed discussion on publisher and subscriber function blocks, interested readers can refer to [29]. We use the traditional parallel symbol $||$ to represent composition for the delayed synchronous pattern (*delayed composition*). For example, the composition for Figure 5.13 is $Resource_{conveyor} \ || \ Resource_{arm}$.

Unlike, the ordered synchronous pattern, the order in which resources or devices are composed does not matter. The resources and devices only communicate after all resources and devices in the system reach their respective ends of each tick. Therefore, all communications between resources and devices are delayed by a single tick. For example, an output event from $Resource_{conveyor}$ is only available to $Resource_{arm}$ in the next tick. This delay decouples any input and output dependencies between the resources and devices. Since the resources and devices do not communicate during each tick, they can execute independently (during each tick).

For an analogy of the delayed synchronous design pattern, consider the shift register presented in Figure 5.12. The shift register is constructed using three D flip-flops, each connected in parallel. Each D flip-flop is driven by its own clock coming from the same clock source. The shift register *executes* in parallel, and each flip-flop executes independently at every clock edge.

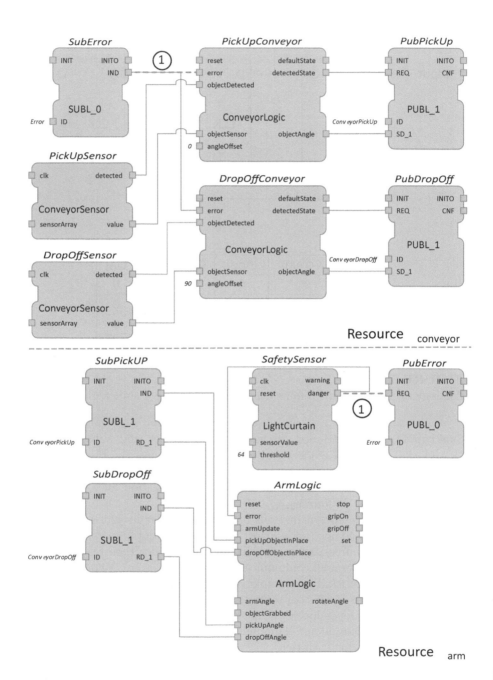

FIGURE 5.13
Distributed robotic arm.

Communication is delayed, that is, it occurs at the next clock edge. The interconnection between the flip-flops represents the function block network in which communication between function blocks is delayed to the next tick. The clock pulses are analogous to the logical ticks of synchronous programming. The D flip-flops operate similar to function block resources or devices under delayed composition ($D1 \parallel D2 \parallel D3$). For example, at the start of each tick, the three flip-flops sample their inputs in parallel. The values they sample are from the previous tick (see the truth table in Figure 5.12). This is the same as delayed composition where each resource and device operates independently during a tick and the communication between resources and devices is delayed to the next tick.

Let us consider a concrete example to illustrate the behavior of function blocks using this design pattern. Figure 5.14 illustrates how the function blocks are executed for the distributed robotic arm example. Across the resources, the function blocks execute by using the delayed synchronous pattern. Within each resource, the function block executes according to the ordered synchronous pattern as presented in the previous section. Therefore, the composition of function blocks for the distributed robotic arm example may be represented as:

{SubPickUp ⊗ SubDropOff ⊗ SafetySensor ⊗ ArmLogic ⊗ PubError} ∥ {SubError ⊗ PickUpSensor ⊗ DropOffSensor ⊗ PickUpConveyor ⊗ DropOffConveyor ⊗ PubPickUp ⊗ PubDropOff }.

Like all synchronous systems, the delayed synchronous pattern also exe-

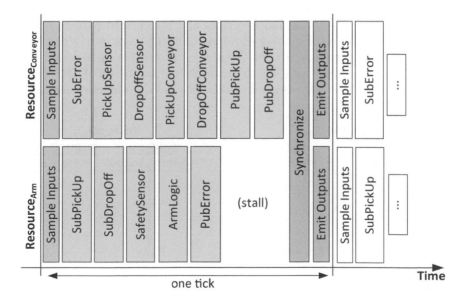

FIGURE 5.14
Distributed execution of robotic arm.

cutes in ticks. During each tick, each resource samples the inputs from the environment, executes the function blocks it encapsulates independently, synchronizes, and emits the outputs to the environment. Synchronization is required because there is no physical clock driving the start of each tick. Therefore, resources must negotiate when to start the next logical tick. Synchronization begins once all resources have executed the function blocks they encapsulate. If a resource completes execution early, it must wait (stall) until all the other resources have completed their execution. During synchronization, the resources exchange the events and data through publisher and subscriber blocks. Once synchronization is completed, the resources and devices emit their respective outputs and begin their next tick.

To illustrate the timing property of the delayed design pattern, Figure 5.15 depicts how events and data are propagated between $Resource_{conveyor}$ and $Resource_{arm}$. The presented scenario is the same as that in Figure 5.11a where the SafetySensor detects an intrusion in the workspace. The event chain is highlighted in dashed connections ① in Figure 5.13. During the first tick, SafetySensor detects the intrusion and emits the error event to PubError. Next, the two resources synchronize and the data is sent from $Resource_{conveyor}$ to $Resource_{arm}$ via PubError and SubError. In the second tick, SubError emits the signal received from PubError to PickUpConveyor, which sends a signal to stop the conveyor. From this example, we can see the differences in timing between the ordered synchronous pattern and delayed synchronous pattern. The ordered synchronous pattern takes only a single tick to respond (see Figure 5.11a), whereas the delayed synchronous pattern takes two ticks to respond. To generalize, the ordered pattern always has a response time between one and two ticks and the delayed pattern has a response time equal to the number of resources in the event chain. The delayed synchronous design pattern is ideal for time predicable distribution in two ways. First, the longest tick length is simply the maximum WCRT between all resources and devices. Second, unlike the GALS semantics for function

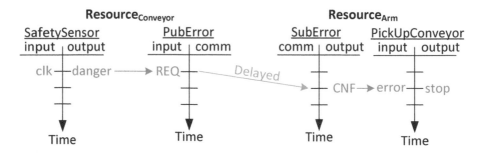

FIGURE 5.15
Response time for robotic arm example under the delayed synchronous pattern.

blocks [29], determinism is preserved while still executing in a parallel fashion. The limitation of the delayed synchronous design pattern is the need to synchronize at the end of each tick.

5.7 Timing Analysis

In the previous section, we presented logical time in terms of ticks. For practical applications, real time values are required. To convert logical time to real time, timing analysis is used to compute the WCRT of the system. The computed WCRT is the duration of one tick in real time. The WCRT is the worst-case execution time (WCET) for all possible ticks within the system.

Timing analysis techniques can be classified into two categories: *measurement-based timing analysis* and *static timing analysis* [27]. Measurement-based techniques execute a given application with a large set of input vectors to compute the WCET. The key lacuna of this approach is the lack of guaranteed soundness. The measured WCET may be an underestimate since the inputs required to elicit the worst case may not be used [23]. Static analysis techniques create models of a given application using sound abstractions. The WCET of the application is determined using path enumeration algorithms to compute the longest execution path. These techniques are exhaustive in nature and always provide a safe over-approximation of the actual worst case. Figure 5.16 illustrates the distinction between actual execution times and the measured execution times of a typical program. The area shaded in light gray depicts the distribution of actual execution times of a typical program. The area shaded in dark gray depicts the distribution of measured execution times.

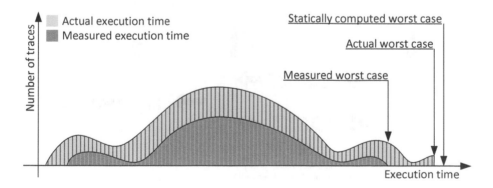

FIGURE 5.16
A graph showing generalization of the observed timing values compared with the actual timing values.

From the figure, we can see that measurement-based analysis usually only samples a subset of all the possible program traces. Therefore, the measured WCET is usually an underestimate of the actual WCET of the program. In comparison, static timing analysis provides a sound and safe overestimation of the WCET of the program.

5.7.1   Static Timing analysis

Static timing analysis is the process of computing the WCET of a system to ensure that all timing requirements are met. Figure 5.17 illustrates a block diagram overview of the static timing analysis process. For IEC 61499 function blocks, the process begins with the XML description of the system. Once the system has been designed and is ready for deployment, the target implementation compiles the XML to a separate binary file (machine code) for each device within the system. Timing analysis is usually performed separately for each device and then for the entire system.

The static timing analysis of a device is split into three processes: *control flow graph (CFG) construction*, *timing analysis of instructions* and *longest path analysis*.

- The *CFG construction* process analyzes the instructions in a binary file and creates a CFG depicting the possible paths through the program. A CFG consists of nodes and transitions. Nodes represent sequential blocks of instructions and transitions represent branching between these blocks of instructions.

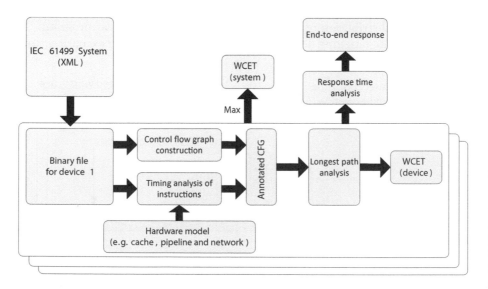

FIGURE 5.17
Overview of static timing analysis.

- The *timing analysis of instructions* determines the execution cost of each instruction in the binary from datasheets and/or hardware models, that is, cache models, pipelines models and network models.

- The *longest path analysis* uses the combined data from the CFG and instruction execution times to determine the path contributing to the WCET of the device.

Once the WCETs of all devices are computed, the WCET of the system can be easily derived if the delayed synchronous design pattern is used. The WCET of the system is simply the maximum WCET of all the devices. Aside from the WCET, another timing property of interest to designers is the end-to-end response time of the system. Defined as the time required to emit a set of outputs in response to a particular combination of inputs. A response time value is computed for each pair of input and output ports using response time analysis. For the design patterns proposed in this chapter, the end-to-end response is simply the WCET multiplied by the number of resources in the event chain from the input events to the output responses. More details appear in [29].

5.8   Conclusion

Design patterns are reusable solutions that solve recurring design problems by using a well defined structure [6, 12]. When presented with a new task, it is typical practice for designers to reuse and adapt proven solutions. While there are many design patterns designed to aid the functional design of IEC 61499 systems [10, 18, 25], there are limited approaches [9, 33] that act as guidelines for the systematic design of safety-critical time-predictable systems.

In this chapter, we proposed new design patterns to aid the design of time-predictable systems. The proposed design patterns provide guidelines for choosing the appropriate semantics based on the application requirements when designing a time-predictable system. We categorize function block models into two categories: intra-resource and inter-resource function blocks. Table 5.3 presents a comparison between the ordered synchronous design pattern and the delayed synchronous design pattern. Both design patterns are time-predictable and provide determinism and reactivity. Formal semantics and associated proofs appear in [19].

Intra-resource function blocks, for example, basic and composite function blocks, are within resources. They are centralized systems; the function blocks shared hardware resources such as the processor and the associated memory. We proposed an *ordered synchronous design pattern* for intra-resource function blocks. The design pattern was inspired by the PRET-C synchronous programming language [3]. This pattern provides time-predictable concurrent

TABLE 5.3

Comparison of Ordered Synchronous Design Pattern and Delayed
Synchronous Design pattern

Property	Ordered Synchronous Design Pattern	Delayed Synchronous Design Pattern
Time predictability	Yes	Yes
Determinism	Yes	Yes
Reactive	Yes	Yes
Suited for	Intra-resource function blocks	Inter-resource function blocks
Communication between function blocks	Within a tick	Delayed by a tick
Response time	1 or 2 ticks	n number of ticks, where n is the number of resources in the event chain
Parallel execution	No	Yes

execution of function blocks and is ideal for centralized systems. Communication between function blocks is instantaneous (within a single tick). The response times are one tick for non-cyclic function block networks and two for cyclic networks that have feedback loops. The duration of a tick is the WCRT of the system. Parallel execution for this pattern is not ideal because a strict execution order is mandatory for the ordered synchronous design pattern.

Inter-resource function blocks are across resources and devices. They are distributed systems in which function blocks are deployed on separate virtual or actual devices. For inter-resource function blocks, we propose the *delayed synchronous design pattern* that provides time-predictable parallel execution of function blocks and is ideal for distributed systems. This design pattern is inspired by [28]. Communication between function blocks is delayed to the next tick. This decouples the communication dependencies between resources and devices. The response time is equal to the number of resources that the input event has to propagate through to compute the corresponding output event. Parallel execution is possible because the resources and devices do not communicate within the same tick.

Although we have developed two different patterns, a combination of both is needed for the execution of distributed function blocks, as highlighted in Figure 5.14. Both design patterns are time-predictable while preserving determinism and reactivity.

In the future, our approach can be extended to match other application requirements to the appropriate execution semantics through new design patterns.

Bibliography

[1] JPL Institutional Coding Standard for the C Programming Language, Mar. 2009.

[2] MISRA-C: 2012: Guidelines for the Use of the C Language in Critical Systems, 2013.

[3] S. Andalam, P. S. Roop, A. Girault, and C. Traulsen. Predictable Framework for Safety-Critical Embedded Systems. *IEEE Transactions on Computers*, Vol. 99, No. 1, 100–114, Feb. 2013.

[4] A. Benveniste, P. Caspi, S. A. Edwards, N. Halbwachs, P. Le Guernic, and R. de Simone. The Synchronous Languages 12 Years Later. *Proceedings of the IEEE*, Vol. 91, No. 1, 64–83, January 2003.

[5] M. Boldt, C. Traulsen, and R. von Hanxleden. Worst Case Reaction Time Analysis of Concurrent Reactive Programs. *Electronic Notes in Theoretical Computer Science*, Vol. 203, No. 4, 65–79, June 2008.

[6] F. Buschmann, R. Meunier, H. Rohnert, P. Sommerlad, and M. Stal. *Pattern-Oriented Software Architecture: A System of Patterns*. John Wiley & Sons, New York, 1996.

[7] G. C. Buttazzo. *Hard Real-time Computing Systems: Predictable Scheduling Algorithms And Applications Second Edition*. Springer, 2005.

[8] J. H. Christensen. Design patterns for systems engineering in IEC 61499. In *Verteilte Automatisierung - Modelle und Methoden für Entwurf, Verifikation, Engineering und Instrumentierung (VA2000)*, pages 63–71, Otto-von-Guericke-UniversitÃđt Magdeburg, Germany, 2000.

[9] G. Doukas and K. Thramboulidis. A Real-Time-Linux-Based Framework for Model-Driven Engineering in Control and Automation. *Industrial Electronics, IEEE Transactions on*, Vol. 58, No. 3, 914–924, March 2011.

[10] V. N. Dubinin and V. Vyatkin. Semantics-Robust Design Patterns for IEC 61499. *Industrial Informatics, IEEE Transactions on*, Vol. 8, No. 2, 279–290, May 2012.

[11] S. A. Edwards and E. A. Lee. The Case for the Precision Timed (PRET) Machine. In *Proceedings of the 44th Annual Design Automation Conference*, DAC '07, pages 264–265, New York, 2007. ACM.

[12] E. Gamma, R. Helm, R. Johnson, and J. Vlissides. *Design Patterns: Elements of Reusable Object-Oriented Software*. Addison-Wesley Longman, Boston.

[13] G. Gebhard, C. Cullmann, and R. Heckmann. Software Structure and WCET Predictability. In *Bringing Theory to Practice: Predictability and Performance in Embedded Systems*, volume 18 of *OpenAccess Series in Informatics (OASIcs)*, pages 1–10, Dagstuhl, Germany, 2011. Schloss Dagstuhl–Leibniz-Zentrum fuer Informatik.

[14] J. Herter, P. Backes, F. Haupenthal, and J. Reineke. CAMA: A Predictable Cache-Aware Memory Allocator. In *Real-Time Systems (ECRTS), 2011 23rd Euromicro Conference on*, pages 23–32, Porto, Portugal, July 2011.

[15] G. J. Holzmann. The Power of 10: Rules for Developing Safety-Critical Code. *IEEE Computer*, Vol. 39, No. 6, 95–97, 2006.

[16] D. E. Knuth. *The Art of Computer Programming, Volume 2 (3rd Ed.): Seminumerical Algorithms*. Addison-Wesley, Boston, 1997.

[17] H. Kopetz. *Real-Time Systems: Design Principles for Distributed Embedded Applications*. Kluwer, Dordrecht, 1997.

[18] G. E. Krasner and S. T. Pope. A Cookbook for Using the Model-View-Controller User Interface Paradigm in Smalltalk-80. *Journal of Object-Oriented Programming*, Vol. 1, No. 3, 26–49, August/September 1988.

[19] M. M. Y. Kuo. *Precision Timed Industrial Automation*. PhD thesis, University of Auckland, New Zealand, 2015.

[20] X. Li and R. von Hanxleden. Multithreaded Reactive Programming - the Kiel Esterel Processor. *Computers, IEEE Transactions on*, Vol. 61, No. 3, 337–349, March 2012.

[21] F. Maraninchi and Y. Rémond. Argos: an Automaton-Based Synchronous Language. *Computer Languages*, pages 61–92, 2001.

[22] M. Masmano, I. Ripoll, P. Balbastre, and A. Crespo. A Constant-Time Dynamic Storage Allocator for Real-Time Systems. *Real-Time Systems*, Vol. 40, No. 2, 149–179, 2008.

[23] A. Roychoudhury. *Embedded Systems and Software Validation*. Morgan Kaufmann, Boston, 2009.

[24] Z. A. Salcic, P. Roop, M. Biglari-Abhari, and A. Bigdeli. REFLIX: A Processor Core for Reactive Embedded Applications. In *Proceedings of the Reconfigurable Computing Is Going Mainstream, 12th International Conference on Field-Programmable Logic and Applications*, pages 945–945. Springer, 2002.

[25] F. Serna, C. Catalan, A. Blesa, and J. M. Rams. Design Patterns for Failure Management in IEC 61499 Function Blocks. In *Emerging Technologies and Factory Automation, IEEE Conference on*, pages 1–7, Bilbao, Spain, 2010.

[26] V. Vyatkin. IEC 61499 as Enabler of Distributed and Intelligent Automation: State-of-the-Art Review. *Industrial Informatics, IEEE Transactions on*, Vol. 7, No. 4, 768–781, 2011.

[27] R. Wilhelm, J. Engblom, A. Ermedahl, N. Holsti, S. Thesing, D. Whalley, G. Bernat, C. Ferdinand, R. Heckmann, T. Mitra, et al. The Worst-Case Execution Time Problem - Overview of Methods and Survey of Tools. *ACM Transactions on Embedded Computing Systems (TECS)*, Vol. 7, No. 3, 36, 2008.

[28] L. Yoong, P. S. Roop, V. Vyatkin, and Z. Salcic. A Synchronous Approach for IEC 61499 Function Block Implementation. *Computers, IEEE Transactions on*, Vol. 58, No. 12, 1599–1614, 2009.

[29] L. H. Yoong, P. S. Roop, Z. E. Bhatti, and M. M. Y. Kuo. *Model-Driven Design Using IEC 61499 A Synchronous Approach for Embedded and Automation Systems*. Springer, 2015.

[30] S. Yuan, S. Andalam, L. H. Yoong, P. S. Roop, and Z. Salcic. STARPro - A New Multithreaded Direct Execution Platform for Esterel. *Electronic Notes in Theoretical Computer Science*, Vol. 238, No. 1, 37–55, 2009.

[31] G. Zhabelova and V. Vyatkin. Multiagent Smart Grid Automation Architecture Based on IEC 61850/61499 Intelligent Logical Nodes. *Industrial Electronics, IEEE Transactions on*, Vol. 59, No. 5, 2351–2362, May 2012.

[32] Y. Zhao. A Model-Driven Approach for Designing Time-Critical Medical Devices . M.E. thesis, Electrical and Computer Engineering Department, University of Auckland, New Zealand, 2014.

[33] A. Zoitl. *Real-Time Execution for IEC 61499*. International Society of Automation eBooks, 2009.

6

Automatic Reengineering of IEC 61131-Based Control Applications into IEC 61499

Monika Wenger

fortiss GmbH

Alois Zoitl

fortiss GmbH

Georg Schitter

Vienna University of Technology

CONTENTS

6.1 Introduction

As we progress toward personalized-production and lot size one, the interconnection level [3] of industrial applications will rise significantly. There will be increased demand for the handling of distributed systems [35] with cooperating devices from different vendors, caused by the integration of more functionality into powerful, embedded microcomputers, attached to different types of objects [3]. Actual production systems already reach the border of programmability [4, 5] caused by the steadily increasing complexity of control systems [24]. In many industrial maintenance efforts, the efforts for the integration of new functionality and the efforts for changes within existing systems are becoming uneconomical [4]. Therefore future systems require better adaption to the ever-changing environment conditions, as well as an easier handling of the increasing complexity and the higher levels of interconnections. Hardware aging, additionally functionality and the need for more flexibility are forcing the companies to change their systems. But new components are not always compatible with current control systems and therefore increase overhead in managing the different control software systems. To counteract the increasing management of different control software systems and increase the flexibility of the current International Electrotechnical Commission (IEC) 61131 standard-based solutions, the control software could be replaced as a whole. The great disadvantage of a replacement as a whole is the fact that the companies already invested lots of money on developing their current IEC 61131-based control systems and therefore usually do not want to start again from scratch. For current systems to remain competitive, an automated reengineering of existing control applications into systems based on new technologies such as distributed control systems [44] is required. Within this chapter, an automatic reengineering approach is presented, which is realized for IEC 61499 as the target technology and is applied to the control code of a sorting station.

6.2 IEC 61131 versus IEC 61499

Currently most control applications are based on the IEC 61131 standard. The standard was developed for centralized systems, which is reflected by its language structure, especially its cyclic execution schema, its communication infrastructure and the support of global variables. Of course it is possible to represent distributed systems also with the methods of IEC 61131 and use global variables for synchronization and communication but at the cost of reduced configurability, flexibility and maintainability. For flexible automation systems more and finer modularity is required that can hardly be managed by the modeling capabilities of IEC 61131. The object-oriented (OO) concepts introduced by the third edition of IEC 61131 did not address these topics, but mainly caused a shift from a domain-specific language for the automation domain to a general-purpose programming language (GPL). Instead the development of richer and more powerful programming languages only leads to mechanically and mentally unmanageable languages [12].

The past shows that "*it would have been better to use really new and improved tools which, incidentally, do not necessarily have to break with all traditions. With technologies which really deserve to be called new technologies we will benefit from slimmer and more reliable software, which is easier to use and to maintain*" [6]. IEC 61499 [30] as a progression of IEC 61131 promises to be such a new technology. Instead of a cyclic execution control and a monolithic design, IEC 61499 provides event-driven execution control for the design of distributed applications. IEC 61499 also promises safe and reusable code by two principles. First, it does not provide global variables and therefore avoids creating hidden dependencies between software modules, which has been identified as some of the most difficult and expensive software defects [14]. Furthermore it provides strong data encapsulation by *algorithms*, which are able to force inevitable range checks of a specific variable [14]. The different function block (FB) types provided by the IEC 61499 standard can be considered as single software components [21], since they conform to a component model and a standardized interface (provided and requested) without hidden interface elements. They can also be deployed independently and be composed without modification and without side effects from the composition, since the component model is able to explicitly describe the context dependencies. IEC 61499 FBs can be reused on any platform and therefore are flexible. Since each component handles its own faults, the system allows easy fault localization and therefore reduces commissioning and maintenance cost.

Besides that IEC 61499 also provides platform independency by offering eXtensible Markup Language (XML) as a storage format for all design artifacts and device management protocols, which allows the development of programmable logic controller (PLC) code that is independent of the hardware platform [30]. Due to the introduction of service interface function blocks (SIFBs), it includes a mechanism for intellectual property protection since instead of supplying the full source code for an FB an XML representation containing only the interface and a behavior description can be passed over [36]. Additionally IEC 61499 offers common administration of devices from different vendors and control software. Due to its architecture and mapping functionality, IEC 61499 allows the development and the maintenance of a whole plant's control software within one tool. The ability to distribute the control software to available devices provides the required flexibility for future plants [17, 8]. Since it supports online reconfiguration [45] it also provides the adaptability required for future systems [3]. These characteristics of IEC 61499 enable it to handle Industry 4.0 ready automation systems.

To support the creation of future plants, IEC 61499 has been chosen as target technology for developing automatic reengineering of IEC 61131-based control applications. After analyzing existing tools and conversion approaches, the reengineering process is introduced. Basically the reengineering process is applicable on any target language. The proposed reengineering process is then applied on the control code of a sorting station, demonstrating the equivalence of the source and target system's execution behavior by dynamic software testing with the physical plant.

6.3   Related Work on Reengineering

At the end of the 20th century companies allocated about 80% of their available programming resources to maintain and reengineer existing code [9]. In the beginning of the 21th century this percentage grew to over 90% in many large companies [28] due to the greater complexity of the software projects and the rapid evolution of new technologies. This led to the development of reengineering tools and conversion concepts within the software and automation domains.

6.3.1   Existing Reengineering Tools

Refine^{TM} was a commercial tool that could be configured for a desired reengineering task in cooperation with its vendor, Reasoning Software, Inc. in the United States. Its language processing system supported Refine, Ada, C, COBOL, IBM JCL, SQL, and SDL to build software management and reengineering tools. Refine has been used as a basis for some blackbox commercial automated migration projects [2], like a prototype for an automated Natural 1.2 to Natural 2.0 conversion that handled several key incompatibilities between the two language versions [9].

Akers et al. [2] introduced Design Maintenance System (DMS) from Semantic Designs Inc., a commercial program analysis and transformation system. They used DMS to construct a tool for modernization of a large C++ industrial avionic system from 1990 to one based on a proprietary variant of the Common Object Resource Broker Architecture (CORBA), preserving functionality but introducing regular interfaces for inter-component communication.

Heidenreich et al. [20] introduced Java Model Parser and Printer (JaMoPP) for reengineering of Java code. JaMoPP allows the representation of Java code in the modeling space where all statements and arguments are present. Reengineering operations, like model analysis and transformation approaches, can therefore be performed also in the modeling space.

Within the automation domain, Itris Automation, a the french company, provides a PLC Converter able to convert a single project from Schneider-

Electric (SMC, Orphée, Série 7), Rockwell Automation (SLC500, PLC5) or Siemens (S5) to a specific PLC (Quantum, Premium , M340, M580) for Schneider Electric Unity (effective 2014). For unknown artifacts it uses a knowledge base, which allows a manual matching with available elements.

Most of the available commercial tools recommend support, since there is a *"significant learning curve in building transformation-based tools"* [2]. These tools do usually not inherently support domain-specific languages of the automation domain or are restricted to special types of PLCs as input and output platforms.

6.3.2   IEC 61131 to IEC 61499 Migration Approaches

For the replacement of monolithic PLC applications with IEC 61499 applications, Hussain et al. [22] present a manual migration of PLC control applications written in signal interpreted petri nets (SIPN) to IEC 61499. The SIPN description of the application is used to generate IEC 61131 programs and is supposed to be extended by an IEC 61499 generator based on their migration rules. Usually control applications are not available as SIPNs but already exist as IEC 61131 applications, which require a previous migration from IEC 61131 to SIPN.

Riedl et al. [29] provide an approach to represent sequential function charts (SFCs) by IEC 61499 FB networks. To rebuild SFC applications as FB networks the authors introduce two FBs, FB_Step and FB_Split_2. FB_Step represents a transition with one subsequent step whereas FB_Split_2 represents two alternative transitions executed successively. The control flow of an SFC is therefore represented by an event flow. The algorithms executed after activation of a specific step are not taken into consideration which implies their re-implementation in their IEC 61499-based runtime environment (RTE) distributed object model environment (DOME) language or C++.

To provide a first step toward manufacturing independence of medium-sized companies, Gerber et al. [15] present six rules for the transformation of IEC 61131 FBs to IEC 61499. The rules focus on a manual transformation of some IEC 61131 FBs since they are the bricks of a distributed control system. The transformation rules derived from the migration of these FBs consider the definition of IEC 61499 FB interfaces and also the separation of algorithms within basic function blocks (BFBs). These rules do not cover the transformation of a whole project and are not detailed enough to support an automatic transformation for a whole project.

Shaw et al. [32] present a semi-automatic reengineering of program organization units (POUs) implemented in a Rockwell Automation ladder diagram (LD) that enables industrial automation manufacturers to remain competitive and produce flexible and re-configurable systems. The POUs are automatically translated into RTE specific C code, which results in several header and source files containing data types, variable declarations and LD routines as C function equivalents. The generated C code has to be encapsulated manually into a proper FB. POUs with state machine behavior therefore have to be manually embedded into the execution control chart (ECC) of a BFB. For more complex state-based LD routines, they extended the ECC definition of IEC 61499 by hierarchical ECCs that cover subroutine calls and have to be developed manually as well. Composite function blocks (CFBs) are used to encapsulate related behavior. The interfaces of these FBs as well as the FB network are not generated but derived manually from system-level understanding achieved by studying a specification document. Global variables and variables scoped to a specific FB are integrated by header files that are specified in the `CompilerInfo` tag of an FB. As suggested by Vyatkin et al. [37], they also encapsulate directly represented variables in SIFBs to offer an easy replacement for hardware changes. The main drawbacks are its restriction to a particular RTE due to the C-code generation, the lack of compliance with the IEC 61499 standard due to the use of hierarchical ECCs, and the amount of manual rework due to the lack of an automatic IEC 61499 system structure generation.

Dai et al. [11] introduced a migration concept to manually transfer centrally controlled baggage handling systems based on IEC 61131 to distributed control systems, according to IEC 61499, to overcome the insufficient PLC performance in an airport baggage handling system, when trying to deploy distributed control and providing sorting and malfunction detection functionality for each PLC. They suggest an *object-oriented conversion approach* for processing control. Each device is represented as an FB and therefore the original PLC code is translated into a state machine which subsequently is represented by an ECC of a BFB. For processing control code where state machines cannot be easily recovered and control code parts retain sequential logic they suggest an *object-oriented reuse approach* which inserts the behavior of the state machine together with the sequential logic as structured text (ST) code within one algorithm of a BFB's ECC. For handling logistics systems or systems with heavy data processing they suggest a *class-oriented approach* where BFBs represent a single service of a plant.

None of these migration approaches supports an automated migration of whole IEC 61131 control applications into IEC 61499. They are either limited on a specific RTE, and only consider a few but very specific conversion rules or very general rules sufficient for manual migrations.

6.4 Developed Reengineering Process

The intended reengineering process is supposed to produce a vendor-independent format for distributed systems according to the IEC 61499 standard. The independent format concerns engineering tools as well as different types of PLCs. The process is also supposed to combine different IEC 61131 input projects within one target project.

To automate language migration, *"program transformation is often the most effective technology because the necessary changes are structural (rather than textual) in nature, and must be made pervasively across large bodies of application source code"* [9]. For a transformation, the application has to be prepared so that the structural information is accessible by a transformation engine. The representation of the subject system in a structured form is the primary task of reverse engineering.

To improve the interoperability between PLC vendors, PLCopen provided an XML [34] interchange format in 2005, which is still not supported by every IEC 61131-3 vendor and therefore cannot be used to reduce the number of IEC 61131-3 dialects. There are as many IEC 61131-3 dialects as there are structure descriptions. The introduction of the third edition of the IEC 61131-3 standard in 2013 tries to address the increasing complexity for programming automation systems by language evolution and introduces OO extensions like inheritance, references and classes. Since all features defined within the IEC 61131-3 standard are optional [27] vendors usually only support subsets of the standard definitions, which in turn increases the number of IEC 61131-3 dialects. To simplify the programming according to their means, vendors also add their own definitions or features. All of these vendor-specific definitions create new IEC 61131-3 dialects which make portability of control applications even more difficult [46].

Due to the different IEC 61131-3 dialects, we propose an intermediate step instead of a direct migration from the source to a specific target language. This also allows us to combine several control applications within one target project. To finally migrate applications, a type of analysis and filtering of valuable information to establish *"the structural relations is indispensable"* [20]. Wright [43] noted that the library elements provided by IEC 61131-3 tools are not always present on all IEC 61131 platforms and are not implemented equally, resulting in different behavior. This circumstance is valid between IEC 61131-3 tools and also affects the libraries of IEC 61131 and any desired target language. In order for the library differences to be resolved, a mapping between the library elements of both the source and the target platform is needed. The previously described requirements result in the reengineering process illustrated in Figure 6.1.

The reengineering process consists of three major steps. The first step contains the reverse engineering, which is divided into a code-to-model (C2M) step that provides the structural information in terms of a vendor-specific model and a model-to-model (M2M) step which converts the vendor-specific model to an intermediate, vendor independent model. The second step considers the required library match that maps the source project's library elements to those of the target library. The last step concerns the language migration which converts the intermediate model to the desired target model and makes use of the library match model before it serializes the target model.

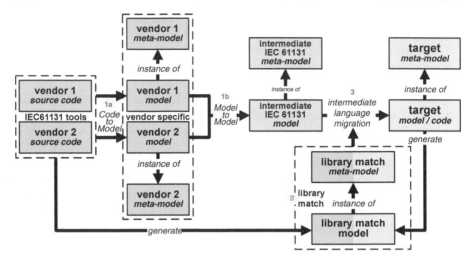

FIGURE 6.1
Reengineering process for the automatic migration of legacy control systems
based on the direct migration presented in [42].

This model-driven approach is chosen since the technologies provide support
for element search and element collection without the requirement to iter-
ate through all elements of a model. This simplifies the implementation and
maintenance of the transformation rules. Besides that it enables the use of
language development frameworks, like Xtext [13] or EMFtext [19], which
provide a complete language infrastructure from parsers over linker, compiler
or interpreter, and access classes generated from models. The second step of
the reverse engineering process and the language migration have been realized
by transformation rules implemented in the transformation language Xtend of
the former openArchitectureWare (oAW) [16]. The three reengineering steps
and their corresponding models are explained in the following section.

6.4.1   Reverse Engineering Process

The two-stage reverse engineering process reads the legacy control codes of
different vendors to generate a vendor-independent, intermediate IEC 61131
model. Since there are many vendors using their own dialects of the IEC 61131
languages, the C2M step results in several vendor-dependent IEC 61131 models. Due to this vendor dependency, it is necessary to select the IEC 61131
vendors that are supposed to be used for demonstration issues rather early.
Among the various vendors and their IEC 61131 tools, the Smart Software
Solutions GmbH (3S), that provides CoDeSys V2.3 [1], has been chosen for
its integration into the reengineering process since it is a common tool within
the domain and provides a pure textual source. For CoDeSys V2.3, a pro-

prietary meta-model based on the structure of the intermediate model has been defined and extended to integrate the hardware structure which consists of several modules and module specific-parameters. The hardware model has been used to reconstruct the required hardware connections between the application and the hardware access developed for each module. The vendor-specific meta-models define the structure of the code and are used within a language development framework to generate parsers and printers and also to define and process the transformation rules of the M2M step.

Vendor-specific transformation rules are processed to convert a vendor-specific model into the intermediate model. The semantics of the different models are therefore defined by the combination of the meta-model with the transformation rules. Since there are several IEC 61131-3 dialects there are as many vendor-specific rules as the number of dialects that are supposed to be integrated into the process. All sets of rules are supposed to compensate variations of IEC 61131 standard elements. Within CoDeSys V2.3 control applications, for example, the handling of network variables is of major interest. Network variables are CoDeSys V2.3-specific variables, which are used to communicate between two PLCs. They are defined as global variables and are declared within the body of any POU that uses them. The IEC 61131-3 standard does not envisage network variables but instead provides access variables. Therefore one CoDeSys V2.3-specific transformation rule is the mapping of its network variables to IEC 61131-3 standard access variables.

The M2M step's result is the intermediate IEC 61131 model that represents the language definitions of the IEC 61131 standard and is supposed to provide a vendor-independent format. It is based on the extended Backus Naur form (EBNF) [23] description of the IEC 61131 standard, which provides the basic architecture of IEC 61131 and contains definitions for letters, digits, identifiers and literals, all possible data types and variable types, as well as definitions for *configurations* and *POUs*. The EBNF description also contains three of five IEC 61131-3 languages namely ST, instruction list (IL) and SFC whereas it leaves out the graphical languages function block diagram (FBD) and LD. Within this work a meta-model is realized for the basic architecture as well as for IL, SFC, ST and FBD, based on the results presented in [39, 40]. The resulting meta-model has been extended to enable reuse of its elements within the library match model. The FBD elements missing in the IEC 61131 standard's EBNF definitions have been added as well. These FBD elements are based on the XML interchange format provided by PLCopen and also on the definitions of an FB network provided by IEC 61499.

6.4.2 Library Match

Existing IEC 61131 control applications often use standard library blocks or restricted vendor blocks. Usually there is no source code accessible for these blocks as C/C++ code or in any of the IEC 61131 languages. Such restricted blocks, which cannot be decomposed any further and where no source code is available, are at the end of the call hierarchy. Also directly represented variables of IEC 61131 belong to these types of blocks since they perform hardware access by sending or receiving signals, whereas they hide their actual implementation for hardware access. To assign any of these blocks to a corresponding block of the target system, a library match model has been designed based on [41].

The library match model assigns each IEC 61131 library element to a corresponding IEC 61499 one while it also serves as lookup table for name differences of corresponding library elements and for interface differences. In contrast to the use of transformation rules, the library match model is able to compensate non-generalizable differences. Type differences between the source and the target library elements also do not have to be calculated by searching in the call history since the interface types are contained within the library match model. Undefined POUs on the target side can also be detected during the migration process, while the library match model matches the library elements of different tools.

The library elements of each source vendor and those of the target platform are described within their own library files, whereas these elements are reused within the library match model in terms of references to the blocks of the different library files. The library match model is currently built manually and therefore requires some initialization effort. But this initialization effort is only required if library elements are missing or a new vendor with its own library elements is supposed to be integrated. In this case, the missing library elements have to be added to the vendor-specific library model and referenced within the library match model. In addition they have to be reimplemented for the desired target tool. The library match model eventually is used as input for the language migration.

6.4.3   Language Migration

The language migration is the last step of the reengineering process. It transforms the intermediate model into the desired target model. Since we chose IEC 61499 for demonstration issues, its meta-model represents the language definitions provided by the IEC 61499 standard. The standard provides an EBNF and a document type definition (DTD) for the design of the meta-model. The DTD can be converted to a corresponding XML schema definition (XSD) which can directly be used as meta-model. It therefore provides a faster integration into the proposed reengineering approach while it also directly supports the XML persistence recommended by the standard. Therefore the provided XSD has been used to generate an Ecore [26] meta-model and its corresponding model access classes. Basically neither the conversion of the XSD to Ecore nor the generation of the model access classes, also called static implementation, is necessary, since the chosen transformation language supports XSD models and is able to work with dynamic implementations. Even if the static implementation is not required and has to be maintained as the model evolves, it is preferred for two reasons. First, it needs less memory and provides faster access to the data [25] while it also simplifies the definition of the transformation rules by supporting code completion within the Xtend editor and by providing the model code as programming reference. Second it is preferred since the target meta-model has to be extended to support the library match meta-model and to implement an IEC 61499 sufficient meta-model part for the five languages of the IEC 61131 standard. This is because the IEC 61499 standard only refers to these languages but does not provide specific descriptions of all required meta-model elements. On the one hand these extensions require the use of non-containment references that are supported by Ecore models and not by XSD models. On the other hand the definition of such a linking mechanism requires equal meta-model types. In order to perform the language migration, general and POU transformation rules are needed. Additionally model queries are used to support the transformation process in terms of the library match, analysis functionality and design decisions.

6.4.3.1 General Conversion Rules

The general conversion rules address the transformation of the basic IEC 61131 architecture to elements of the basic IEC 61499 architecture. Sünder et al. [33] present two general conversion approaches, the name-driven and the execution-driven conversion, as shown in Table 6.1. The use of the execution-driven or the name-driven concept depends on the elements provided by the vendor and on the concrete structure of the control application. While the execution-driven concept is aligned on the executing elements, the name-driven concept is aligned on a name consistency of the IEC 61131 and the IEC 61499 standard.

Since the execution-driven concept is aligned on the executing elements, its main weakness concerns the handling of configurations as they are summed within one system and therefore disappear during the conversion. Each configuration represents one PLC, while in IEC 61499 PLCs are represented by devices. Since all resources$_{61131}$[1] are converted to separate devices, this leads to several devices containing equal Internet Protocol (IP) addresses, even if a device is defined as an *independent physical entity* [30]. As a consequence, the execution-driven concept better suits vendors that support only configurations with one resource or for control applications that use just one resource but lots of different tasks.

The name-driven concept is aligned on a name consistency. Its main weakness is the handling of tasks since several tasks might be mapped into one application, which might result in execution conflicts. As a consequence, the name-driven concept is better suited for control applications that contain several configurations and resources$_{61131}$ but only one or very few tasks per program.

---

[1]Resource$_{61131}$ denotes an IEC 61131 resource.

TABLE 6.1
Name-Driven and Execution-Driven Element Mapping Based
on [33]

IEC 61131-3	IEC 61499 Name-Driven	IEC 61499 Execution-Driven
Projects	System	—
Configurations	Devices	System
Resources	Resources	Devices
Tasks	Applications	Resources
Programs	Applications	
Functions, Function Blocks	Function Blocks	

Besides the general architecture, Sünder et al. [33] also present transformation rules for tasks, access variables, global variables and configuration variables. Each task is represented by an FB network consisting of an E_CYCLE FB for periodic execution and an E_SWITCH for starting and stopping the cycle, as well as an SIFB reacting on the rising edge of a Boolean variable for event-triggered execution. The desired Task FB network triggers the entire control algorithm, whereas at the beginning of every cycle any hardware input is read and at the end of every cycle every hardware output is written. Within this work we split the FB network into three different FB networks, one for each type of task, cyclic, periodic and event triggered, to reduce the number of additional FBs within an application.

Access variables provide inter PLC communication. In IEC 61499 the interface of each FB that internally accesses such an access variable is extended. The interface elements that represent the former access variable are then connected through PUBLISH and SUBSCRIBE SIFBs. Global variables and configuration variables are represented as SIFBs, whereas former configuration variables have to be initialized during the initialization phase of the application and former global variables have to be updated whenever their value changes.

6.4.3.2 POU Conversion Rules

The interface of any POU supports input variables, output variables, internal variables, temporary variables, external variables, input-output variables and access paths. Since IEC 61499 uses the data types of IEC 61131, input and output variables can be transferred to every FB type, whereas internal variables are only provided by BFBs without further effort and temporary variables have to be handled like internal variables and must be reset before FB execution. External variables, input-output variables and access paths all require input and output interface elements that allow them to update the values on each change.

POUs can be programmed within two textual languages, ST and IL, as well as three graphical languages FBD, LD and SFC. The textual languages are preferable be integrated into algorithms of BFBs or, in case of internal

FB calls, as a combination of BFBs and CFBs, since BFBs and SIFBs are not able to instantiate other FB types internally.

Since the demonstrator used within this work is implemented with ST, only rules for this language are summed up. Based on the structure of the ST, code a BFB or a CFB is created. The ST code is therefore added to an algorithm of the BFB's ECC. POUs that internally invoke other POUs or update the value of any access or global variable need to be split into several Algorithms, as illustrated in Figure 6.2.

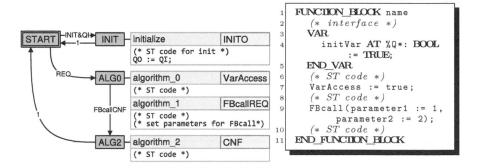

FIGURE 6.2
Execution control chart (left) for structured text code with access variables and function block call (right).

The interface of the BFB is therefore extended by additional elements for updating access or global variables and invoking other FBs. Read and/or write access of any access or global variable is achieved by an additional event data input and event data output pair, respectively. For every FB invoked a request and confirm event is added to the interface as well as the mirrored interface of this POU instance. The mirrored interface allows us to set and get the parameters of the FB to be called. For ST code that indicates initialization, an initialization part is added to the ECC, as shown in Figure 6.2 at the top. The interface is extended by initialization events connected to relevant data inputs and outputs.

6.4.3.3   Model Queries

To collect information for the existence or generation of a specific pattern, as described in [40], or decide whether a target pattern is supposed to be generated, model queries are defined. Two types of model queries are distinguished within this work. *Information collecting queries* check for the existence of a specific pattern. There are internal and external *information collecting queries* which internally perform an information collection or call other *information collecting queries*. *Decision making model queries* are based on metrics to automatically decide how to generate the target model. That way the target model can be optimized in terms of smaller and easier readable FB networks for example. These queries consist of an information collecting part and a comparison part. The first part collects the information from the source model according to the related metric by external or internal *information collecting queries*. The second part compares the collected information to a predefined and empirically determined threshold of the corresponding metric. Based on the result of this comparison, a specific model transformation rule is chosen. *Decision making model queries* are used to check whether initialization is required or ST has to be represented by a BFB or a CFB, for example. The calculation for this decision is achieved by comparing the lines of code (LOC) of the different statement types. To make a distinction between BFB and CFB the following metric is used:

$$(LOC_{for} + LOC_{while} + LOC_{repeat} + LOC_{case} > 0) \vee$$
$$(LOC_{statement} - LOC_{FBinvocation} > LOC_{FBinvocation}).$$

6.5   Proof of Concept

To show the applicability of the proposed reengineering process a sorting station as illustrated in Figure 6.3 (left) is used. The machine sorts three types of objects that are transported on pallets by a conveyor belt. It consists of an identification station and two handling units. Each station is equipped with a sensor that detects the presence of a pallet and a stopper which has to be deactivated for releasing a pallet.

6.5.1   Sorting Station Structure

The identification station scans the transported part by a simulated color sensor (manual set of a specific color input). After identification, the pallet is released. If both handling units are inactive it is assumed that both subsequent stations are free. Based on the identified color, the part passes through the system or is sorted out by one of the two handling units. Each inactive handling unit deactivates its stopper so that unsuitable parts can pass through the system, whereas an active handling unit activates its stopper as long as the manipulation is processed. The manipulation is performed by two Festo pneumatic drives that allow horizontal and vertical end-to-end position movements, indicated by four position sensors as well as a vacuum gripper.

Each station is controlled by one Festo CPX-CEC PLC, programmed with CoDeSys V2.3 in ST. The CPX-CEC PLC has been used since it supports both an IEC 61131 RTE and there also exists an IEC 61499 RTE porting. CoDeSys V2.3 has been used as an IEC 61131 engineering tool, since it is supported by the CPX-CEC PLC. Since each of these projects contains a program PLC_PRG, they do not require the specification of a task and are

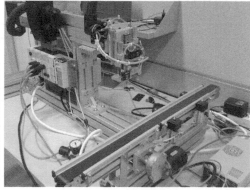

FIGURE 6.3
Structure of the sorting application based on [18] (left) and hardware implementation (right).

all executed cyclically. All three projects, the identification station and both handling units, are reengineered by the proposed method. The sorting station has been used within the Odo-Struger-Lab at the Automation and Control Institute (ACIN) of the Technical University of Vienna for demonstration within a funded project. Due to teaching activities within this lab, components of the sorting station, like the color sensor, were needed for other stations and the available space was limited. Therefore only the first handling unit was available for test, as illustrated in Figure 6.3 (right). The second handling unit is therefore reengineered but not downloaded on a PLC for testing purposes. Since the second handling unit behaves as the first one with the exception of sorting another color, it can be assumed that the behavior of the reengineered result equals the source behavior if the behavior of the first handling unit equals the source behavior.

6.5.2   Reengineering Result

The projects of the sorting station have been processed by the proposed reengineering method. Table 6.2 lists the elements of the source projects related to the corresponding target elements that are automatically generated or added during the language migration process.

   The target elements are generated according to the name-driven transformation rules to better retain the hardware structure. The target model therefore contains one system with three devices, one for each of the source project's configurations and tree applications, one for each program.

   Each application is mapped to one resource$_{61499}$, which has been added to one of the devices. All applications contain an FB network that represents the cyclic execution behavior of the task and reads the hardware inputs, indicated by the PLC modules in Table 6.2, executes the desired algorithm

TABLE 6.2
Reengineering Results[3]

IEC 61131-3 Elements	IEC 61499 Elements
Configurations: *not named*	Devices: *Configuration_0, Configuration_1, Configuration_2*
Resources: *not named*	Resources: *Resource_0, Resource_1, Resource_2*
Cyclic task	FB network
Programs: *PLC_PRG*	Applications: *PLC_PRG_0, PLC_PRG_1, PLC_PRG_2*
Access Variables: *SORT1_start, SORT2_start, SORT1_complete, SORT2_complete*	Interface variables connected through PUBLISH/SUBSCRIBE
User defined FBs: *Identification, HandlingUnit*	BFBs: *Identification, HandlingUnit*
Library FBs: *AND, OR, TOF*	Library FBs: *F_AND, F_OR, FB_TOF*
PLC modules: *DI8, DO4, MPA1S*	Library SIFBs: *DI8, DO4, MPA1S*

afterward, and writes the hardware outputs before it again triggers the read of the hardware inputs. The original execution order is therefore represented by a corresponding event chain. The user-defined programs and FBs within the source projects are identified to be best implemented as BFBs by the model queries. Since the programs all call subprograms in terms of user defined FBs and also use several access variables, it is not possible to just insert the program's code into one algorithm but to split them into several algorithms, as described by the POU conversion rules. Figure 6.4 shows the resulting IEC 61499 application for one of the two handling units, consisting of communication (SUBSCRIBE, PUBLISH), hardware access (DI8, DO4, MPA1S) and entire control application (PLC_PRG_0, HandlingUnit) FBs, triggered in a loop to represent the cyclic Task of the IEC 61131 application. The structures of the other two stations are similar, since they also contain communication, hardware access and entire control application FBs, executed in a loop.

6.5.3 Source and Target Behavior Comparison

From the currently available industrial automation software compliant to IEC 61499 [10], the framework for distributed industrial automation and control (4DIAC) [47] was chosen as target tool, since it has been developed by fortiss with Profactor, ACIN of TU Vienna and AIT. The reengineering result was imported into the IEC 61499 tool, 4DIAC-IDE, and the 4DIAC runtime environment (FORTE) was started on both of the Festo CPX-CECs. *Configuration_2*, for the identification station was downloaded to PLC 2 and *Configuration_0*, for the first handling unit, to PLC 1. Based on the sorting

---

[3]Note: Names and types are italicized.

```
1   PROGRAM PLC_PRG
2   VAR
3     active : BOOL;
4     gripper : HandlingUnit;
5   END_VAR
6   VAR
7     sensorPresence AT %I*: BOOL;
8     sensorStopper AT %I*: BOOL;
9     stopper AT %Q*: BOOL;
10  END_VAR
11  SORT1_start;
12  SORT1_compl;
13  IF SORT1_start=TRUE AND SORT1_compl=FALSE THEN
14    active := TRUE;
15    stopper := TRUE;
16  END_IF
17  IF active=TRUE AND sensorPresence=TRUE AND sensorStopper=FALSE THEN
18    gripper(active := active);
19    IF gripper.complete=TRUE THEN
20      SORT1_compl := TRUE;
21      active := FALSE;
22      stopper := FALSE;
23    END_IF
24  END_IF
25  IF SORT1_start=FALSE AND SORT1_compl=TRUE THEN
26    SORT1_compl := FALSE;
27  END_IF
28  END_PROGRAM
```

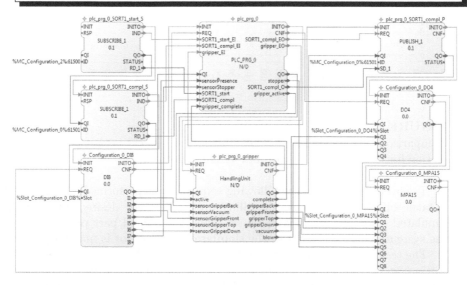

FIGURE 6.4

Original handling unit program in structured text (top) and IEC 61499 reengineering result of the handling unit (bottom), consisting of communication (SUBSCRIBE, PUBLISH), hardware access (DI8, DO4, MPA1S) and entire control application (PLC_PRG_0, HandlingUnit) function blocks.

functionality, several test cases have been defined, similar to the approach presented in [38].

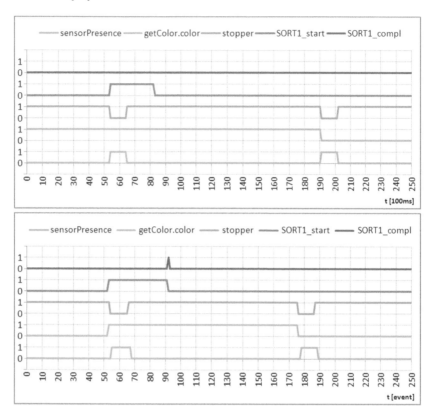

FIGURE 6.5
Comparison of the variable changes during the test case execution of the ID station for CoDeSys V2.3 (top) and 4DIAC (bottom).

To cover all test cases the following sequence was performed: the first part is taken off the pallet by the handling unit, where the desired input of the simulated color sensor is manually set, then the second part passes through the system, where the input of the simulated color sensor is manually reset.

The test cases were performed for the CoDeSys V2.3 version of the sorting station as well as for the reengineered 4DIAC version. During these tests relevant inputs and outputs of the ID and the HU station were measured. The reengineered version has been extended by a CVS_WRITER FB, triggered by the stations' main FBs (former programs), to record the desired values. These measurements are shown in Figure 6.5 and Figure 6.6.

The detection of the pallets is represented by the two peaks of the sensorPresence. The first of the two pallets contained a part to be sorted out of the system, whereas the second one is passed through the system together

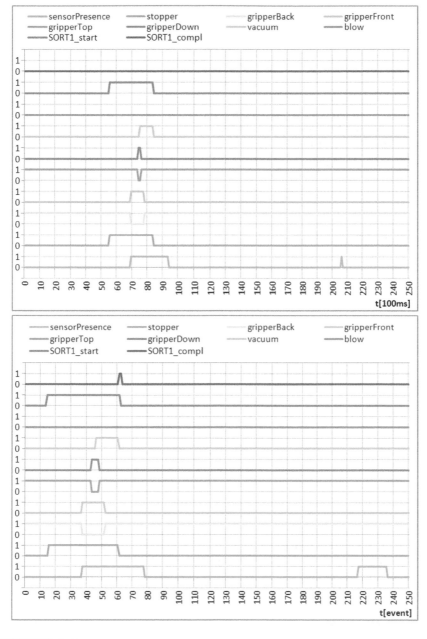

FIGURE 6.6
Comparison of the variable changes during the test case execution of the HU station for CoDeSys V2.3 (top) and 4DIAC (bottom).

with its part. The different test cases can be identified by the movement of the gripper, illustrated by the values of gripperBack, gripperFront, gripperTop and gripperDown and also by its drag and drop through the vacuum output and the pallet stopper by the stopper output (Figure 6.6). The time span of the peaks differs between the CoDeSys V2.3 and the 4DIAC implementations since they are recorded at a sampling rate of 100 ms and also recorded according to the number of asynchronous CNF events sent by the main FBs. This is especially reflected by the second detection of a pallet through the sensorPresence (Figure 6.6) where the HU's main FB is executed about 18 times until the pallet has passed the presence sensor.

In CoDeSys V2.3 the blow output (Figure 6.6) never gets true even if it is set and reset within the application, which also holds for the network variable SORT1_compl. This behavior results from the PLC's ability to execute one statement in about 5 μs. Within 4DIAC no peak for the blow output can be observed since between its set and reset no CNF event is sent, whereas the change of the variable SORT1_compl can be observed since a value change of this variable is associated with an event.

When comparing the results of the identification station, a difference of the getColor.color value can be observed. For the source application, the value of getColor.color is true before a pallet is detected, whereas for the target application it is not true until a pallet is detected. This behavior of the source application occurs due to the simulated color sensor and since the application is initialized by the values of the hardware inputs and outputs (IOs) directly after download, whether the application is started or not. Even if this does not change the observable behavior of the plant, it shows a difference between the initialization of the source and the target application.

Within the target application, the output values of an FB, affected by hardware initialization is not updated during the initialization phase. To resolve this difference, proper assignment statements need to be added to the initialization algorithm. This behavior difference caused no visible difference in the system behavior, and since it was caused by the simulated color sensor it cannot be observed for a real color sensor. In summary the proposed reengineering approach resulted in the successful automated transformation of the IEC 61131-3-based sorting station into an IEC 61499-based implementation in 4DIAC that shows the same behavior.

6.6 Conclusion

In this chapter a model-driven reengineering process for an automatic conversion of control applications implemented in IEC 61131 into a control application implemented in IEC 61499 is presented for CoDeSys V2.3 ST. The ST code is analyzed by model queries to either generate a CFB with an FB network or a BFB with an ECC, dependent on its structure. The reengineering results are evaluated by the control application of a sorting station, which is executed in 4DIAC. The whole process is performed automatically with exception of the library match, which is built manually and therefore requires some initialization effort. For the sorting station, neither the library match model nor the target tool's library had to be changed, since it could be reused from a previous test application.

The proposed reengineering approach currently supports CoDeSys V2.3 ST projects and logi.CAD FBD projects. Additional languages and vendors can be integrated by adding a corresponding reverse engineering step, which is two- or one-phased based on the source code. More research to automate the initial library match and to improve the model analysis as well as the corresponding model queries is worthwhile.

Another aspect concerns the scalability. With about 1200 elements generated from about 1275 LOCs, including information about the hardware configuration, the demonstration example contains relatively few objects, but large software models can consist of up to 10^9 elements [31]. It has been shown that EMF is able to handle model sizes with $3.6 \cdot 10^6$ objects for tasks such as create, modify, traverse, query, and partial load, very well [7]. For larger model sizes it could be considered to reduce the model elements by redesigning the meta-model and/or using a database to (un)load single objects on demand together with fragmentation techniques to improve time efficiency [31].

To the best of our knowledge the proposed reengineering process is currently the only one able to convert a whole IEC 61131 project from its configuration though implementation of its POUs into an equivalent IEC 61499 project. Furthermore it produces a tool vendor-independent and platform-independent format usable for any PLC supporting the execution of an IEC 61499 RTE with a functionally adequate implementation of vendor-specific library elements and providing the required number of IOs as well as a proper communication infrastructure. It is even able to combine several IEC 61131 projects. Therefore the resulting control application is no longer composed of several single projects but only one IEC 61499 project for the whole system that distributes its applications to a corresponding number of PLCs.

Bibliography

[1] 3s Smart Software Solutions GmbH. Codesys v2.3.

[2] R. L. Akers, I. D. Baxter, M. Mehlich, B. J. Ellis, and K. R. Luecke. Re-engineering C++ component models via automatic program trans-

formation. In *12th IEEE Working Conference on Reverse Engineering (WCRE)*, *Pittsburgh*, pages 7–11. IEEE Computer Society, 2005.

[3] Arbeitskreis Industrie 4.0. *Umsetzungsempfehlungen für das Zukunftsprojekt Industrie 4.0*. Forschungsunion im Stifterverband für die Deutsche Wirtschaft e.V., Berlin, Oct. 2012.

[4] Arbeitskreis 'TB Konzept Normung zu Industrie 4.0' des DKE-Fachbereichs 'Leittechnik' (FB 9). *Die Deutsche Normungs-Roadmap Industrie 4.0*. DKE Deutsche Kommission Elektrotechnik Elektronik Informationstechnik im DIN und VDE, Frankfurt am Main, 1.0 edition, Dec. 2013.

[5] F. Basile, P. Chiacchio, and D. Gerbasio. On the implementation of industrial automation systems based on PLC. *IEEE Transactions on Automation Science and Engineering*, Vol. 10, No. 4, 990–1003, 10 2013.

[6] M. Bautsch. Cycles of software crises. *European Network and Information Security Agency (ENISA)*, Vol. 3, No. 4, 3–5, 2007.

[7] A. Benelallam, A. Gómez, G. Sunyé, M. Tisi, and D. Launay. Neo4EMF, a scalable persistence layer for EMF models. In J. Cabot and J. Rubin, editors, *European Conference on Modeling Foundations and Applications*, volume 8569, pages 230–241. Springer, 2014.

[8] G. Black and V. Vyatkin. Intelligent component-based automation of baggage handling systems with IEC 61499. *IEEE Transactions on Automation Science and Engineering*, Vol. 7, No. 2, 337–351, April 2010.

[9] S. Burson, G. B. Kotik, and L. Z. Markosian. A program transformation approach to automating software reengineering. In *IEEE International Computer Software and Applications Conference*, pages 314–322, Chicago, 1990.

[10] J. H. Christensen, T. Strasser, A. Valentini, V. Vyatkin, and A. Zoitl. The IEC 61499 function block standard: Software tools and runtime platforms. presented at ISA Automation Week, 2012.

[11] W. Dai and V. Vyatkin. Redesign distributed PLC control systems using IEC 61499 function blocks. *IEEE Transactions on Automation Science and Engineering*, Vol. 9, No. 2, 390–401, 2012.

[12] E. W. Dijkstra. The humble programmer. *Communications of the Association for Computing Machinery (ACM)*, Vol. 15, No. 10, 859–866, 10 1972.

[13] S. Efftinge and M. Völter. oAW xText: A framework for textual DSLs. In *Eclipsecon Summit Europe*, Esslingen, Germany, 2006.

[14] K. A. Fischer. Concepts and applications of the IEC 61499 PLC programming standard for naval engineers. In *American Society Naval Engineers - Electric Machines Technology Symposium (EMTS)*, Villanova, PA, 2014.

[15] C. Gerber, H. Hanisch, and S. Ebbinghaus. From IEC 61131 to IEC 61499 for distributed systems: A case study. *EURASIP Journal on Embedded Systems*, 2008.

[16] A. Haase, M. Völter, S. Efftinge, and B. Kolb. Introduction to openArchitectureWare 4.1.2. In *MDD Tool Implementers Forum (Part of the TOOLS Conference)*, 2007.

[17] V. Hajarnavis and K. Young. An assessment of PLC software structure suitability for the support of flexible manufacturing processes. *IEEE Transactions on Automation Science and Engineering*, Vol. 5, No. 4, 1545–5955, Oct 2008.

[18] I. Hegny, M. Wenger, and A. Zoitl. IEC 61499-based simulation framework for model-driven production systems development. In *15th IEEE International Conference on Emerging Technologies and Factory Automation*, pages 1–8, Bilbao, 2010.

[19] F. Heidenreich, J. Johannes, S. Karol, M. Seifert, and C. Wende. Model-based language engineering with emftext. In R. Lämmel, J. a. Saraiva, and J. Visser, editors, *Generative and Transformational Techniques in Software Engineering IV*, volume 7680 of *Lecture Notes in Computer Science*, pages 322–345. Springer, Heidelberg, 2013.

[20] F. Heidenreich, J. Johannes, M. Seifert, and C. Wende. Construct to reconstruct: Reverse engineering Java code with Java model parser and printer (jamopp). In *International Workshop on Reverse Engineering Models from Software Artifacts*, Lille, 2009.

[21] G. T. Heineman and W. T. Councill, editors. *Component-based Software Engineering: Putting the Pieces Together*. Addison-Wesley, Boston, 2001.

[22] T. Hussain and G. Frey. Migration of a PLC controller to an IEC 61499-compliant distributed control system: Hands-on experiences. In *IEEE International Conference on Robotics and Automation*, pages 3984–3989, Barcelona, 2005.

[23] IEC SC65B. IEC 61131-3 Programmable Controllers. Part 3: Programming Languages, 2003. Geneva.

[24] K. H. John and M. Tiegelkamp. *SPS-Programmierung Mit IEC 61131-3*, volume 4. Springer, Heidelberg, 2009.

[25] T. LiB. Learning EMF. Online, 2013.

[26] W. Moore, D. Dean, A. Gerber, G. Wagenknecht, and P. Vanderheyden. *Eclipse Development using the Graphical Editing Framework and the Eclipse Modeling Framework*. 2004. IBM Redbooks.

[27] H. Otto. Aktuelles zur IEC 61131 Teil 3. *ATLAS Themenreihe Steuerungstechnik*, pages 45–47, 2012.

[28] Reasoning Systems, Inc. Reengineering and rewriting legacy software systems (component-based reengineering technology), 1 2002. Advanced Technology Program (ATP) funded project 94-06-0026, project duration 1.1.1995 to 31.12.1997.

[29] M. Riedl, C. Diedrich, and F. Naumann. SFC inside IEC 61499. In *IEEE International Conference on Emerging Technologies and Factory Automation*, pages 662–666, Prague, 2006.

[30] I. SC65B. IEC 61499-1 Function Blocks for Industrial Process Measurement and Control Systems. Part 1: Architecture, 2005. Geneva.

[31] M. Scheidgen, A. Zubow, J. Fischer, and T. H. Kolbe. Automated and transparent model fragmentation for persisting large models. In *15th International Conference on Model Driven Engineering Languages and Systems*, pages 102–118, 2012.

[32] G. D. Shaw, P. S. Roop, and Z. Salcic. Reengineering of IEC 61131 into IEC 61499 function blocks. In *8th IEEE International Conference on Industrial Informatics*, pages 1148–1153, Osaka, 2010.

[33] C. Sünder, M. Wenger, C. Hanni, I. Gosetti, H. Steininger, and J. Fritsche. Transformation of existing IEC 61131-3 automation projects into control logic according to IEC 61499. In *IEEE International Conference on Emerging Technologies and Factory Automation*, pages 369–376, Hamburg, 2008.

[34] Technical Committee 6. XML formats for IEC 61131-3. Technical report, PLCopen, 2009.

[35] VDI/VDE-GMA-Fachausschuss. Industrie 4.0 auf dem Weg zu einem Referenzmodell, 2014.

[36] V. Vyatkin. IEC 61499 as enabler of distributed and intelligent automation: State-of-the-art review. *IEEE Transactions on Industrial Informatics*, Vol. 7, No. 4, 768–781, 11 2011.

[37] V. Vyatkin, M. Hirsch, and H. Hanisch. Systematic design and implementation of distributed controllers in industrial automation. In *IEEE International Conference on Emerging Technologies and Factory Automation*, pages 633–640, Prague, 2006.

[38] M. Wenger, R. Hametner, and A. Zoitl. IEC 61131-3 control applications vs. control applications transformed in IEC 61499. In *IFAC Workshop on Intelligent Manufacturing Systems*, pages 30–35, Lisbon, 2010.

[39] M. Wenger and A. Zoitl. IEC 61131-3 model for model-driven development. In *38th IEEE Conference on Industrial Electronics Society*, pages 3744–3749, Montreal, 2012.

[40] M. Wenger and A. Zoitl. Re-use of IEC 61131-3 structured text for IEC 61499. In *IEEE International Conference on Industrial Technology*, pages 78–83, Athens, 2012.

[41] M. Wenger, A. Zoitl, C. Sünder, and H. Steininger. Semantic correct transformation of IEC 61131-3 models into the IEC 61499 standard. In *IEEE International Conference on Emerging Technologies and Factory Automation*, pages 1–7, Mallorca, 2009.

[42] M. Wenger, A. Zoitl, C. Sünder, and H. Steininger. Transformation of IEC 61131-3 to IEC 61499 based on a model driven development approach. In *IEEE International Conference on Industrial Informatics*, pages 715–720, Cardiff, 2009.

[43] T. Wright and C. Powers. Re-engineering legacy controls systems using IEC 61131. In *IEE Colloquia on the Application of IEC 61131 in Industrial Control: Improve Your Bottom-Line Through High Value Industrial Control Systems*, pages 6/1–6/24. Institution of Electrical Engineers (IEE), 2000.

[44] M. B. Younis. *Re-Engineering Approach For PLC Programs Based On Formal Methods*. PhD thesis, Technische Universität Kaiserslautern, 2006.

[45] A. Zoitl. *Real-Time Execution For IEC 61499*. International Society of Automation, 2009.

[46] A. Zoitl, T. Strasser, C. Sünder, and T. Baier. Is IEC 61499 in harmony with IEC 61131-3? *IEEE Industrial Electronics Magazine*, Vol. 3, No. 4, 49–55, 2009.

[47] A. Zoitl, T. Strasser, and A. Valentini. Open source initiatives as basis for the establishment of new technologies in industrial automation: 4DIAC a case study. In *IEEE International Symposium on Industrial Electronics*, pages 3817–3819, Bari, 7 2010.

7

Unit Test Framework for IEC 61499 Function Blocks

Reinhard Hametner

Thales Austria GmbH

Ingo Hegny

Vienna University of Technology

Alois Zoitl

fortiss GmbH

CONTENTS

7.1   Introduction

The functional requirements of modern industrial production systems rapidly increase. Additionally, there is the need to shift functional implementations from hardware components to software components in order to have flexible reusable components. Such components are encapsulated functional parts which are necessary to increase the engineering efficiency for reusable application parts for upcoming projects [33]. Therefore software components (i.e., industrial automation code) become more important and must cope with the increasing complexity of modern applications. Szyperski [31] coins the term "software components" for software modules with properties like unit of independent deployment and the possibility of third-party composition which have no external observable states. IEC 61499 function blocks (FBs) have been compared to software components and similar properties have been identified by Sünder et al. [29]. The development of applications for industrial production systems becomes more time-consuming and has to deal with adaptations of the control code on short notice.

An important factor for an efficient development process is the immediate reaction to changing requirements during the development phase. Changes of the mechanical system also infer changes of the electrical and control software parts of the control application [16, 22]. Low software quality and the increased use of untested software in automation systems bear high failure risks in fulfilling the overall system quality requirements. Furthermore, the reuse of untested software components in multiple automation projects spreads the risk. Thus software testing helps to increase and maintain a high software quality level over several projects.

A common technique to increase the software quality in the field of software engineering is test-driven development, which is also known as test-first development (TFD). TFD breaks the traditional development process work flow of "coding then testing". The test specification for a component cannot only used for the test process. It also acts as a design and development guide to support the developer in better understanding the implementation problems [8]. Thus, TFD facilitates automatic test case execution, improves the interactions between various stakeholders from different disciplines, and decreases the number of defects during the design and implementation phase [20].

First, all test cases for a system under test have to be specified. Then the engineers execute the specified test cases for the developed product. The control engineers should not need to implement the specified test cases for the target automation system. A fully automatic test case execution starting from the test case specification is required [34]. It helps to compensate for the lack of required skills (e.g., extensive knowledge of the testing methodology) as well as shorten the implementation and testing cycles.

In this chapter a testing framework for IEC 61499 FBs and the feasibility of the proposed approach are demonstrated through an example implementation.

7.2 Related Work

This section summarizes work on testing processes based on the TFD methodology, presents related work on unit testing, and explains work related to test case specification methods. Due to the potentially complex behavior of software components, complete tests must be numerous and comprehensive. Tests have to be structured hierarchically to increase the comprehensibility.

A common way for test structuring is to introduce the following layers [14, 34]: The lowest layer consists of *test cases*. These are contained in *test scenarios* in restricted order based on dependencies defined in the behavior specification. *Test scenarios* are independent from each other. *Test suite*—the topmost layer—consists of one or multiple *test scenarios*.

7.2.1 Test Process Overview

Applying strategies from software engineering can help to increase the quality of automation systems through improving testability and enhancing the development efficiency. Use cases drive sequences of actors' interactions with the system and lead to usage-based testing approaches. Models describe the functional system behaviors and can be used for deduce test cases and code (i.e., model-driven testing [8, 34]). Early definition of test cases (e.g., based on models) leads to TFD, a well-established software engineering practice in agile (business) software development projects [2]. Based on system specifications and customer requirements, test cases are specified prior to the implementation of the system or component. Early test case specification enables early defect detection in the system's specification during test case construction, for instance, inconsistencies between requirements, and missing and/or unclear requirements. It represents the foundation for frequent development iterations to increase the software quality. TFD-based development needs test strategies [10], which ensure fast feedback on the current development state.

Shull et al. [27] present a study based on a systematic literature review of the evidence of TFD's effectiveness. The influence of TFD on the delivered quality, internal quality, and productivity is shown. Madeyski [20] presents an overview of the majority of empirical studies which have investigated TFD and pair programming versus test-last and solo programming. Mattesson et al. [21] identify TFD and an applicable method for the development of software projects that provides a good improvement of implementation efficiency with limited additional effort. These works support our attempt for improving the software development process for control applications.

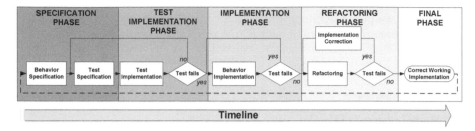

FIGURE 7.1
Testing and implementation workflow of the test-first development process.

Figure 7.1 presents the basic workflow of TFD consisting of five phases: specification phase, test implementation phase, implementation phase of the functional behavior, refactoring phase, and the final phase. In the *specification phase* the behavior and test cases for the software component will be specified. In the next phase, *test implementation phase*, the test cases will be implemented, based on the existing specification. Starting from this phase, test cases are executed, but due to the missing behavior implementation the test case execution has to fail [20]. In the third phase, called *implementation phase*, the functional behavior will be implemented and the test cases can be executed repeatedly until the desired behavior is reached.

If the test suite (i.e., all specified test cases) passes, the *refactoring phase* can be started. At this step the functionality is already implemented but the quality and structure of the software may not be sufficient. Therefore the implemented behavior has to be optimized without changing the functionality of the code. After a successful validation of the refactored code with test cases, the *final phase* is reached. Finally, if additional functionality is needed, the process can be started over from the beginning of the five phases.

7.2.2   Unit Testing

Software artifacts without external dependencies are used to provide the basis for unit testing. Their fitness for use is assessed with appropriate interface-oriented test cases which check the units' functional behaviors. These units (units under test) must have unambiguous testable behavior.

The tested units form the basis for the application development. Hence, the testing of as many units as possible shall be achieved. Therefore an automatic test process implemented in a TFD process is desirable.

In the software engineering community, software testing is a well established technique. Several unit testing techniques are evaluated and classified in [35]. Since several years agile programming methods (e.g., eXtreme Programming) are heavily utilizing unit testing and unit test frameworks (e.g., JUnit) are constantly improved. These unit test frameworks typically require that the tests are implemented in the same language as the code to be tested.

7.2.3 Test Specification Methods

For testing software components the procedure or strategy, test scenarios, and the test cases have to be specified. Test cases can be classified as: normal/-positive test cases, special test cases, and error/negative test cases. Normal/-positive test cases are necessary to test the components' intended (specified) behavior. Special test cases check the components' behavior in the border area of intended execution behavior (e.g., upper or lower limit). Error/negative test cases are used for verifying the implementation in case of error states as well as the behavior on wrong input.

There are several approaches to specify a test case. A common testing methodology in the field of software engineering is model-based testing, see [4]. Formal or semi-formal models are used for the specification. The Unified Modeling Language (UML) [3], as the dominant software modeling language, is a potential tool for describing and specifying functional and test behaviors of complex control systems. The UML state diagram, which is part of the UML diagram family, is mostly used for structural test processes; see [14, 18, 23]. A selection of useful and applicable UML diagrams for the industrial automation domain is presented in [26] and [15]. Chow [7] describes a method for checking the correctness of control structures at the design level by finite-state machines. Hence, state charts and automata are used in the software testing process. Swain et al. [30] use the combination of UML state chart diagrams and activity models named state-activity-diagram for test case specification and generation. The combination provides both control flow and event-oriented state change information.

Thramboulidis [32] proposes a hybrid approach that integrates UML with IEC 61499 function block constructs. Therefore a model transformation process is used from UML diagrams to a FB design model.

7.2.4 Testing versus Verification

A different approach for ensuring software quality and its correctness is formal verification (e.g., model checking [17]). By using a verification method, the control software is represented in form of mathematical models. These can be analyzed and checked if all requirements are fulfilled at all times. Thus the correct behavior of the verified component can be certified. Several approaches for performing formal verification on IEC 61499 applications have been presented in [6, 9, 25].

Critical parts of executing control applications are the execution environment and the control hardware. Gerber and Hanisch [11] showed that the execution environment can change the execution behavior and added it to their verification models. However, [28] showed that a full mathematical model of the whole execution environment and the control hardware leads to highly complex models (i.e., inefficient calculation effort). In order to check the components' behavior on real control devices, software testing can complement model checking methods. Furthermore, methods for specifying tests (i.e., the components' intended behaviors) can support the modeling phase for model checking approaches [24].

7.3   Requirements for IEC 61499 Unit Tests

Unit testing is a technique that helps to ensure the correctness of a functional software unit (e.g., module, component, class) according to its specification. Therefore the first step is to define units in IEC 61499. This includes their interfaces, their behaviors, and the specifications of the behavior for the test cases.

7.3.1   Software Units in IEC 61499

The greatest advantage of unit testing can be gained if the units can be used in different contexts without changed behavior, as the tests guarantee that false behavior and resulting errors can only stem from the combination of the units. Software components as defined in [31] feature this characteristic. There a software component is defined as a unit of independent deployment, of third-party composition, and with no (externally) observable state. Another aspect of the component-orientated concept is that components shall be usable without any knowledge of their internals. In order to facilitate this aspect this work focuses on black box testing of control software units for IEC 61499 control applications. That means that for the test specification and process no knowledge and assumptions on the internals of the unit are required. Only the publicly specified interface is used for the specification and execution of tests, according to the black box testing method.

Sünder et al. [29] show that FBs can in general be considered as software components when applying the software component concept to IEC 61499. This holds true especially for basic FBs (BFBs), as these perfectly encapsulate the internal data. The contained algorithms and the execution control chart (ECC) are only allowed to access internal or interface data. For service interface FBs (SIFBs) the situation is different. SIFBs contain hidden interfaces to the underlying device specific services, which, first violates the definition of software components. Second, these services need to be controlled by and interact with performing tests. This is typically done during device development by the device vendor and needs special knowledge of the device or the service represented by the SIFB.

Finally IEC 61499 provides a third choice known as the composite FB (CFB). A CFB encapsulates a set of other FBs (BFB, CFB, or SIFB) and provides this combination as a new unit to the application developer. Nevertheless the type of CFB is indistinguishable from any other FB for the application engineer (i.e., provided FB interface). Therefore CFBs can also be units for the unit test framework developed in this chapter.

7.3.2 Observable Behavior of Function Blocks

IEC 61499 defines FBs as passive elements whose execution can be requested by input events. Associated data inputs will be sampled at the occurrence of the triggering input event. The encapsulated functionality will take the data inputs, internal data and data outputs and perform its calculations generating output data and zero or more event outputs, each again with associated data outputs (see Figure 7.2). This behavior is also the starting point for the targeted black box unit tests. With these, the observable behavior of the FB's interface is tested. The behavior can be formalized to a transformation function FB. It takes an input vector $\vec{I} = (ei_x, di_1 \cdots di_r)$ and produces an output vector set O. ei_x represents an input event and $di_1 \cdots di_r$ represents all data inputs [13].

$$FB \quad : \quad \vec{I} \mapsto O. \tag{7.1}$$

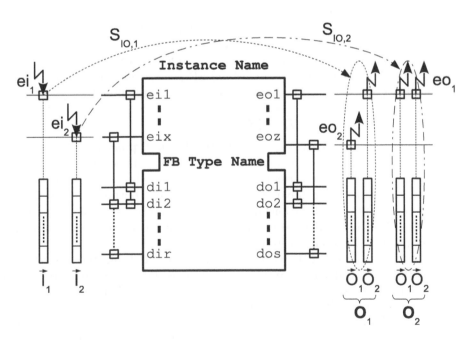

FIGURE 7.2
Observable behavior of IEC 61499 function block as basis for unit tests.

O consists of zero or more output vectors $\vec{O}_x$

$$O = (\vec{O}_1 \cdots \vec{O}_y), \qquad O = \emptyset \; : \; y = 0, \tag{7.2}$$

$$\vec{O}_x = (eo_z, do_1 \cdots do_s). \tag{7.3}$$

Similar to the input vector, eo_z represents an output event and $do_1 \cdots do_s$ represent all data outputs.

The mapping in Equation 7.1 is only bijective, if the FB has no internal state (e.g., a simple sine function). y represents the number of output vector elements. If $y = 0$, no output event and no output data exist. However as FBs may have internal state (i.e., ECC states, internal variables, and the values of data outputs), the history of previously applied input vectors $(\vec{I}_1, \ldots, \vec{I}_t)$ is required to determine the according output vector set O. Therefore a sequence of $\vec{I}, O$ tuples for describing an FB's observable behavior is needed:

$$S_{IO} = \{\{\vec{I}_1, O_1\}, \ldots, \{\vec{I}_t, O_t\}\}. \tag{7.4}$$

7.3.3   Unit Test Specification for Function Blocks

Typically to perform unit tests, several test cases are required. Each test case checks a certain aspect of the component under test. This separation into test cases has the advantage that tests are easier to develop and maintain, and are reusable. Furthermore it is easier to identify the reason for a failed test. Applying this for testing IEC 61499 FBs, a test sequence will be S_{IO} as defined in Equation 7.4. The full test suite U_T is represented by a set of these sequences:

$$U_T = \{S_{IO,1}, \ldots, S_{IO,u}\}. \tag{7.5}$$

A unit test execution system called "Test Runner" processes this data. For each test sequence entry, the input data and input event to the FB under test (FBuT) is applied. Furthermore the Test Runner will monitor the outputs (events and data) and compare them to the provided output expectations. In order to simplify the development of test cases, each test sequence assumes that the FBuT is in a defined initial state. The unit test execution system has to ensure that this prerequisite is met.

For more complex FBs with many inputs and outputs (i.e., events and data) these test sequences can become very complicated and hard to define. Therefore means are necessary to allow the description of the behavior of FBs in a way more intuitive for humans that can be transformed into the unit test representation according to Equation (7.5). The unit test specification methods strongly depend on the developer of the unit tests. Two main target audiences can be distinguished. The first are the control application developers who develop and maintain specific FBs for their applications. The second are FB library developers who provide general FBs to the first group. Their work is mainly focused on maintenance, refactoring, and functional improvement of FBs, which are elaborate tasks [36]. As both groups work with IEC

61499 models, the test specification method should be a technique they are familiar with. Taking this further, Baker et al. [1] state that the test specification should utilize the same means as the target application in order to be effectively usable. For our target domain that means that test specifications should be IEC 61499 models or derived models of IEC 61499 at best. Based on this, existing unit test frameworks and methods (e.g., JUnit or UML testing profile [1]) cannot be used directly but they can serve as the bases for the required features and functionality.

7.4 Modeling Unit Tests According to IEC 61499

Based on the identified requirements, three potential methods for a unit testing process for IEC 61499 [12, 13] are detailed and explained in the following. The main difference between these methods is the specification method that is supported by IEC 61499 or an external method. An important point is the location of the test specification relative to the specification of the FBuT. This is a further distinguishing element.

7.4.1 Separated Test Component

The first natural approach for a unit test framework is to apply approaches from the software engineering domain. This means that the tests, for example for Java classes, are specified utilizing Java classes containing the test code. It could be shown that this is a natural way for software developers to specify their unit test cases in the same language as the development or implementation language. Applying this approach to test cases for IEC 61499 FBs, unit test cases should be implemented as separated and related test FBs. These test FBs would feature the mirrored interface of the FBuT (i.e., inputs will become outputs and vice versa). For performing the tests, both FBs would be connected and the tests executed. The whole development and test process can be done in the original IEC 61499 development environment (e.g., 4DIAC [37]), which would be convenient for automatic control.

With this approach, all IEC 61499 FB models are available and can be used for the test implementation. The ECC provides a powerful means for defining the test cases for the stateful behavior of the FBuT. The combination of the ECC with several algorithms included in a FB can be used for specifying complex test procedures. Furthermore certain commonly needed test procedures can be provided in separate FBs. These can be reused in different test FBs in form of CFBs.

Well established testing frameworks (e.g., JUnit) automatically build test applications. In contrast, there currently exist no such application generators for IEC 61499-compliant FBs. Hence, test applications need to be developed manually and independently from the targeted automation applications. Such test applications need IEC 61499 execution containers (e.g., resources) for deployment and test execution.

Because of the separated components (i.e., test FB and FBuT), it is difficult and time-consuming to maintain the changes in both FBs (e.g., interface changes in FBuT). A further disadvantage is that IEC 61499 currently has no means for fulfilling all requirements necessary for test specification, test result interpretation, and results reporting.

7.4.2   Model-Based Test Specification

The second approach is based on the model-based testing methodology commonly used in software engineering. The model-based method can also be adapted for testing industrial control components like IEC 61499 FBs. In this approach a separated test specification in the form of a software model which can be derived from the system requirements specifications, and enables automated code and test case generation is used [34]. The aim of this method is to automatically implement executable test cases for IEC 61499 FBs from their test specifications.

For modeling test specification, the UML diagram family is an appropriate solution. It is a powerful specification language and is widely used and accepted in the software engineering domain. Several UML diagram types are applicable for the test specification [12]. A restricting UML profile could provide the needed clarity for the test case implementation. Hametner et al. [14] present a model transformation process which uses the test case information from UML state chart diagrams and automatically generates an IEC 61499 test FB. Similar to the approach in Section 7.4.1, this FB includes all executable test cases. The ECC and algorithms in the Structured Text language are used for the test implementation.

An advantage compared to the previous testing approach is that changes of the interface and/or changes of the functional behavior of the FBuT can be handled more easily, as the executable test cases are generated automatically. This approach supports regular changes of the test specification (i.e., UML models). Hence, the effort for changing and maintaining the test specification parallel to the behavior specification is reduced. Furthermore no programming skills are necessary for specifying and maintaining the unit tests, which will increase the acceptance of TFD in the industrial automation domain.

One of the shortcomings of the previous approach, separated maintenance for FBuT and the related test specification, is a drawback of this approach.

7.4.3   Utilizing Service Sequence Diagrams for Test Specification

Service sequence diagrams allow the specification of sequences of input events and the resulting output events. Additionally, input data values can be used for FB test specification. Furthermore, IEC 61499 allows the specification of multiple service sequences, each describing a certain execution aspect of the FB. This can be utilized for defining several sequences which will test different aspects of the interface behavior of the FBuT in a predefined order. Service sequences resemble the general FB unit test description as defined in Equation (7.5). Furthermore positive, special, and negative test cases can be specified and used for the testing process.

The main advantage of this approach is that the test specification is directly included in the type definition of the FBuT. Therefore the dedicated test specification is always part of the FB definition and no inconsistencies between different specification versions can occur. Furthermore the test specification method is a native IEC 61499 method and only limited additional IEC 61499 training is required for learning test case specification.

For FBs with complex interfaces, the service sequence diagrams can become elaborate but service sequences can help to focus on specific aspects of the interface behavior. Furthermore, similar tasks needed in multiple test sequences (e.g., bringing the FBuT to a predefined state) have to be included in each test sequence. IEC 61499 does not allow the invocation of a sequence diagram from another one. Here extensions like sequence calls, loops, or conditions introduced to UML sequence diagrams in the second edition, may be helpful for IEC 61499 service sequence diagrams too.

7.4.4 Evaluation of Test Specification Approaches

The requirements identified in Section 7.3 are fulfilled by the three approaches. Several criteria have been identified as relevant for the industrial automation domain. A comparison of the approaches is presented in Table 7.1 which shows the evaluation results for the six selected properties. The three different approaches are estimated by three values (i.e., +, 0, −) for the comparison.

- \+ indicates that the property is fully supported.

- 0 means that the property is only partly supported. An implementation of the testing technique is possible with some effort.

- − shows that the property is not supported. Therefore testing is not possible.

Comparing all the approaches, the first one from Section 7.4.1 is identified as having the greatest *flexibility* regarding the test specification. This approach resembles current unit test methods applied in the software engineering domain. The model-based approach using UML diagrams (as described in Section 7.4.2) is powerful. However, restrictions (i.e., a UML profile) are needed for an unambiguous test case execution which also limits the flexibility. With respect to flexibility, the utilization of service sequence diagrams in Section 7.4.3 are definitely the most limited methods.

TABLE 7.1

Evaluation results for Separated Test Component
(7.4.1), Model-Based Test Specification (7.4.2), and
Utilizing Service Sequence Diagrams (7.4.3)

Evaluated Property	7.4.1	7.4.2	7.4.3
Flexibility in test specification	+	0	−
Portability	0	−	+
Readability	−	+	0
Integrity	−	0	+
Expressiveness	+	0	−
Test maintenance	−	0	+

An advantage of the first and third approach is that the artifacts needed for testing (i.e., test FB and test service sequence diagrams) can be accessed by all IEC 61499-compliant engineering tools. *Portability*, exchange of engineering data (e.g., FBs) between tools, is an important property defined in the standard. However, exchanging UML diagrams between different engineering tools is problematic because of prevalent data format incompatibilities.

UML has been specified for the implementation of software. Therefore *readability* is important for a good analysis of the components' behavior specification. Service sequence diagrams are easy to grasp. However, an overview on all test cases can hardly be gained for complex FBs. Test-relevant properties (e.g., ECC, algorithms, FB interface) are distributed within the test FB. Therefore, neither a single test case nor the full test suite can easily be understood at the first glance.

In the third approach in Section 7.4.3, the test-related information is part of the type definition of the FBuT. Therefore it cannot get lost or forgotten on transfer. Limited *integrity* is offered by the approach in Section 7.4.2, if the same UML model is also used for the specification of the FB behavior proposed in [32]. A major drawback of the first approach is that the test specification and implementation are completely separated from the FBuT which may lead to inconsistencies.

All means for the specification and implementation of FBs can be used in the first approach. The second and third approaches need a generic interpreter for the test specification. For an unambiguous interpretation of the test case specifications, the *expressiveness* has to be limited. IEC 61499 service sequence diagrams used in the third approach provide only limited expressiveness. Hence, the Test Runner can easily reach an unambiguous interpretation of the specified test cases.

Integrated modeling methods as proposed in the third and second approaches (if the FB behavior is also specified in UML) enable easy checking for inconsistencies between FBuT and the test specification (e.g., interface changes). These *test maintenance* tasks are harder to achieve with a complete separation of tests and FBuT as proposed in the first and second approaches.

After assessing the evaluation criteria, the third approach of Section 7.4.3 has been chosen as the basis for the IEC 61499 unit test framework. The usability of service sequence diagrams can be increased by potential improvements in IEC 61499 (e.g., loops, sequence calls, conditions). The benefits of this approach—portability, integrity, and the easy test maintenance—outweigh its disadvantages.

7.5 Resulting Test Framework

Based on the service sequence diagram test specification approach, the test framework as depicted in Figure 7.3 is implemented. The tests in the form of service sequence diagrams are part of the FB type file, as are all other FB type specific definitions.

The Test Runner (see Figure 7.3(a)) is the core element in the presented test framework and handles parsing, selection, execution, and evaluation for the test cases. The Test Runner is implemented and included in the Eclipse-based IEC 61499 engineering environment 4DIAC-IDE [37] as an additional open-source plug-in. The tests are specified by service sequence diagrams (see Figure 7.3(b)) which are stored in the FB type file together with the interface and behavior implementation. The test cases and the FB interface are parsed by the Test Runner. The gathered information is interpreted as a set of test cases, each with an input vector and a set of expected output vectors.

Based on the interface definition of the FBuT, a type-specific test application is automatically created and instantiated at the (remote) test device for the procedure. The IEC 61499-compliant runtime environment 4DIAC-RTE [37] is used. The test application consists of the FBuT, FBs for the communication with the Test Runner (receiver, sender), as well as multiplexing and de-multiplexing FBs (MuX, DeMuX). MuX and DeMuX FBs are generic FBs; that means the runtime environment is able to instantiate FBs with a variable number of event inputs and outputs, according to the interface of the FBuT. The encoding of the events as data allows a synchronous transmission of the input set on a unidirectional connection from the Test Runner to the test device. Also the synchronous transmission of the output vector sent back from the test device to the Test Runner is facilitated.

For the communication between the Test Runner and the test device, the connection-based TCP/IP protocol is chosen. The numbers of inputs and outputs of the sender and receiver functionality are derived from the interface of the FBuT. For the transmission of the encoded events, an additional data input is added for the sender and an additional data output for the receiver, respectively. The necessary interfaces are shown in Figure 7.3(c).

The test process is executed as described in the following:

1. The type specific test application is instantiated at the test device.

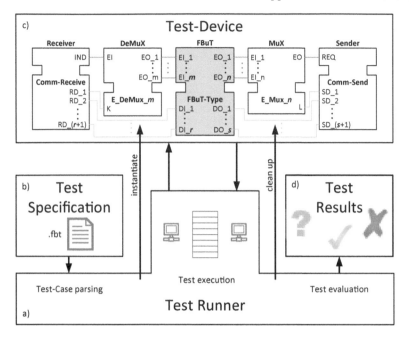

FIGURE 7.3
Overall unit test framework. a) The Test Runner. b) The tests are specified in the FB type file together with the interface and behavior. c) FB type specific test application is instantiated and deployed to a target control system. d) Test results are reported and visualized.

2. The Test Runner starts the evaluation of the specified test cases.

3. The Test Runner transfers the given input set to the test device.

4. The execution of the FBuT is triggered by the specified input event.

5. The resulting output events and data are transferred back to the Test Runner.

6. The Test Runner removes the test application from the device (i.e., clean-up phase), analyzes the replies from the test application, and evaluates the test results of the unit test (see Figure 7.3(d)).

A specific test sequence passes the test if the set of the received output vectors equals the expected output vectors. Both are extracted from the test specification. As soon as one of the following criteria is met, the test sequence is evaluated as failure:

- Number of received output vectors is not equal to the specified number

- A wrong event is received in an output vector

- Output data does not match the expected values

- Predefined timeout value of the test execution is reached

Upcoming tests within the same sequence are not evaluated, since preconditions might not be fulfilled. After the execution of all tests, the test application is removed from the device and the results are reported and visualized by the engineering tool.

The separation of Test Runner and test device bears the following advantages:

1. The tests can be executed on the deployed runtime environment. Errors and changes during the porting to new devices can be tested.

2. For the instantiation of the test application on the test device, only standard FBs are needed along with the FBuT.

3. Any kind of FB (including SIFBs) can be tested on devices for which an implementation exists.

4. The network-based test execution enables tests on platforms that do not provide visualization capabilities or file system access (for report saving). The test evaluation on the engineering system further reduces the need to provide memory to gather the results on the test device; thus tests can be executed even on small devices.

5. As long as standard compliant communication interfaces are provided by the devices, tests on devices from various vendors are possible. The proposed test framework also accomodates the influence of different execution semantics on different target platforms which arise from different interpretations of the standard [6, 5].

7.6 Application Examples

For the validation of the function block test framework representative but comprehensible FBs are selected. The selection of the FBs is performed on the basis of typical execution features and FB properties that have to be considered for testing. A comprehensive taxonomy of the execution features and properties is presented in Figure 7.4.

As a result of the classification and evaluation process, two well known FBs are chosen, which are defined in Annex A of IEC 61499-1 [19]. The E_PERMIT and E_CTU FBs are apt to provide insights into the specification and execution of unit tests for FBs.

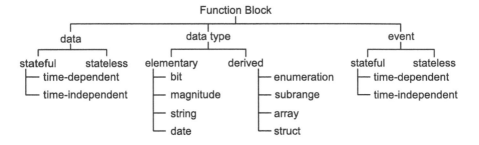

FIGURE 7.4
Taxonomy of observable function block execution features and properties.

7.6.1   Testing Stateless Function Block: E_PERMIT

The E_PERMIT FB provides a permissive event propagation. As can be seen in Figure 7.5(a), the FB has one event input *EI* and an associated Boolean data input *PERMIT*. The behavior of the E_PERMIT FB is specified as stateless with respect to data and events. The observable output behavior resulting from an input event EI depends only on the value of *PERMIT*. As this is Boolean data, the following two outputs are valid:

- *PERMIT* is *false*: no output event should be triggered.

- *PERMIT* is *true*: one output event *EO* should be triggered.

Since the two test cases are independent from each other, they are specified as separate test sequences. An additional test (i.e., *negative test*) that should fail on execution validates the functionality of the test environment.

The specification of the test cases is edited with the service sequence editor of the 4DIAC-IDE; see Figure 7.5(b). All three test cases with the respective input and output vectors are visible in the figure. As the service sequences are included in the FB type definition and only IEC 61499-compliant description methods are used, any FB type editor may be used to specify or change the test cases.

7.6.2   Testing Stateful Function Block: E_CTU

The second FB is the E_CTU FB. Its interface includes two input events (*CU*, *R*) and two output events (*CUO*, *RO*), as well as one data input (*PV* of a 16-bit unsigned integer type UINT) and two data outputs (*Q* of Boolean type, *CV* of type UINT) as shown in Figure 7.6. The counter value *CV* is increased at the occurrence of a count-up event *CU*, and reset to 0 on the occurrence of the reset event *R*. The behavior of the FB depends on the counter value; therefore the E_CTU FB is stateful and time-independent with respect to data.

The use of magnitude data types as input as well as output data highly

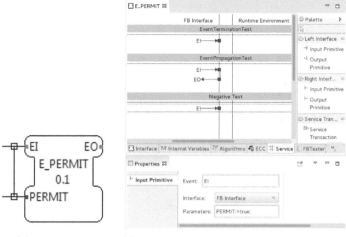

(a) E_PERMIT   (b) Service sequence diagrams for the test of E_PERMIT

FIGURE 7.5
IEC 61499 standard library FB **E_Permit** and its test specification with 4DIAC.

increases the number of potential test cases. There exist 2^{17} possible input vectors (i.e., 2 non-concurrent event inputs $+ 1 \times$ 16-bit data input). Furthermore this comprehensible example already has 2^{18} output interface states. Therefore, exhaustive testing of the **E_CTU** FB would require 2^{35} (i.e., $2^{17} \times 2^{18}$) test cases.

The requirement for a well-considered test case selection is evident. A representative set of test cases (including special tests and negative tests) has been specified for the **E_CTU** FB type.

The test cases are parsed by the Test Runner and run when tests are triggered by the user in the user interface (plug-in in the 4DIAC-IDE). Within the same user interface, test results are reported both textually (showing detailed

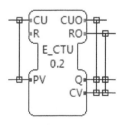

FIGURE 7.6
Interface specification of the IEC 61499 standard library FB E_CTU.

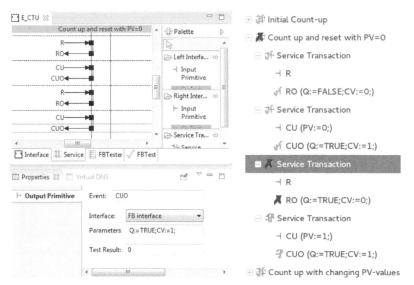

(a) Test specification with sequence diagram   (b) Results of unit test execution
of three scenarios

FIGURE 7.7
E_CTU FB: (a) Count up and reset with PV=0; Test sequence diagram. (b)
Test results: scenarios 1 and 3 successful; scenario 2 failed due to failure of
scenario 3; scenario 4 not executed as a result.

information on failed tests) as well as graphically (providing a good overview
on test results), as presented in Figure 7.7(b). At the occurrence of the first
error (see negative test) a test sequence is stopped and marked as failed (see
cross mark in Figure 7.7(b)). Further test cases within the failed test sequence
are marked as untested (see question mark in Figure 7.7(b)).

7.7   Conclusion and Future Work

The increasing use of software in industrial automation solutions demands
methods to increase the quality of the software while at the same time reducing
development costs. The TFD approach has proven a valuable solution for this
problem in the domain of software engineering.

A unit test framework suitable for industrial automation is presented as
basis for a TFD approach. FBs are identified as the software units to be tested
in IEC 61499. Based on the formal specification of their external observable
behavior, a unit test specification method and a relevant unit test execution
framework are derived. The main advantages of the presented solution are that

it is fully IEC 61499 standard compliant and can be utilized in any IEC 61499 standard compliant environment. With two representative FBs taken from the IEC 61499 standard library, the usage is shown and the applicability of the proposed approach are proven.

The presented unit test framework has been added to 4DIAC and is available with all current releases of the 4DIAC-IDE. Furthermore, also the two test examples presented in this chapter are also available in the 4DIAC FB library. This allows interested researchers to test and evaluate the unit test framework for it suitability.

The limitations of service sequence diagrams in case of FBs with complex interfaces are identified. The sequences become very large and only a subset can be efficiently specified. A solution for such FBs could be to model the test cases with UML and automatically generate the resulting service sequence diagrams as input for the unit test framework. This has the advantage that the test cases can be developed more easily and that the tests are stored as part of the FB.

To overcome these limitations, a set of extensions to the IEC 61499 service sequences is proposed that reduces the effort for specifying tests and analyzing FB behavior by increasing their usability. A first step will definitely be the introduction of sub-service sequences and a calling mechanism, after which common tasks needed for several test cases (e.g., FB preparation), can be defined once and reused multiple. Further extensions are control structures, conditional execution, and loops in order to implement repetitive test sequences. Finally, as control applications are real-time constrained the timing behavior is an important property and quality feature. Therefore it should be possible to specify the timing behavior in test sequences. These requirements have been brought to the standardization committee of IEC 61499 and will be considered in the development of IEC 61499-5. The presented test framework is one possible solution for requirement changes on short notice in the development phase to keep the quality of industrial control software at a high level.

Bibliography

[1] P. Baker, Z. Dai, and J. Gabrowski. *Model-Driven Testing: Using the UML Testing Profile.* Springer, 2007.

[2] K. Beck. *Test-Driven Development by Example.* Addison-Wesley, 2002.

[3] G. Booch, J. Rumbaugh, and I. Jacobson. *The Unified Modeling Language Reference Manual.* Addison-Wesley, 1999.

[4] M. Broy, B. Jonsson, J.-P. Katoen, M. Leucker, and A. Pretschner. *Model-Based Testing of Reactive Systems*. Springer, 2005.

[5] G. Čengić and K. Åkesson. On formal analysis of IEC 61499 applications, Part B: Execution semantics.

[6] G. Čengić and K. Åkesson. On formal analysis of IEC 61499 applications, Part A: Modeling. *Industrial Informatics, IEEE Transactions on*, Vol. 6, No. 2, 136–144, May 2010.

[7] T. Chow. Testing software design modeled by finite-state machines. *Software Engineering, IEEE Transactions on*, Vol. SE-4, No. 3, 178–187, May 1978.

[8] L. Crispin and J. Gregory. *Agile Testing: A Practical Guide for Testers and Agile Teams*. Addison-Wesley, 2008.

[9] V. Dubinin, V. Vyatkin, and H.-M. Hanisch. Modelling and verification of IEC 61499 applications using Prolog. In *IEEE Conference on Emerging Technologies and Factory Automation*, pages 774–781, Sept. 2006.

[10] M. Duvall, S. Matyas, and A. Glover. *Continuous Integration: Improving Software Quality and Reducing Risk*. Addison-Wesley, 2007.

[11] C. Gerber and H.-M. Hanisch. Does portability of IEC 61499 mean that once programmed control software runs everywhere? In *10th IFAC Workshop on Intelligent Manufacturing Systems*, pages 29–34, Lisbon, July 2010.

[12] R. Hametner. *Test-Driven Software Development for Improving the Quality of Control Software for Industrial Automation Systems*. PhD thesis, Vienna University of Technology, Oct. 2013.

[13] R. Hametner, I. Hegny, and A. Zoitl. A unit-test framework for event-driven control components modeled in IEC 61499. In *IEEE Conference on Emerging Technology and Factory Automation*, pages 1–8, Barcelona, Sept 2014.

[14] R. Hametner, B. Kormann, B. Vogel-Heuser, D. Winkler, and A. Zoitl. Test case generation approach for industrial automation systems. In *5th International Conference on Automation, Robotics and Applications*, pages 57–62, Wellington, New Zealand, Dec. 2011.

[15] R. Hametner, D. Winkler, T. Östreicher, N. Surnic, and S. Biffl. Selecting UML models for test-driven development along the automation systems engineering process. In *IEEE Conference on Emerging Technologies and Factory Automation*, pages 1–4, Bilbao, Sept. 2010.

[16] R. Hametner, A. Zoitl, and M. Semo. Component architecture for the efficient development of industrial automation system. In *6th IEEE Conference on Automation Science and Engineering Proceedings*, pages 156–161, Toronto, Aug. 2010.

[17] H. M. Hanisch, A. Lobov, J. L. Martinez Lastra, R. Tuokko, and V. Vyatkin. Formal validation of intelligent automated production systems towards industrial applications. *Intl. J. of Manufacturing Technology and Management*, Vol. 8, No. 1, 75–106, 2006.

[18] T. Hussain and G. Frey. UML-based development process for IEC 61499 with automatic test-case generation. In *IEEE Conference on Emerging Technologies and Factory Automation*, pages 1277–1284, Prague, Sept. 2006.

[19] IEC 61499-1. *Function blocks. Part 1: Architecture*. International Electrical Commission, 2.0 edition, 2012.

[20] L. Madeyski. *Test-Driven Development: An Empirical Evaluation of Agile Practice*. Springer, 2010.

[21] A. Mattsson, B. Lundell, B. Lings, and B. Fitzgerald. Linking model-driven development and software architecture: A case study. *IEEE Trans. on Software Engineering*, Vol. 35, No. 1, 83–93, Jan./Feb. 2009.

[22] M. G. Mehrabi, A. G. Ulsoy, and Y. Koren. Reconfigurable manufacturing systems: Key to future manufacturing. *J. of Intelligent Manufacturing*, Vol. 11, No. 4, 403–419, Aug. 2000.

[23] S. Mouchawrab, L. Briand, Y. Labiche, and M. Di Penta. Assessing, comparing, and combining state machine-based testing and structural testing: A series of experiments. *IEEE Trans. on Software Engineering*, Vol. 37, No. 2, 161–187, Mar./Apr. 2011.

[24] S. Preusse and H.-M. Hanisch. Specification and verification of technical plant behavior with symbolic timing diagrams. In *3rd International Design and Test Workshop*, pages 313–318, Monastir, Dec. 2008.

[25] S. Preusse and H.-M. Hanisch. Verifying functional and non-functional properties of manufacturing control systems. In *3rd International Workshop on Dependable Control of Discrete Systems*, pages 41–46, Saarbruecken, June 2011.

[26] C. Seidner and O. Roux. Formal methods for systems engineering behavior models. *IEEE Trans. on Industrial Informatics*, Vol. 4, No. 4, 280–291, 2008.

[27] F. Shull, G. Melnik, B. Turhan, L. Layman, M. Diep, and H. Erdogmus. What do we know about test-driven development? *IEEE Software*, Vol. 27, No. 6, 16–19, Nov./Dec. 2010.

[28] C. Sünder and V. Vyatkin. Functional and temporal formal modelling of embedded controllers for intelligent mechatronic systems. *Intl. J. of Mechatronics and Manufacturing Systems*, Vol. 2, No. 1/2, 215–235, 2009.

[29] C. Sünder, A. Zoitl, J. Christensen, H. Steininger, and J. Fritsche. Considering IEC 61131-3 and IEC 61499 in the context of component frameworks. In *6th IEEE International Confernece on Industrial Informatics*, pages 277–282, Daejeon, July 2008.

[30] S. K. Swain, D. P. Mohapatra, and R. Mallc. Test case generation based on state and activity models. *J. of Object Technology*, Vol. 9, no. 5, 1–27, 2010.

[31] C. Szyperski. *Component Software: Beyond Object-Oriented Programming.* ACM Press, 2nd ed., 2002.

[32] K. Thramboulidis. Using UML in control and automation: A model driven approach. In *2nd IEEE Intlernational Conference on Industrial Informatics*, pages 587–593, June 2004.

[33] K. Thramboulidis. Model-integrated mechatronics: Toward a new paradigm in the development of manufacturing systems. *IEEE Trans. on Industrial Informatics*, Vol. 1, No. 1, 54–61, Feb. 2005.

[34] M. Utting and B. Legeard. *Practical Model-based Testing.* Morgan Kaufmann, 2007.

[35] S. Vegas, N. Juristo, and V. Basili. Maturing software engineering knowledge through classifications: A case study on unit testing techniques. *IEEE Trans. on Software Engineering*, Vol. 35, No. 4, 551–565, July 2009.

[36] V. Vyatkin and V. Dubinin. Refactoring of execution control charts in basic function blocks of the IEC 61499 standard. *IEEE Trans. on Industrial Informatics*, Vol. 6, No. 2, 155–165, May 2010.

[37] A. Zoitl, T. Strasser, and A. Valentini. Open source initiatives as basis for the establishment of new technologies in industrial automation: 4DIAC a case study. In *IEEE International Symposium on Industrial Electronics*, pages 3817–3819, Bari, July 2010.

8

Verifying IEC 61499 Applications

Petr Kadera

Czech Technical University in Prague

Pavel Vrba

Czech Technical University in Prague

CONTENTS

8.1 Introduction

Previous chapters have described basic principles of using the IEC 61499 standard for development of distributed control systems. Although certainly not trivial, the task of producing a working IEC 61499 application is relatively straightforward. Nevertheless, producing an application that **always** works is much more challenging. How can we be sure that the developed application will meet our requirements in all cases? How can we actually describe these requirements? Another important aspect is related to re-usability of IEC 61499 fragments. How can we assess whether a piece of code developed earlier can be safely re-used?

The biggest challenges of embedded system design have been defined by Henzinger and Sifakis [15] as follows:

1. Encompassing heterogeneous execution and interaction mechanisms for system components.

2. Providing abstractions that isolate the design subproblems requiring creativity from those that can be automated.

3. Scaling by supporting compositional, correct-by-construction techniques.

4. Ensuring the robustness of the resulting systems.

What is system verification? Generally speaking, system verification checks whether a system meets the qualitative requirements that have been identified. Frequently, **verification** and **validation** are mistaken terms, although their meanings are different. Definitions presented in [9] say:

Validation is the process of evaluating software during or at the end of the development process to determine whether it satisfies specified business requirements.

Verification is the process of evaluating workproducts (not the actual final product) of a development phase to determine whether they meet the specified requirements for that phase.

In other words, the goal of validation is to demonstrate that the product fulfills its intended use when placed in its intended environment, i.e., confirm that we are building the **right** product. The goal of verification is to ensure that products meet their specified requirements to confirm that we are building the product **right**. This chapter addresses the problems of verification.

The need for verification can be illustrated from four well known technological incidents that would have been avoided if proper verification methods had been applied.

Therac-25 Radiation Overdosing: A radiation device for treatment of patients suffering from cancer incorrectly measured the produced radiation and at least six significant overdoses were experienced from 1985 to 1987, which resulted in deaths of three affected patients. The accidents occurred when the high-power electron beam was activated instead of the intended low power beam, and without the beam spreader plate rotated into place. Previous models had hardware interlocks in place to prevent this, but Therac-25 removed them, depending instead on software interlocks for safety. The software interlock could fail due to a race condition. The defect occurred because a one-byte counter in a testing routine frequently overflowed; if an operator provided manual input to the machine at the precise moment that this counter overflowed, the interlock would fail.

AT&T Telephone Network Outage: In January 1990, large parts of the US telephone network were affected by a 9-hour outage. The estimated losses caused by this incident exceeded US$ 100 million. A surprising reason for the outage was a wrong interpretation of a break statement in the C programming language.

Pentium FDIV Bug: In 1994, Intel introduced a new version of Pentium chips. Unfortunately, problems in the floating point division unit were experienced. Under certain circumstances, this unit produced incorrect results. The overall losses related to this issue reached US$ 500 million.

Ariane 5 Disaster: The European Space Agency (ESA) experienced a well-publicized accident in 1996. The reason for the crash has not been certainly proven, but the most probable reason for the control system failure was an incorrect datatype conversion from a 64-bit floating point to a 16-bit signed integer.

Testing and debugging of IEC 61499 is acknowledged to be inherently complex due to the nature of its applications, which are naturally distributed and parallel systems that can behave in a non-deterministic way due to the autonomous scheduling of control actions over the individual components of the distributed system. Therefore, testing is both challenging and necessary.

In this chapter we address these important questions. We will provide an overview of methods and approaches that have been already developed for general software verification. Then, we will briefly mention some techniques developed for verification of IEC 61131 control code and finally we will describe the current achievements of IEC 61499 verification.

8.2   General Software Verification

An alternative to extensive tests of newly developed software and hardware solutions is utilization of verification methods. With growing complexity of computer systems, the attention is continuously shifting from construction to verification. Software model checking is a powerful approach for the formal verification of software.

8.2.1   Dynamic Verification

Dynamic verification investigates correctness of the software execution. Depending on the scope, tests are organized into following categories:

Unit tests: tests of smallest isolatable program fragments

Module tests: tests of interconnected units

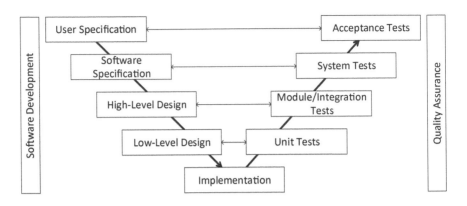

FIGURE 8.1
The V-model of the software engineering process.

Integration tests: tests of interconnected modules

System tests: tests of the entire system

Tests can be assigned to analyze the following: functional, reliability, performance, operability, security, compatibility, maintainability and transferability.

Test levels and their relation to the quality assurance of the final software product can be well illustrated on a V-model (see Figure 8.1). Currently, this is the dominating software verification method [1] and [8].

8.2.2   Static Verification

Formal methods represent verification approaches that do not require code execution. The source code is transformed into an abstract form which is then verified in a formal way. Two main representaives of static verification are model checking and theorem proving. More details and comparison of these two approaches can be found in [20].

8.2.2.1   Model Checking

The authors of [2] define model checking as follows: "Model checking is an automated technique that, given a finite-state model of a system and a formal property, systematically checks whether this property holds for (a given state in) that model." Model-based software verification was introduced in the early 1980s by Clarke et al. [5] and developed into a complex methodological framework. The achievements of Clarke's group are summarized in [6].

Various model checking algorithms have been developed that share a com-

mon property: before the verification can be conducted, the piece of software
to be verified has to be transformed into form of finite state automata.

The main challenges related to model checking are as follows:

Model Construction: The first problem is the construction of the model.
There is a significant semantic gap between a program expressed in terms
of a programming language (methods, inheritance, exceptions, etc.) and a
model of such a program that is described as a set of states connected by
transitions.

State Explosion: If a model contains too many states, it can be impractical.

Property Specification: Formalization of a functional requirement can be
clumsy by means of temporal logic.

Output Interpretation: An error trace can be very long and difficult to
map to a model and consequently to the original source code. This problem is related to the same semantic gap that makes model construction
difficult.

8.2.2.2 Theorem Proving

The utilization of theorem proving for software verification was introduced
by Hoare in 1969 who built his approach on the fact that: "Computer programming is an exact science in that all the properties of a program and all
the consequences of executing it in any given environment can, in principle,
be found out from the text of the program itself by means of purely deductive reasoning" [16]. Deductive reasoning is equivalent to theorem proving and
means that valid rules are applied to sets of axioms. Hoare proposed a calculus
that describes behavior of a program in terms of pre- and post-conditions in
form of triples in the following form:

$$P\{Q\}R, \tag{8.1}$$

which can be translated as: "If the pre-condition P is true before initiation
of a program Q, then the post-condition R will be true on its completion."
This concept was further developed by Dijkstra into the concept of "predicate
transformers" that given the program Q and post-condition R identifies the
set of pre-conditions that have to hold to meet the post-condition [7].

The main challenges of the theorem proving approach described in detail
in [17] are as follows:

Invention of lemmas and new concepts: Perhaps the biggest issue is the
first step: the construction of a given formula from the axioms by rules of
inference.

Examples and counterexamples: It is difficult to transform a set of examples into a logical proof and also it is difficult to automatically produce a counterexample from a failed proof.

Analogy, learning, and data mining: Users frequently blame theorem provers for repeating old proofs with new symbols instead of finding a generalized lemma. Unfortunately, generalization is a tricky problem that requires a cautious approach and thus learning and data mining for theorem proving are not easy.

An open architecture: It is challenging to design a prover that behaves autonomously (i.e., tries various proving strategies on its own) and can be easily reconfigured, so that it eliminates some proving techniques.

Parallel and collaborative theorem provers: Parallelization is a highly requested feature for provers working with large use cases. It is important for (i) decomposition of a large problem into smaller sub-problems and (ii) integration of already proven theorems into new solutions.

User interface and interactive steering: Finding a suitable form of presenting results of a prover to its users is another challenge, since the form needs to present all important facts and not bother the users with a lot of mindless details.

Education: The availability of people trained in the use of automatic reasoning is limited. The usual undergraduate courses on predicate calculus and logic do not provide sufficient understanding of all aspects of theorem proving.

A verified theorem prover: It is hard to guarantee that a prover always provides correct outputs.

8.3 Verification of IEC 61131-3

Ladder logic (LL) is one of the most used notations for programing PLCs defined by IEC 61131-3. It is a graphical notation whose basic elements are based on an analogy to physical relay diagrams. This made LL accessible for engineers who were not familiar with computer programming. However, LL programs are difficult to debug and verify because their graphical representation of switching logic obscures the sequential, state-dependent logic inherent in the program design [10]. A method for verification of programs developed in LL is proposed by Bender et al. [3]. This method represents a model-driven approach for formal verification of LL programs through model checking. The model is designed according to the LL metamodel which is able to model a subset of the LL language that is transformed to the time Petri nets that are then formally verified. The authors focused on the verification of generic, i.e., model independent, properties, especially the absence of race conditions in LL programs. However, the applicability of this approach is limited because only the controller is modeled while a plant and its model are not considered. As stated by the authors, taking the plant into account could eventually lead to a state space explosion, which would make the analysis infeasible.

Verification of programs written in Instruction List (IL), which is another textual PLC programming language standard defined by IEC 61131-3, is addressed in [13]. In order to be able to apply a Petri net-based verification framework on IL programs, the authors introduced Petri net semantics for the IL notation. Neither the previous LL verification nor this work models the plant behavior. Therefore, the authors extended their work and proposed an expanded version of their method in [14].

The results of these experiments have shown that the model-based verification is applicable for small and mid-size control applications. Large applications were transformed into giant models with huge numbers of states and state transitions that were computationally unmanageable.

8.4 Dynamic Verification of IEC 61499

This section provides an overview of approaches designed for testing and verification of correct behavior of a running IEC 61499 application. First, a monitoring and debugging framework is introduced. Second, adoption of a unit test for IEC 61499 is described.

8.4.1 Monitoring and Debugging

Monitoring and debugging are necessary features of any industrial automation system, because they enable prompt responses to unexpected situations and outages and determine the roots of the detected problems. According to [26], the key aspect of designing a proper monitoring system is the sound selection of the observed information inputs; because if too much information is harvested, performance issues arise. On the contrary, if too little information is gathered, the system behavior cannot be fully reconstructed and may lack determinism.

A monitoring and debugging infrastructure for distributed control systems according to IEC 61499, which is provided as part of the open source implementation of the 4DIAC [24], is introduced in [34]. Its important characteristic is that it provides insight into the control program run on the application level, i.e., a centralized view of the control system, which can be spread across multiple computational resources. It requires receiving data from various resources

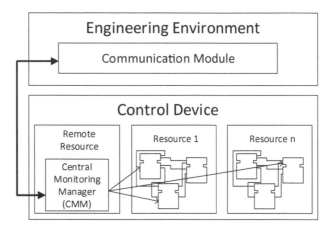

FIGURE 8.2
Monitoring manager as a single point of communication.

and their synchronization to prevent data inconsistency. The authors solve the issues related to distributed monitoring by introduction of the central monitoring manager (CMM). The CMM is the only connection point for the tool for monitoring and debugging actions. Thus, the number of communication channels is bound to one connection per device. Furthermore, the device central monitoring manager can be executed in an independent execution context (i.e., process or task), reducing the disturbance to the control application parts within the device. The architecture of this solution is shown in Figure 8.2.

The monitoring system consists of several extensions of the standard IEC 61499 management interface as follows:

Add watch: subscription of listening to changes of an event or data port

Remove watch: an inverse operation to *Add watch*

Read watches: update of all watched ports

Write with force flag: write command that ensures that the written data port value will not be overwritten

Write on event port: triggers execution of an event

The list of extensions illustrates that the functionality of this solution is based only on increasing the awareness of the current event and data port values. Breakpoints that are common parts of debuggers for procedural languages as Java or C++, are not supported.

8.4.2   Unit Tests

Since the tests of the final complex applications are very challenging, an approach based on early tests of individual software components has been introduced. Generally speaking, the process starts with unit testing, where the basic units, i.e., individual function blocks, are tested in isolation to guarantee that each one behaves correctly.

The application of the test first development (TFD) approach improves the quality of the program code units to detect possible problems as soon as possible. This approach has been widely adopted by software engineers and now it finds its way into the control domain. Its main principle is that the development process starts with definition of the requested behavior of a particular application fragment. This is followed by the development and the code is continuously tested against the specified requirements. Each fragment of the application is finished when all requirements are met.

The application of this approach for IEC 61499 development has been introduced by Hametner et al. who developed a unit test framework for event-driven control components modeled in IEC 61499 [11]. Within the development of the framework, the authors had to answer the following research questions:

1. What are testable software units in IEC 61499?

2. What is a suitable way for specifying test cases and unit tests in IEC 61499 so that the input for a unit test execution system can be generated?

3. How can we execute the tests and automate the test execution independently from the targeted control device?

Answer to Question 1: A testable software unit is a general concept going beyond the borders of IEC 61499. Szyperski defines a software component as a unit of independent deployment, of third-party composition, and with no (externally) observable state [25]. These characteristics are well aligned with the concept of function blocks and thus a function block is a testable software unit.

Answer to Question 2: The authors have developed a method for automated generation of event sequences for testing function blocks with many inputs. Sub-service sequences and calling mechanisms have been introduced to define several test cases (e.g., FB preparation) and reuse them many times.

Answer to Question 3: The authors developed a stand-alone unit test framework, which is suitable for industrial automation as the basis for a TFD approach. Moreover, the unit tests and the reports on the results can be generated automatically.

The detailed description of the infrastructure for IEC 61499 unit tests can be found in Chapter 7.

8.4.3 Keyword-Driven Testing

An alternative to the well-known verification approaches is keyword-driven testing (KDT) [35]. Its utilization for testing IEC 61499 applications has been proposed [12]. The principle of KDT is to describe each test case as abstractly as possible while making it precise enough to interpret and execute the test cases by a test tool [27].

KDT clearly separates the software development and definition of a test case. This enables us to define test cases during early development phases as expected in the TFD approach. The term "keywords" refers to words that control processing of a particular application. Thus, sets of keywords dedicated for testing a desktop application and an industrial control application will be different.

The test framework introduced in [12] contains following components:

Test Case Definition represents individual test scenarios and test cases and a basic set of keywords according to implementation standards.

Test Management coordinates and controls test case execution.

Fixture communicates between the test management system and the system under test.

Analysis and Reporting analyzes the individual test runs including coverage report generation and aggregation of test results.

The fixture component provides flexibility that enables us to compare behaviors of two systems developed in different ways. Authors of [12] illustrate this capability on parallel testing of two systems developed according to the standards IEC 61134-3 and IEC 61499.

8.5 Static Verification of IEC 61499

If the function blocks behave as expected when separated, it is possible to carry out integration testing or module testing in which the units within a module are tested together.

An attempt to model behavior of a IEC 61499 control system to prove the controller's safety was made by Schnakenbourg [22] who proposed a formal method based on translation of the function block code into synchronous SIGNAL language [19]. The method requires expression of the expected behavior of the control code in terms of SIGNAL language in order to use automated verification methods to confirm that the control application meets the specified properties.

Stanica proposed a methodology for the modeling of IEC 61499 function block behavior using timed automata in [23]. His work provides a method that is able to associate basic timed automata with each functional block and compose them to verify the execution of a network of functional blocks.

A framework for control system verification has been developed by Vyatkin and Pang [32]. The engineering methodology proposed is based on the component design of automated manufacturing systems from intelligent mechatronic components. The authors call their approach closed-loop modeling as it takes the controlled system into account. The main goal of this framework is to integrate heterogeneous modeling notations and tools (e.g., Unified Modeling Languages or Simulink) and its key feature is the inherent support of formal verification using automated transformation among different system models.

An alternative verification approach proposed in [33] is based on translation of the IEC 61499 control code to Esterel language [4]. A basic building block of an Esterel program is a module which has defined its input and output signals. Each signal may consist of a status and/or a value. Esterel programs are executed in discrete time steps called ticks and execution within a tick is instantaneous. The mapping between function blocks' and Esterel concepts is summarized in Table 8.1.

Vyatkin and Hanish developed a method for modeling behavior of function blocks and connected plants (closed-loops) using Net Condition/Event Systems (NCES) [31] modeling language. NCES reuses the basic concepts of Petri nets and extends them to capture events and data flow. Thus, a network of simple and composite function blocks can be mapped to a structure of net condition/event modules (NCEMs). This modeling approach does not take into account the communication delays caused by the communication infrastructure interconnecting individual function blocks. This is an important limiting factor which decreases the potential application of this method for real-world industrial use where the communication delays cannot be omitted from the model.

Lapp et al. [18] proposed a solution to overcome this issue. Their solution is based on bringing more temporal properties regarding run times and execution times into the formal model to consider them during verification. The key improvement is introduction of transformation rules that convert into NCEMs the function blocks and also service interface function blocks (SIFBs) that do not contribute directly to the control, but provide communication across multiple computational resources. Behavior of SIFBs is specified by service primitives according to ISO-IEC-10731. The service primitives intuitively describe the dependencies of communication signals and thus they are

TABLE 8.1
Mapping between Function Block and Esterel Concepts

Function Block Element	Esterel
Function block	Module
Event	Pure interface signals
Data	Value-only interface signals
Internal variables	Value-only local signals
Algorithms	Instantaneous modules
Function block network	Parallel composition of modules

also mapped to the corresponding NCEMs. The created NCES models have to be refined, i.e., extended with quantitative information such as execution times of algorithms in function blocks or response times of communication interfaces. Consequently, the corresponding parts of the models have to be quantitatively extended, e.g., with capacities, weighted arcs, initial markings or time intervals. The values for these parameters should be determined from hardware specifications and reasoned estimations regarding safety constraints.

A comprehensive work on formal analysis of IEC 61499 was developed by Čengić and Åkesson and described in two articles covering modeling [29] and execution semantics [30]. The former contribution provides a formal framework of IEC 61499 for mathematical modeling and comparison of different execution semantics and the latter illustrates how three execution semantics, the buffered sequential execution model, non-preempted multi-threaded, and the cyclic buffered execution model, can be mathematically formalized. Together, the models can be used to analyze and compare how an application would behave when executed using different semantics.

The introduced formalism can be integrated with a model of a controlled plant to obtain a closed-loop model. In such case, the model of the plant provides inputs for the controller and receives control signals, respectively. Moreover, the formalisms and mathematical models introduced within this work have been used as a basis for implementation of a runtime environment of Fuber [28] and a software tool of Moger as part of Supremica [21] that generates models suitable for formal verification.

8.6   Conclusion

The wide range of notations, methods and tools mentioned within this chapter illustrates the complexity of the software verification domain. Some require users with specialized skills which limits their adoption in industrial practice. This is not the case of the monitoring and debugging systems that are currently integrated within a development environment for IEC 61499 and therefore are directly usable. The unit tests represent promising approaches for verification of control code fragments, but their nature does not enable their use for modeling and verification of complex behavior that emerges from interaction of multiple function blocks. On the contrary, the static verification methods are able to do so, but require creation of an underlying model. Whereas there have been proposed several methods transforming the function block schema into a model, automated modeling of the hardware and the control plan remains an open problem. Currently, these models have to be provided by the control engineer and their use may be a challenging and time consuming task requiring specialized skills.

Bibliography

[1] P. Ammann and J. Offutt. *Introduction to Software Testing.* Cambridge University Press, 2008.

[2] C. Baier and J.-P. Katoen. *Principles of Model Checking.* MIT Press, 2008.

[3] D. Bender, B. Combemale, X. Cregut, J. Farines, B. Berthomieu, and F. Vernadat. Ladder metamodeling and PLC program validation through time Petri nets. In *Model Driven Architecture: Foundations and Applications*, pages 121–136. Springer, 2008.

[4] G. Berry and G. Gonthier. The Esterel synchronous programming language: Design, semantics, implementation. *Sci. Comput. Program.*, Vol. 19, No. 2, 87–152, 1992.

[5] E. Clarke, E. Emerson, and A. Sistla. Automatic verification of finite state concurrent system using temporal logic specifications: A practical approach. In *Proceedings of 10th ACM SIGACT-SIGPLAN Symposium on Principles of Programming Languages*, pages 117–126, New York, 1983.

[6] E. Clarke, Jr., O. Grumberg, and D. A. Peled. *Model Checking.* MIT Press, 1999.

[7] E. Dijkstra. Guarded commands, nondeterminacy and formal derivation of programs. *Commununications of the ACM*, Vol. 18, No. 8, 453–457, 1975.

[8] A. Engel. *Verification, Validation, and Testing of Engineered Systems.* John Wiley & Sons, 2010.

[9] M. Fisher. *Software Verification and Validation: An Engineering and Scientific Approach.* Springer, 2007.

[10] A. Guasch, J. Quevedo, and R. Milne. Fault diagnosis for gas turbines based on the control system. *Engineering Applications of Artificial Intelligence*, Vol. 13, No. 4, 477–484, 2000.

[11] R. Hametner, I. Hegny, and A. Zoitl. A unit-test framework for event-driven control components modeled in IEC 61499. In *Proceedings of IEEE Conference on Emerging Technology and Factory Automation*, pages 1–8. IEEE, 2014.

[12] R. Hametner, D. Winkler, and A. Zoitl. Agile testing concepts based on keyword-driven testing for industrial automation systems. In *Proceedings of 38th Annual Conference of IEEE Industrial Electronics Society*, pages 3727–3732, 2012.

[13] M. Heiner and T. Menzel. Instruction list verification using Petri net semantics. In *Proceedings of IEEE International Conference on Systems, Man, and Cybernetics.*, volume 1, pages 716–721, 1998.

[14] M. Heiner and T. Menzel. Time-related modelling of PLC systems with timeless petri nets. In R. Boel and G. Stremersch, editors, *Discrete Event Systems*, pages 275–282. Springer, 2000.

[15] T. Henzinger and J. Sifakis. The discipline of embedded systems design. *Computer*, Vol. 40, No. 10, 32–40, 2007.

[16] C. Hoare. An axiomatic basis for computer programming. *Commununications of the ACM*, Vol. 12, No. 10, 576–580, Oct. 1969.

[17] M. Kaufman and S. Moore. Some key research problems in automated theorem proving for hardware and software verification. *In Revista de la Real Academia de Ciencias Exactas, Físicas y Naturales. Serie A: Matemáticas*, Vol. 98, No. 1, 181, 2004.

[18] H. Lapp, C. Gerber, and H. M. Hanisch. Improving verification and reliability of distributed control systems design according to IEC 61499. In *Proceedings of IEEE Conference on Emerging Technologies and Factory Automation*, 2010.

[19] P. Le Guernic, T. Gautier, M. Le Borgne, and C. Le Maire. Programming real-time applications with signal. *Proceedings of IEEE*, Vol. 79, No. 9, 1321–1336, 1991.

[20] M. Ouimet and K. Lundqvist. Formal software verification: Model checking and theorem proving. Technical report, Massachusetts Institute of Technology, 2008.

[21] K. Åkesson, M. Fabian, H. Flordal, and R. Malik. Supremica: An integrated environment for verification, synthesis and simulation of discrete event systems. In *Proceedings of 8th International Workshop on Discrete Event Systems*, pages 384–385, 2006.

[22] C. Schnakenbourg, J.-M. Faure, and J.-J. Lesage. Towards IEC 61499 function blocks diagrams verification. In *IEEE International Conference on Systems Man and Cybernetics*, pages 1–6, 2002.

[23] M. Stanica and H. Guéguen. Using timed automata for the verification of IEC 61499 applications. In *Proceedings of 7th IFAC Workshop on Discrete Event Systems*, pages 375–380. Elsevier, 2005.

[24] T. Strasser, M. Rooker, G. Ebenhofer, A. Zoitl, C. Sunder, A. Valentini, and A. Martel. Framework for distributed industrial automation and control (4DIAC). In *Proceedings of 6th IEEE International Conference on Industrial Informatics*, pages 283–288, 2008.

[25] C. Szyperski. *Component Software: Beyond Object-Oriented Programming*. Addison-Wesley, 2002.

[26] H. Thane. *Monitoring, Testing and Debugging of Distributed Real-Time Systems*. PhD thesis, Royal Institute of Technology, Stockholm, 2000.

[27] M. Utting and B. Legeard. *Practical Model-Based Testing: A Tools Approach*. Morgan Kaufmann, 2010.

[28] G. Čengić and K. Åkesson. Definition of the execution model used in the Fuber IEC 61499 runtime environment. In *Proceedings of 6th IEEE International Conference on Industrial Informatics*, pages 301–306, 2008.

[29] G. Čengić and K. Åkesson. On formal analysis of IEC 61499 applications. Part A: Modeling. *IEEE Transactions on Industrial Informatics*, Vol. 6, No. 2, 136–144, 2010.

[30] G. Čengić and K. Åkesson. On formal analysis of IEC 61499 applications. Part B: Execution Semantics. *IEEE Transactions on Industrial Informatics*, Vol. 6, No. 2, 145–154, 2010.

[31] V. Vyatkin and H. M. Hanisch. A modeling approach for verification of IEC1499 function blocks using net condition/event systems. In *Proceedings of 7th IEEE International Conference on Emerging Technologies and Factory Automation*, pages 261–270, 1999.

[32] V. Vyatkin, H. M. Hanisch, C. Pang, and C.-H. Yang. Closed-loop modeling in future automation system engineering and validation. *IEEE Transactions on Systems, Man, and Cybernetics, Part C: Applications and Reviews*, Vol. 39, No. 1, 17–28, 2009.

[33] L. Yoong and P. Roop. Verifying IEC 61499 function blocks using Esterel. *IEEE Embedded Systems Letters*, Vol. 2, No. 1, 1–4, March 2010.

[34] A. Zoitl, G. Ebenhofer, and M. Hofmann. Developing a monitoring infrastructure for IEC 61499 devices. In *Proceedings of 18th IEEE Conference on Emerging Technologies and Factory Automation*, pages 1–6, 2013.

[35] A. Zylberman and A. Shotten. Test language: Introduction to keyword driven testing. *Methods & Tools*, Vol. 19, , 15–21, 2010.

9

Fault-Tolerant IEC 61499 Applications

Mario de Sousa

University of Porto

CONTENTS

9.1 Introduction

Anyone reading this book from the beginning is certainly more than aware that IEC 61499 was created with the intention of supporting distributed automation and control applications. Non-distributed IEC 61499 applications are possible and may even make sense in many situations. This chapter will nevertheless focus on applications whose executions are distributed among multiple execution devices, whether PLCs (programmable logic controllers), embedded platforms, or full-fledged computers. A data communication network linking these execution devices and allowing the distinct sub-applications distributed among the execution devices to exchange data and synchronization events is assumed.

When deploying a distributed application the designer should be aware of the possibility of partial failures. What is meant by partial failures is when one of the execution devices simply stops to execute for some unexpected reason, while the remaining execution devices continue their normal processing. This is a scenario that never happens in a centralised application running on a single execution device, where a typical failure of the execution device will result in the complete shut-down of the control application. Distributed applications with dependability constraints must therefore take these new failure modes into account.

Note that this failure mode is not specific to IEC 61499 applications, but is actually common to all distributed applications no matter the underlying framework that is being used. Even distributed IEC 61131-3 applications are subject to this occurrence, but in this case the cyclic execution semantics of IEC 61131-3 provides a straightforward means of detecting a partial failure. A typical distributed control application based on IEC 61131-3 consists of many individual applications, each running on an individual PLC, along with the specific configuration of the communication network used for exchanging data. All local programs are constantly running an event loop. Likewise, the communication network is typically configured to periodically exchange well defined data points amongst the PLCs. The periodic nature of the data exchange over the network allows any PLC to determine that a remote PLC is no longer active—be it from a failure of the PLC itself, or due to a network problem—when the remote PLC fails to communicate during one of the communication cycles. This is the mentioned method for failure detection, and procedures for handling these failures are typically either implemented directly in each of the locally executing programs, or are taken into account in the communication network configuration (for example, fall-back values for when a Remote Terminal Unit stops receiving updated values for its outputs).

On the other hand, IEC 61499 control applications have completely distinct execution semantics—FB activation is based on passing of events, and the transfer of data and/or events between execution devices does not necessarily happen periodically. In this scenario detection of the failure of a remote execution device cannot be based on the failure to receive an event and/or data from that device. However, we have seen that IEC 61499 applications are also subject to partial failures, and if they are to be used in situations with high dependability requirements then a solution to tolerate these partial faults is required.

In this chapter a design pattern for tolerating partial faults in IEC 61499 applications is presented. This design pattern is an adaptation of traditional approaches already applied in the field of high dependability applications, and to understand its properties and for ease of presentation, an overview of the commonly used expressions in this domain is presented in the first section of this chapter. The following sections present the design pattern and show how it can be implemented on an existing IEC 61499 execution environment, and how some of the dependability properties of the resulting IEC 61499 application can be determined.

9.2 Background

In the introduction to this chapter the words *dependability* and *failure* were used. For the community dealing with safety-critical systems these words and many others (the syntax) have very well defined meanings (the semantics)—this makes the exchange of ideas much easier between people who have agreed to common syntax and semantics and also provides a framework on which to build the ideas and approaches for building dependable applications. In this subsection we will explain some of these terms and their meanings, as well as the common approaches of achieving dependable systems.

9.2.1 System Dependability

The dependability of a system has been defined by the IFIP Working Group 10.4 on dependable computing and fault tolerance as "[..] the trustworthiness of a computing system which allows reliance to be justifiably placed on the service it delivers [..]". Another similar but slightly simpler definition is given by Avizienis et al. [2] as "the ability to deliver service that can justifiably be trusted".

These definitions are, however, very generic, and can mean many things to different people. Additionally they are not of much help unless these properties can be quantifiably measured. Once we can put a number to the trustworthiness of a system, meaningful comparisons can be made between competing solutions. To simplify this task, dependability has been further defined to mean a system that has, to varying degrees, some or all of the following properties:

1. Availability

2. Reliability

3. Safety

4. Maintainability

5. Integrity

6. Confidentiality

Availability is used to measure the proportion of time that the system is available to execute its functions. This is obviously a numerical value, and 99% would be considered typical for relatively mundane completely non-critical applications. Notice how 99% availability that is sometimes touted as very good by common service providers (e.g., Internet service providers, and electrical distribution companies) is actually not very high since it allows for more than 3 full days of non-service every year.

More formally, availability may be defined as the "probability that an item will perform its required function [..] at a stated instance of time" [3], and is given as a fixed probability value.

The next attribute, reliability, is defined as the "probability that an item will perform its required function [..] for a stated time interval" [3]. This is actually a measure of how long a system is able to work continuously without interruptions, but is formally given as a probability function for each time interval. If, for example, a hard disk has a reliability of 50% over 5 years, this would mean that it has only a 50% chance of working continuously without interruption over a 5-year period. The probability of continuous operation would decrease even further if longer intervals were considered. If we were to plot the probability of continuous operation of this same disk for a specific period t on a chart, we would get its reliability function $R(t)$.

Notice that availability and reliability measure distinct properties. An embedded system that automatically reboots every hour and takes 100 ms for each reboot would have an availability higher than 99.99%, but a very low reliability. On the other hand, a production machine that can work continuously throughout the year but is always stopped for maintenance for 1 week over the new-year period would have a much higher reliability, but an availability not much better than 98%.

Safety is defined as "the absence of catastrophic consequences on the user(s) and the environment"[2]. What is considered catastrophic must be defined for each specific system and use context. Notice again that neither availability nor reliability is synonymous with safety. A system with an availability of 0% can be perfectly safe (for example, an airplane that never leaves the ground). A system with a low reliability can also be safe if it is built so it always fails in such a way as never to produce catastrophic consequences (for example a barrier for a railway crossing that will always fall shut upon the occurrence of any failure).

Maintainability measures how easy it is to repair any failures in the system. Confidentiality is the absence of unintended disclosure of information, and integrity the absence of unintended changes to the system. Security is often viewed as a property of computing systems, and is considered as simultaneously providing the availability, confidentiality and integrity properties.

9.2.2 Faults, Errors, and Failures

In the context of dependable systems, *failure, error* and *fault* have very well defined and distinct meanings.

A failure occurs when a system does not fulfill its mission, i.e., when it does not complete the device's desired function. For example, a PLC fails to de-activate an actuator at the appropriate time.

A failure is said to be caused by an error. The error is usually explained as an incorrect internal state of the system that may lead to a failure. In the previous example of the PLC, the error would be the PLC's output relay that is stuck in the closed position. Notice that this error can exist when no one is aware of its presence. The error only causes a failure of the PLC when that specific output needs to be de-activated.

A fault is the cause of an error. The sticking of the output relay contacts could have been caused by excessive current on that output.

This sequence is summarized by the following sequence of events:

Fault $\rightarrow$ Error $\rightarrow$ Failure.

Notice how the classification of a specific situation as a fault, error or failure will depend on the boundaries of the system considered. For example, a PLC's power LED may have a failure (stop lighting up), but this same occurrence is merely considered an error for the PLC if it does not affect the execution of the PLC's program.

9.2.3 Achieving Dependability

There are two main approaches of achieving dependability:

1. Fault prevention

2. Fault tolerance

The fault prevention approach consists of building a system in such a way as to prevent the occurrence of faults. On a software project, this would mean following strict development rules that will reduce the probability of inserting bugs (software faults) into the software. On a hardware project, this could be achieved by using higher quality parts or by designing the system with higher margins.

The second approach, fault tolerance, assumes that it is not credible to predict all possible fault scenarios at design time, and accepts that faults will eventually occur in the system when it is in use. In this situation, the system has to be built in such a way that some faults will be tolerated, and the system therefore continues to provide correct service even in the presence of faults. This is the approach taken by the design pattern presented in this chapter.

Fault tolerance will be achieved through the introduction of redundancy. This means that some specific components of the system will be duplicated (or even tripled, quadrupled, etc.), so the failure of one sub-component does

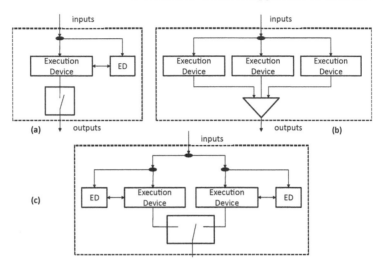

FIGURE 9.1
Fault-tolerant architectures.

not result in the failure of the whole system. An airplane with two engines that requires only only engine to fly is said to be able to tolerate the failure of one engine.

Architectures for introducing fault-tolerance may be classified into three main approaches, which are best explained in a diagram (see Figure 9.1). The two approaches based on the use of error detectors (EDs) suffer from the drawback that it is not always possible to identify the existence of an error, and any change from one processor to another commonly requires an additional reconfiguration delay that may not be acceptable in control applications with hard real-time deadlines. For this reason, the design pattern proposed in this chapter follows the architecture (c) that uses a voter to determine the consolidated output of the redundant replicas.

9.3 Replication in IEC 61499 Applications

As mentioned in the introduction, distributed IEC 61499 control applications are subject to failures of one (or more) of the execution devices running the distributed application. In centralized applications running on a single PLC, the traditional approach of tolerating faults is to introduce a second (or more) PLC that is identical (both in hardware and in software) to the original PLC. Doing the same for every execution device running a distributed application may be considered overkill, as the number of execution devices to be replicated may be rather large.

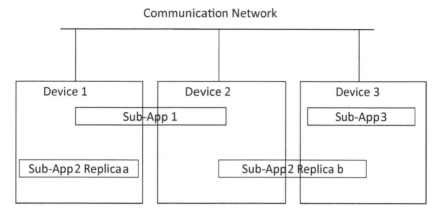

FIGURE 9.2
Distribution of replicated FBs among execution devices.

A possible solution is to exploit the distributed nature of IEC 61499 itself—each execution device may be configured to run replicas of the sub-applications running on other execution devices of the same system. For example, every execution device may be paired with one other execution device, and each execution device pair may be configured to execute all sub-applications of the pair. This would result in a situation similar to the replicated PLCs, where each replica runs an identical set of software as its pair.

This approach is still rather awkward and unnecessarily constrains the liberty of design of the application designer and/or developer. A better approach is to allow the application designer to decide which software components of the distributed application need to be replicated. The criteria for this choice are not relevant for now, but an obvious consideration will be to replicate the sub-components that have a higher influence on the required properties for the overall system (reliability, availability, safety, ...), and execute on the execution devices that are more prone to failure.

The proposed replication design pattern therefore allows for the replication of only parts of the IEC 61499 application, and places no restrictions on how these are allocated to other new or existing execution devices. For this fault-tolerant design pattern, the unit of replication is the FB. This means that the internal parts of a composite FB may not be partially replicated—either the entire composite FB is replicated completely, or not at all. This is due to the restriction that composite FBs may not be distributed among several execution devices, which in turn arises from the fact that composite FBs maintain internal state (copies if there input and output data ports) that must be stored in a single centralized location for each FB. IEC 61499 sub-applications may be partially replicated, as they do not have this same restriction.

Figure 9.2 shows an example of how parts of a distributed IEC 61499

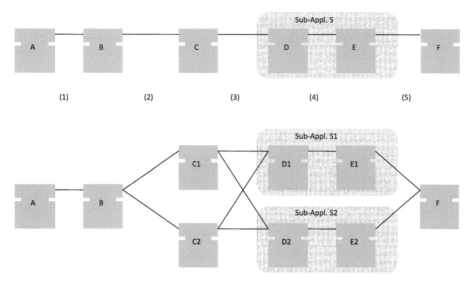

FIGURE 9.3
Interaction scenarios with replicated FBs.

application may be replicated and the replicas distributed among execution devices. Notice how replicas do not necessarily need to reside in a single execution device. It is acceptable to have one replica of a specific sub-application executing in one device, and another replica of that same sub-application distributed among two or more execution devices. In this case, however, some restrictions apply.

It is now possible to identify several interaction scenarios (see Figure 9.3), depending on whether the sender (2), the receiver (5), both(3), or none (1) of the FBs exchanging data and/or events are replicated. Notice that another possible interaction scenario arises when a sub-application is replicated in full. A sub-application contains data and events connections within itself, and these connections are also replicated when the complete sub-application is replicated. This constitutes another distinct interaction scenario, identified as (4).

In each of these scenarios, voters will be placed at the input of any FB that receives data from a replicated FB. Considering each of the interaction scenarios in turn, one voter will be required before FB F of the interaction scenario (5), and one voter before each of the replicas of sub-application S of the scenario (3) (see Figure 9.4). Special note should be made of the replicated voter in scenario (3). When a FB is replicated, any voters *supplying* data to that FB should also be replicated, and each of these voters should be allocated to the same execution device as each of the replicas. The reasoning for this architecture is to remove single points of failures (mentioned again in Section 9.3.3).

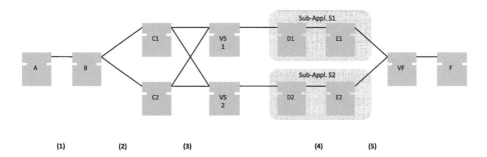

FIGURE 9.4
Interaction scenarios and placement of voters.

If the original non-replicated application contains a situation in which a single FB processes events and/or data coming from two upstream FBs (FBs G, H and I of Figure 9.5) or sends data and/or events to two downstream FBs (FBs J, K and L of Figure 9.5), the replication scenarios are still the same as those identified in Figure 9.3, but each of them will occur several times in the same application.

9.3.1 Message Propagation

The use of replicas usually makes sense only if all replicas behave in the exact same way when presented with the same set of input values. In other words, each replica of the same FB must be guaranteed to make the same decisions upon receiving the exact same input data and/or events. The internal algorithms may not, for example, depend on the use of a random variable that takes on a distinct value at each of the replicas. Assuming that none of the replicas has an internal failure, this will allow the replicas to always present the same values on their respective outputs, and therefore provide the voters with consistent input data to work with.

Notice that for the replicas to maintain the same output data and events, it will also be necessary for their internal variables to do the same. If an internal variable is mismatched between replicas, this may lead to a possible mismatch of the outputs in the future, even if the outputs are the same when the internal variable becomes mismatched. The set of current values of all the outputs and internal variables is often termed the FB's internal state. The property of keeping this internal state synchronized between replicas is termed replica determinism.

To maintain replica determinism, it is also necessary to guarantee that all replicas receive the same input data and events in the exact same order. Taking interaction scenario (2) of Figure 9.3 as an example, the data sent by FB B must arrive in the same order to each of the replicas of C. This may

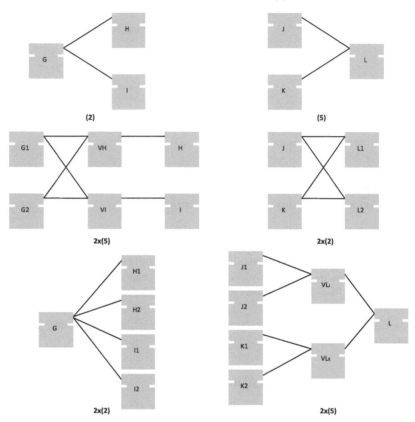

FIGURE 9.5
Replicating FBs with multiple upstream and downstream connections.

seem trivial, but we must also take into account that FB C may have other
input data sources other than FB B. For example, when FB L of Figure 9.5 is
replicated (L1 and L2), both replicas must receive the same inputs in the same
order, from both FBs J and K. This must be guaranteed to occur whatever
the underlying communication network is used.

Networks do not usually provide these ordering guarantees, and therefore a
strategy is required to guarantee that all messages arrive in the same order at
all replicas. Several communication algorithms have been developed that can
provide these guarantees, but they commonly require several messages to be
exchanged between all FBs so they can come to a common agreement of which
order to consider (see [4] for an overview). These solutions are not sufficient to
fulfill the common tight timing requirements of control applications. For this
reason we suggest the use of the timed-messages protocol [7]. This approach is
based on time, and assumes that the clocks of all the execution devices used
to run the replicated application are kept synchronized.

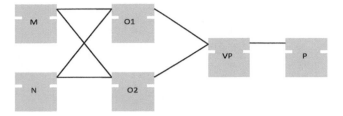

FIGURE 9.6
Synchronization with timed-messages.

Using this protocol, each message between two FBs (where at least one of them is replicated) carries with it a constant time in the future, a time at which the message is considered to become valid. The sending FB calculates this future time as the current time at which the message is being sent, plus a fixed time offset specific to the message. The fixed offset is determined off-line, and should be a value greater than all possible delays that the message may suffer while being sent to the destination FBs.

Consider, for example, FBs M and N sending messages to the replicas O1 and O2 (Figure 9.6). The fixed offsets used by each of the sending FBs (M and N) do not have to be the same; in fact, if one FB has a network connection with higher delays (for example, because it uses WiFi), it will probably require the use of a higher offset value. What needs to be guaranteed is that each message will arrive at both O1 and O2 before its validity time has expired. Each replica (O1 and O2) will then buffer all the received messages, and only release them at the exact time at which they become valid. This will guarantee that each message will be released in all replicas in the same order and at the same time.

This exact time synchronization provided by the timed-messages protocol is also used to allow the voting FBs to know when to vote. Remember that in the presence of a failure of one of the replicas, that replica will not send any output messages. Once the receiving FB (voter VP in Figure 9.6) has received a message from one of the working replicas, it must know the maximum time it must wait for the arrival of the corresponding messages from the other replicas. Only after this maximum waiting time has expired can it decide that the remaining replica has failed, and can therefore proceed with the voting using the data it has received up to that point in time.

To do this each replica O1 and O2, when sending an output message to the voter, will add to this message the time at which the message can be considered valid. This validity time is determined by adding a fixed offset to the validity time of the event that started the chain of execution. Let us consider, for example, that FB M sends a message to both replicas of FB O (O1 and O2). Each of these replicas will process the message coming from M, which will result in each replica sending a new message to the voter VP. In this case, the validity time of the messages sent by O1 and O2 will be determined by adding a fixed offset to the validity time of the message arriving from M. The value

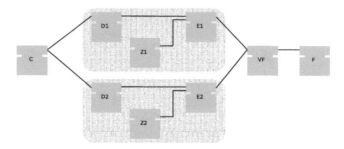

FIGURE 9.7
Synchronization of internal events.

of the fixed offset added by O1 and O2 must be the same, so the validity time of both messages sent by O1 and O2 have the exact same validity time too. The fixed offset will be determined off-line, and must take into account not only the maximum transmission time of the messages sent by O1 and O2, but also the execution time of the O1 and O2 FBs themselves.

The validity time of the messages arriving at the VP voter is also used to group them together. The voter will vote only on messages whose validity time is identical, as only these messages are the results of the same source event. It would not make sense to let the voter VP vote on one message arriving from O1 that originated from the processing of a message from M, together with another message arriving from O2 but originating due to a message from N. Although this occurrence might be considered unlikely, it is not impossible— for example, we can consider the large variation of possible network delays on complex networks using a communication protocol based on UDP.

It should also be noted that IEC 61499 applications do not usually run in isolation, and they commonly require an input to the physical world. In IEC 61499, this is done by using service interface FBs (SIFBs) that provide the value of a physical input connected to the execution device. When introducing replication, it might also become necessary to replicate the physical inputs, especially if the replication was introduced because we cannot place sufficient trust on the execution device to which the input is connected. In this situation, the same physical input will be connected to two or more execution devices, and the SIFBs providing the value of this same physical input on each of the execution devices must be viewed as replicas of the same SIFB—FBs Z1 and Z2 in Figure 9.7.

Note that the connections within the replicated sub-applications of Figure 9.7 constitute an interaction identified as type (4) in Figure 9.3. The exact time synchronization of the timed-messages protocol is also important in interactions of this type. Here it may be necessary to guarantee that the physical input data is read at the exact same time in each replica. We can force the simultaneous reading of the input by using the timed-messages protocol in the event that requests the reading of the physical input (assuming a requester SIFB is used). In this case, the fixed time offset will be determined by the maximum execution time of the FBs that will be executed from the previous synchronization point (in this case, the arrival of a message from C which is released simultaneously in both replicas) until the reading of the physical input (arrival of event at FB Z in Figure 9.7). Since the physical input is always read by every replica at the exact same instant, the corresponding value of the output of FB Z will also always be the same in every replica. The output of Z can therefore be used safely by any other FB in the replica while still guaranteeing synchronization between the replicas.

Note that this synchronization may not be enough to guarantee internal state synchronization between replicas if the exact time at which the physical input is read by the SIFB is not under the control of the application. This is the case if a responder SIFB is used as an interface to the physical inputs, as this FB may read the physical input and generate output events at times that are only controlled by the IEC 61499 execution environment. These situations are best avoided if very strict replica determinism is required. When avoidance is not possible, it becomes necessary to share among the replicas all the input values read by each of the replicas, and let each one vote (using the same voting algorithm) on the value to use. This approach has the drawback of introducing extra messages on the network and more delays in the processing sequence.

9.3.2   Platform Execution Semantics

As stated previously, for the replicas to maintain synchronized internal state, the execution of every replica must be identical. This was explained as replica determinism, and a restriction was imposed that the FB algorithms must be deterministic. However, the FBs are typically executed by an IEC 61499 execution environment, and replica determinism also requires that the execution environments also execute each replica in exactly the same way.

This is especially important when replicas are sub-applications or composite FBs, where the execution of the replica includes the propagation of events between the FBs in each replica's FB network. The order by which the events in a FB network are executed must be the same for every replica. Special care must therefore be taken when one replica executes completely within one execution device, whereas another replica is distributed between second and third execution devices. In this last case it is not possible to guarantee replica determinism.

It should be noted that the time each replica takes to execute is not important—the execution devices do not need to have the same processing capacity, nor do they need to have the same processing load. The timed-messages protocol guarantees that the initiating events are processed at the exact same time at each replica, and when this event is released, the same data is available at the FB's inputs of all replicas. On the other hand, the IEC 61499 execution semantics guarantees that when a FB is activated by an event, it will start by reading all the relevant data provided at the FB's data inputs. Even if the input data changes while the FB is processing that event (because a message transmitting a new data value has reached its validity time), those changes are simply ignored by the executing FB.

These semantics do not apply to sub-applications, that do not keep copies of the values at their inputs when they start processing an input event. In this case, it is no longer possible to guarantee internal state synchronization. Although the data will become valid at the same time in each replica, no guarantees can be made of the time at which the values will actually be read (as the execution devices may have different processing speeds and loads). In summary, state synchronization between replicas is simply not possible when replicating sub-applications.

When state synchronization between replicated FBs is required, it is also important to guarantee that each FB processes only one event at a time. A replica that is already processing one input event and receives a second input event should only process the second event once the first has been completely processed. Basically, the execution of the replica FBs must be serializable. This restriction is required since the replicas are not guaranteed to execute at the same speed, and a situation could occur where one replica executes two events, one after the other, while another replica will execute the two events concurrently.

It is also important to note that an execution device that starts to fall behind in processing the timed-messages that it receives must store those messages and process them in the order by which they reach their validity time. Consider the example of an execution device that receives two events with validity times of 100 and 110 time units and a data value with validity time of 120 time units. At $t=100$ it starts processing the first event, but only finishes this processing at $t=140$. At this time it will start processing the second event, but it must do so without considering the data that became valid at t$=120$.

9.3.3   Voting Function Blocks

Voting FBs are used to consolidate into a single output value the various output values generated by replicas of the same FB. Many voting strategies and algorithms are possible, and choosing which algorithm to use will need to take into account the desired semantics of the application.

Probably the most common voting algorithm is the majority voter that produces as its output the input value that occurs most often (i.e., that represents the majority). This algorithm usually has two possible variations. One alternative proceeds with the voting as soon as it has received the same value from a majority of the replicas, even though some of the replicas have not yet sent their values. Another variation waits until it has received a value from each of the replicas before proceeding with the voting. This last version must also be configured with a time-out, so it does not wait eternally for a value that will never arrive due to a failure of a replica. This second approach does not make sense for the replication design pattern adapted for IEC 61499, as the data arriving at the voters is only considered valid at the validity time contained in the message itself. The voting always proceeds at this exact time, with whatever values may have been received up to that time. Any messages arriving at a voter with a validity time in the past must simply be ignored as they are considered invalid—the replica that sent these messages is considered to have failed.

The majority voting strategy cannot always be used—for example, when voting on floating point values. As is commonly known, the handling of floating point values may differ between hardware platforms in some slight but sometimes important detail, or perhaps the compiler for each platform translates the exact same mathematical expression into slightly different assembly operations, which may cause the values to be voted upon to be not exactly identical. Another possible source of variation is when each replica reads its own physical input of the same analog value, which necessarily must include some conversion error which may result in slightly different input values in each replica. These situations require the use of voting algorithms that do not require the input values to be identical.

Other alternative voting strategies are the averaging voter, which produces as the output the average of all the inputs, and the median voter, that chooses for the output whichever input falls in the middle. A mixture of these two strategies is to throw away the outlier values (maximum and minimum), and determine the average of the remaining values. Yet another alternative is to only consider as outliers those values that differ from the median by more than a configurable constant (for example, any value that differs by more than 30% of the median is considered an outlier).

The averaging voter requires that the input values be numerical, whereas the median voter only requires that the input values can be ordered. When the input values do not satisfy any of these properties, the majority voter can be used, as the only property required of the input values is that they can be compared for equality.

Whatever the voting strategy, note that the voter will reside on the same execution device of the FB receiving the messages. This means that when a FB is replicated, the associated voter is not replicated. The voter will therefore constitute a single point of failure. In the last section of this chapter, it will be shown that the reliability of the resulting replicated application can never be higher than the reliability of any single non-replicated component, including the voter. For this reason, non-replicated voters and the FBs that process their data must have high reliability, or otherwise they too must be replicated. Replicating a voter implies replicating the FB receiving the data of the voter, and will result in communication scenario (3) of Figure 9.3.

A control application will necessarily include some FBs that activate a physical output, and usually it is not considered possible to have two or more execution devices control the same physical output. Accepting this would imply that at least one non-replicated voter must exist at the end of the processing sequence of FBs to write to that physical output, therefore limiting the reliability of the resulting application. Note, however, that in extreme cases, the physical outputs can be replicated, and the final voter can be built in hardware—for example by placing relays in series or in parallel, depending on voting strategy required for the application. In very extreme cases, even the actuators can be replicated, and all actuators connected to the same physical device. In this case it can be said that the final voter is the physical device that connects all actuators, and therefore sums all the forces applied by the actuators.

In summary, no voting strategy is optimal for all situations. The replication design pattern therefore requires that the voters be explicitly inserted by the application designer. The voting FBs can be chosen from a library of FBs that have been thoroughly tested and validated, and even perhaps approved by an external authority. IEC 61499 does not support FBs with variable numbers of inputs, so each voting strategy requires several FBs—each version will have a set number of inputs and be able to handle a fixed number of replicas. The number of replicas must therefore be fixed at design time.

9.3.4　Communication Function Blocks

IEC 61499 applications access the communication network through communication SIFBs. Many different implementations of these FBs are allowed by the IEC 61499 standard, with each implementation providing support for a specific communication protocol stack or network. The IEC 61499 standard does, however, specify a common interface for all these communication SIFBs, so a distributed application can easily be reconfigured to run on a distinct communication network by simply replacing the communication SIFBs.

This adaptability of the IEC 61499 architecture is used to support the replication design pattern. By providing specific implementations of the communication SIFBs that support the timed-messages protocol, the application designer is freed from having to implement this protocol. The design pattern assumes two distinct pairs of publish and subscribe communication SIFBs. One pair is used for many-to-one message propagation, whereas the other pair is used for one-to-many propagations.

The one-to-many publisher and subscriber is used in interaction scenario (2) of Figure 9.3, and is shown as "1tom" in Figure 9.8. The many-to-one pair is used in scenario (5), and is represented as "mto1". Scenario 3 uses a mixture of these two pairs—the one-to-many publisher, with the many-to-one subscriber.

Note that if any time synchronization is required inside a replicated sub-application (e.g., because it reads a physical input, as explained in Sec-

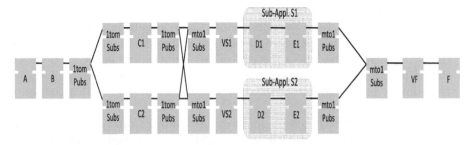

FIGURE 9.8
Interaction scenarios and publish and subscribe SIFBs.

tion 9.3.1), the one-to-many publish and subscribe communication SIFB pair may be used. In this case, each subscriber will be configured to receive from a single publisher as they should not exchange data between the replicas. The publish and subscribe communication SIFB pairs are merely inserted to synchronize the reading of the input data.

9.4 Replication Framework on FORTE

Support for the described replication design pattern has been implemented on the FORTE IEC 61499 execution environment [1].

9.4.1 FORTE Execution Semantics

As explained above, for the replicas to maintain synchronized internal states, the execution semantics of the IEC 61499 execution environment must be deterministic. In other words, for each activation of a source event (i.e., an event that is generated autonomously by a SIFB without having received an input event—for example a subscribe FB or a periodic timer FB), the sequence of activated FBs must always be the same in all replicas. FORTE is relatively well suited for supporting the replication design pattern as its execution semantics almost guarantee a deterministic execution of event sequences.

An instance of the FORTE execution environment implements one or more resources—what we called execution devices up to now. Since each resource in FORTE is handled autonomously, without lack of generality, the following discussion will consider only one resource per FORTE virtual machine.

All events inside a resource are handled by a single background thread, and zero, one or more real-time threads. Each of these threads maintains a queue of outstanding events that have been released and are waiting to be processed. Each source event is explicitly mapped onto one real-time thread (i.e., placed on that thread's event queue) or left to be executed by the single background thread. Each event is handled individually in a first-in-first-out order. While handling any given event, new events that are generated during that execution are added to the end of the outstanding event queue of the thread processing the original event. This means that all other events generated during processing are handled by the same thread in which the source event is handled.

Note that a FB that accepts input events coming from two distinct SIFBs will run in the threads to which the SIFBs are associated. The FB may therefore run in two distinct threads, perhaps even simultaneously (on a preemption-based scheduler, higher priority real-time threads will pre-empt lower priority threads). This will destroy the serializability of FB processing, so care must be taken when configuring the association between SIFBs and execution threads to guarantee that each replica will run inside a single thread.

Although it has not yet been stated explicitly, we can see that the described replication design pattern ensures that all data and events arriving at a replica must come through the timed-messages protocol. Replicas may therefore be freely allocated to any thread in a resource (as long as they always run in only one thread), as any data and events they will receive will be protected by the timed-messages protocol.

9.4.2   Communication Function Blocks

The replication design pattern requires the use of two pairs of publish and subscribe FBs that implement the timed-messages protocol (one-to-many and many-to-one pairs). In principle a specific version of these pairs of FBs would be required for each communication protocol to be supported. However, since all FORTE communication SIFBs are already implemented using a layered design pattern, only two new replication layers had to be added.

When adding a publish or subscribe FB to an application, FORTE requires that these FBs have an input parameter that lists all the communication protocols that this specific FB will use. Standard communication protocols follow a layered approach, and this gives the designer full control of which protocols to use. An application designer who adds replication to an existing replication will therefore merely need to instantiate the new replicas, and add one of the synchronization protocols (one-to-many or many-to-one) to the protocol stack of all communication SIFBs sending and receiving data from the replicas.

Remember that the synchronization protocols must be configured with an appropriate fixed offset value. It is up to the application designer to determine the correct value for this fixed offset as it will depend on the maximum expected network delays of the current setup as well as the maximum execution time of some replicated FBs.

Note too that when using the many-to-one publish and subscribe pair, the subscribe SIFB must provide a set of data outputs for each replica. These outputs are all then connected to the voter that will consolidate them into a single set of output data. For example, if each of the many-to-one publishers sends two data values, and three replicas are used, each publisher will have two data inputs for each data value, while the subscriber will have six data outputs (two per replica).

The implementation of the many-to-one and the one-to-many replication layers both send with the data the time at which the data and associated event will become valid. On the subscribe side, both layers will buffer all received messages until the validity time has been reached. Note that the many-to-one publish FBs need access to the validity time of the original source event, as explained previously with Figure 9.6. This time needs to travel with the events as they propagate in the replicated composite FB until it reaches the publish SIFB. The standard implementation of FORTE merely sends events and is not able to associate this validity time with the propagated event. The revised implementation of FORTE that supports the replication design pattern has been changed so the validity time is automatically propagated with the events in the FB network.

9.4.3   Voting Function Blocks

No significant effort has focused on implementing a full suite of voting FBs for the revised implementation of FORTE that supports the replication design pattern, mostly due to the large number of permutations of data types, voting algorithms, and number of inputs in each voter. Only very simple and basic voters are provided. Implementing new voters is nevertheless relatively trivial—it is the same as implementing any other basic FB in the FORTE and 4DIAC development environment. Any replicated application designer is therefore encouraged to implement voters that cater to the situation at hand, taking into account the failure modes that need to be tolerated.

9.5   Example of Replicated IEC 61499 Application

This section will present an example of how to apply replication in a IEC 61499 application, using the previously described design pattern. The application is very simple and the focus will be on what needs to be added to provide fault tolerance.

9.5.1   Example Application

The application that will serve as an example is taken from a larger application that handles the control of work piece distribution within a plant based on conveyors. To keep the application simple, only two types of conveyors are

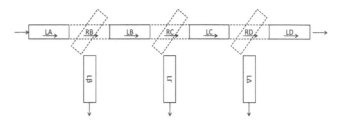

FIGURE 9.9
Conveyor layout.

considered: linear and rotating. Both conveyor types are relatively short, and can only be occupied by a single work piece at any time. Each conveyor has a single sensor that detects a work piece centered on the conveyor. The conveyors are mounted according to the layout in Figure 9.9.

The work pieces arrive on the left and should be transferred to any of the processing stations at the bottom. Each conveyor is controlled by a single FB instance of the appropriate type (linear or rotating controller type). Each instance is also executed by its own independent execution device (or PLC), to which the relevant conveyor's actuators and sensors are connected.

The control algorithms of each FB are very simple. The linear conveyors simply try to push the work piece onto the following conveyor as soon as it becomes free. Rotating conveyors will receive a work piece from the conveyor on their left, and will push it to one of the following conveyors (to the right or the bottom), whichever becomes free first.

Due to possible work piece slippage during transfer from one conveyor to the next, a conveyor is only considered free to receive a new work piece when the current work piece has reached the following conveyor. Every linear conveyor has therefore four basic internal states:

1. Free and stopped

2. Free and running (receiving work piece from previous conveyor)

3. Occupied and stopped

4. Occupied and running (transferring work piece to next conveyor)

The rotating conveyors will have a slightly more complex state diagram as they need to decide which conveyor will receive the next work piece.

The FB network controlling the sequence of conveyors is depicted in Figure 9.10. Note how each conveyor needs to receive and send events to each of the following conveyors and the previous conveyor. The data sent to each of the following conveyors is used to tell that conveyor to accept a new workpiece. The data received from each of the following conveyors is to determine whether that conveyor is free to receive a work piece.

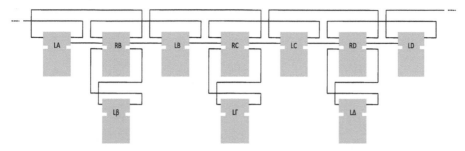

FIGURE 9.10
Example of non-replicated application.

9.5.2 Adding Replication

Let us accept that this application did not achieve the required reliability, and that it was therefore decided to add replication to the controllers of the conveyors along the horizontal line. It was also decided that each execution device would control two conveyors simultaneously, as shown in Figure 9.11.

The replicated control application, including the required publish and subscribe FBs, will become rather complex as in this application each FB needs to simultaneously transmit and receive events to each of the surrounding FBs. Figure 9.12 represents the replicated application, considering only the communication SIFB required to transmit the information forward to the following FB (telling it to start receiving a work piece). The communication channel that goes backward to the previous FB (indicating it is free to receive a work piece) is not represented.

Notice how this situation contains an interaction of type (3) (represented in Figure 9.3) between RB and LB and an interaction type (5) between RB and Lβ. For this reason both the one-to-many and the many-to-one publish

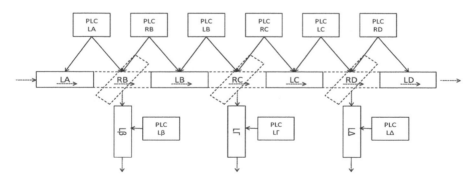

FIGURE 9.11
Allocation of replicas to PLCs.

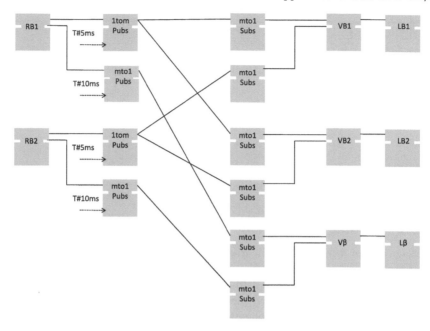

FIGURE 9.12
Example of replicated application.

communication SIFBs are used. All the subscribers will be of the many-to-one type.

Notice too how each of the publishing communication SIFBs need to be configured with an extra delay parameter. This delay parameter must be the same in each of the publishers transmitting to the same replica so the voter can group the arriving messages together.

9.5.3   Handling Voting

Since each FB in this application has only two replicas, the voters used will assume that any message received is correct and will therefore choose the first message to arrive as the message to place on its output. Voting on the physical outputs is also required, as each physical actuator is controlled by two PLCs, instead of only one. In this application, it is assumed that a failed PLC will place all its digital outputs at value false (i.e., each PLC follows a fail-stop failure model). This means that the outputs of both PLCs can be electrically connected in parallel, so that the controlled motor will be activated as long as at least one of the PLCs activates its associated output. This is basically an at-least-one voter built as an electrical circuit.

9.6 Quantifying System Reliability

The reliability of a system $R(t)$ is the probability of correct operation during a specified period of time. The system is assumed to be built from several sub-components whose reliability is known. Determining the reliability of the complete system is therefore a relatively straightforward computation of probabilities. A systematic way of doing this is to use the reliability block diagram (RBD) method [6].

The system is modeled by a diagram of blocks, with an entry on the left and an exit to the right. A block is added for each sub-component, and the connections between the blocks follows the logical function they perform (and not their physical layout). For example, if all sub-components need to work correctly for the system to work, they are placed in series between the entry and the exit nodes. If the system will work correctly if only one of the sub-components is operational, the blocks are placed in parallel.

Assuming that the probability of failure of each sub-component is independent, the reliability of the system composed by blocks in series is determined by the product of the reliabilities of all the sub-components. For a system composed of blocks in parallel, it is easier to do the calculations based on the unreliability of the sub-components (given by $1 - R(t)$). In this case, the unreliability of the system is determined by the product of the unreliabilities of all the sub-components.

The reliability of a system composed of a complex network of series and parallel sub-components can be determined by successively reducing each sequence of blocks in parallel or series to a single block of the equivalent reliability. As an example, consider a trivial system consisting of a single non-replicated input, a triply replicated processing block, followed by a non-replicated voter and output block. Assuming that each block is executed by a distinct execution device and that the voter uses an algorithm that only requires one of the processing replicas to be functioning to produce an output, this system may be modeled by the RBD in Figure 9.13. Looking at this RBD, model it is easy to understand that the reliability of the voter will reduce the overall reliability of the system (as it is placed in series with all the other blocks), so it is important that the voter have high reliability. If the voter is implemented inside the same execution device as the the final output device, it may be removed from the RBD model and it will not affect the overall reliability of the system.

It should, however, be noted that not all systems can be described by a network of independent series and parallel components. If the voter uses majority voting and therefore requires n out of m replicas to remain functioning, simply placing blocks in parallel will not do and special nodes that identify the n out of m semantics (for example 2 out of 3) need to be used in the RBD. Another example in which drawing a RBD is more complex is the example

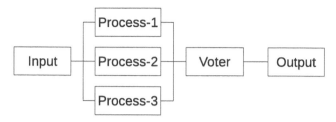

FIGURE 9.13
A simple reliabilty block diagram.

from the previous section (see Figure 9.11) in which the same PLC is used by several replicas and these PLCs cannot be organized into disjoint groups. In this case it becomes necessary to have multiple blocks in the RBD representing the same execution device, invalidating the use of the previously mentioned method of reducing the RBD model and calculating the resulting reliability as that method is only valid as long as all blocks are independent (i.e., the probability of failure of one block in the RBD is completely independent of the probability of failure of another block). More complex algorithms may be used in these cases which, due to their complexity, are not usually feasible for manual calculation [3, 5]. In these cases software tools that perform RBD analysis are preferred.

Another situation that may require software tools is when the reliability of a block is not constant, but varies in time according to some reliability function $(R(t))$. One that is commonly used in this case is the exponential function as it represents a system with a constant probability rate of failing. In other words, the probability of failing in the future is always the same and is independent of how long it has been working correctly without any failure. The exponential function is given by

$$R(t) = e^{-\lambda t} \tag{9.1}$$

where λ is the constant failure rate. In this scenario the multiplication of reliabilities for blocks in series (multiplying exponential functions) results in simply adding the failure rates, but the situation for blocks in parallel is unfortunately more complex.

The reader should also be aware that in the previous examples only the reliability of the hardware is taken into account. In reality the extra software added to support replication may contain unintentional bugs, and this too may be modeled by reliability functions. In this case the RBD will be composed of blocks that represent software functions, instead of hardware.

9.7 Summary

When developing distributed applications, it becomes necessary to take into account the possibility of partial failures. Measures that tolerate the more probable faults may need to be used if the application is required to achieve a certain level of reliability. When the application also has real-time requirements, the fault-tolerant measures used must be limited to those that adopt the forward-recovery approach.

Adding module replication with voting on the outputs of the replicas is one such measure, but many voting algorithms require that all replicas be kept synchronized. In this chapter, an analysis was made of how this may be achieved for IEC 61499 applications and the restrictions that need to be imposed on the IEC 61499 execution environments.

It may be concluded that although the standard design patterns for replicas and voting may be easily adapted for the IEC 61499 applications, care must be placed on the IEC 61499 execution semantics implemented by the execution devices. These must use deterministic algorithms, and event execution must be serializable. A global order on event execution must also be enforced, and the timed message protocol was proposed as a means of achieving this with little messaging overhead. Support for this protocol was added to the FORTE execution environment, but at the time of writing has not yet been introduced into the mainline distribution of this software package.

The author would like to thank the entities that funded this work, namely the ERDF-European Regional Development Fund through the COMPETE Programme (operational programme for competitiveness) as well as Fundação para a Ciência e a Tecnologia (Portuguese Foundation for Science and Technology), Projects FCOMP-01-0124-FEDER-037281 and FCT EXPL/EEI-AUT/2538/2013.

Bibliography

[1] 4DIAC and FORTE project. http://www.fordiac.org/.

[2] A. Avizienis, J.-C. Laprie, B. Randell, and C. Landwehr. Basic Concepts and Taxonomy of Dependable and Secure Computing. *IEEE Transactions on Dependable and Secure Computing*, pages 11–33, 2004.

[3] A. Birolini. *Reliability Engineering: Theory and Practice*. Springer Verlag, 2014.

[4] X. Defago, A. Schiper, and P. Urbanr. Total Order Broadcast and Multicast Algorithms: Taxonomy and Survey. *ACM Computing Surveys*, Vol. 36, No. 4, 372–421, 2004.

[5] S. Distefano and A. Puliafito. Dependability Evaluation with Dynamic Reliability Block Diagrams and Dynamic Fault Trees. *IEEE Transaction on Dependable and Secure Computing*, Vol. 6, No. 1, 4–17, 2009.

[6] M. Modarres, M. Kaminskiy, and V. Krivtsov. *Reliability Engineering and Risk Analysis: A Practical Guide*. CRC Press, 2009.

[7] S. Poldena, A. Burns, A. Wellings, and P. Barret. Replica Determinism and Flexible Scheduling in Hard Real-Time Dependable Systems. *IEEE Transaction on Computers*, Vol. 49, No. 2, 100–111, 2000.

10

Developing IEC 61499 Communication Service Interface Function Blocks in Distributed Control and Automation Applications

Georgios Sfiris

Aristotle University of Thessaloniki

George Hassapis

Aristotle University of Thessaloniki

CONTENTS

10.1   Introduction

Typical industrial control and automation systems for process and manufacturing industries perform a variety of tasks in parallel, such as reading sensor data, computing control algorithms producing outputs to regulate process parameters and updating display and transferring process data to the enterprise information systems. These tasks, depending on the number and complexity, are implemented by hierarchical and distributed computer architectures consisting of a multiplicity of computational nodes. The first such industrial control and automation system was announced by Honeywell in the mid 1970s under the trade name TDC 2000/3000. Since its introduction, this concept was used widely in power plant control and manufacturing systems.

The use of local area networks to interconnect computers and other devices within an industrial control and automation system has become popular since 1980 and today is the norm in all the systems installed in most industrial sectors. There are two implementation approaches. The first utilizes proprietary designs and architectures and the second one considers open distributed architectures with communication over Ethernet type networks or networks built according to publicly available standards. Considering open architectures, any industrial control and automation system is usually structured into several hierarchical levels. Each level has different communication and functionality needs which place different requirements on the communication network of the level. Figure 10.1 shows an example of the hierarchy of an industrial control and automation system.

The lowest level of the hierarchy is the field level which includes such devices as actuators and sensors. The task of the network at this level is to transfer data from the sensors of the manufacturing process to the controllers which perform algorithms for the regulation of the industrial process to desired operating points and transfer the results of the algorithms to actuators. This is a large category of networks characterized by short message sizes and predictable response times in the range of 0.001 to 0.01 sec. In general these networks connect smart devices which offer diagnostic and configuration capabilities at the cost of more intelligence, processing power and price.

At the next network level, the control level in Figure 10.1, the information flow mainly consists of peer-to-peer communication between controllers and computer systems used for human and machine interface (HMI), historical archiving, supervisory control, synchronizing machines and handling

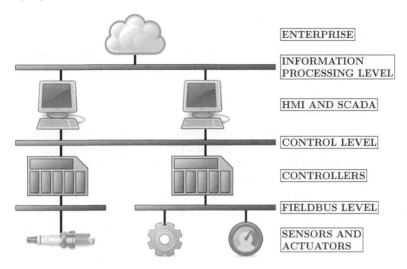

FIGURE 10.1
A hierarchical industrial automation system.

events. For the last tasks, the network requirements differ from those of the first tasks making an adaptable message segmentation necessary. The sizes of the messages and files for supervisory and synchronizing control are relatively large and their transfer response times are in the range of 0.5 to 2 sec. As far as instant display and historical graphs are concerned, the sizes are even larger than those of the supervisory and synchronizing control and for their transfer, response times in the range of 2 to 6 sec are required. Many companies have developed proprietary networks for both the field level and the control level. The field level networks are usually called fieldbuses and they are distinguished by characteristics such as message size and response time. Typical trade names are the FOUNDATION Fieldbus [6], CANopen [2], DeviceNet [33] and Profibus-DP [19]. At the control level, some of the fieldbuses can also be used but networks with higher data transfer rate, better data protection and guarantee time synchronization along with not-so-demanding real time response are preferable in most cases. Typical trade names of networks for this level are the ControlNet [34], Profinet [20] and Modbus [29, 28, 27]. In addition, switched or Industrial Ethernet with TCP/IP [26] is frequently used.

The top network level of an industrial control and automation system is the information level which gathers management information from the control level and manages all the plant automation functions. At this level there exist large scale networks, e.g., Ethernet and WANs for process or factory planning and management information exchange and Ethernet networks can act as gateways to connect other industrial networks.

The increasing deployment of industrial control and automation systems with open distributed computer architecture has placed a high demand on tools and techniques for designing distributed software in a productive and efficient way. Previous standardization efforts which led to the IEC 61131-3 standard [18] proposed programming languages for control and automation with constructs that are independent from the hardware on which their executable codes will run but it did not address the distributed nature of the software. This hardware independence holds only for the control logic. However, the functions of reading values from sensors and setting the actuators and exchanging data over a fieldbus or control level network requires the implementation of specific constructs, depending on the network characteristics and the sensor and actuator technology.

Fieldbuses and control level networks are characterized by protocols modelled according to the OSI standard [21]. However, in the case of the fieldbuses and control networks, a reduced and simplified protocol stack is utilized including only the layers 1, 2, and 7 of the OSI model [35]. The designer must select a fieldbus network protocol that may be appropriate for the application in mind and devices that have the same physical layer characteristics with the considered network. Obviously, programming the data exchanges among the network devices that the control functions require is not a straightforward task, unless there are libraries in the form of services that can be called by the control program through an abstract interface which should be compatible with the constructs of the programming language of the control logic.

The above problems are addressed to a certain extent by the IEC 61499 standard [16] which proposes a component-based approach for the design of distributed software. It seems to be appropriate for developing hardware-independent abstract software designs as far as the control logic is concerned. It provides the template for developing the necessary communication libraries for transferring data to and from the control logic from nodes having different hardware characteristics and requiring this transfer to take place according to the rules of the network protocol considered. These library components make the transfer transparent to the application designer, but must be developed by specialists and become available to the average application designer of the control logic. In what follows, the mechanism offered by the IEC 61499 standard for developing such libraries and including them in the development environment of an application will be explained. Also, a simple example will show how the application software can be written by using a library of a network service. But first we provide a brief overview of the basic concepts of the IEC 61499 standard communication constructs and explain how the provided communication library mechanisms can be integrated with other constructs of software built according to the IEC 61499 recommendations.

10.2 IEC 61499 Programming and Communication Semantics

10.2.1 General Semantics

The IEC 61499 standard introduces a series of modeling notations consisting of the function block, the application, the resource, the device and the system.

The function block (FB) encapsulates local data and algorithmic behavior with a well-defined interface. Such components can be connected by events and data lines and eventually can be distributed on different nodes and devices of a distributed computer system. The application is a software entity consisting of a network of interconnected FBs that solves a control problem. A resource is a functional unit contained in a device. A device is a self-contained hardware item containing a processor, memory and communication interfaces. The system is a collection of devices interconnected and communicating with each other by means of communication networks.

10.2.2 Communication Service Interface Function Block

Every device of an industrial control and automation system with distributed computer architecture must be loaded with software that implements the protocol of the network to which it will be connected. This software must be in the form of executable code designed to run on the specific hardware of the device. This software must also have the interface that will allow its call from the application that will request the reading or transmission of data over the network. A special type of function block called communication service interface function block (CSIFB) is provided by the IEC 61499 to serve the above purpose. Communication mechanisms are abstracted inside these special function blocks hiding the underlying implementation while exposing the communication interface to the application.

A CSIFB can be adjusted to the particular characteristics of the network protocol. To adjust the CSIFB, the network parameters must be known and software implementing the protocol must be available for inclusion in the algorithmic section of the CSIFB. The standard proposes four types of CSIFBs for applying different communication models: PUBLISH, SUBSCRIBE, CLIENT and SERVER FBs (see Figure 10.2). CSIFBs allow the IEC 61499 application to interact with the communication infrastructure of the device either by acquiring access to it or being interrupted by it.

In addition to type specification, IEC 61499 defines another notation to better show the timing and sequential relationships between a CSIFB and the underlying hardware or a CSIFB placed in another device. These notations are called service sequence diagrams and are adopted from communication standards. An example is given in Figure 10.3. We will use these notations to

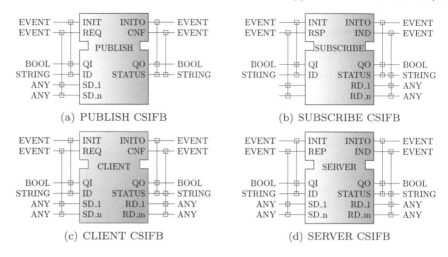

(a) PUBLISH CSIFB

(b) SUBSCRIBE CSIFB

(c) CLIENT CSIFB

(d) SERVER CSIFB

FIGURE 10.2
The four most commonly used IEC 61499 communication service interface funcion blocks.

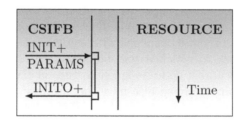

FIGURE 10.3
Service sequence diagram of a CSIFB initialization.

better explain how CSIFBs are initialized and paired to achieve inter-device communication.

10.2.2.1   CSIFB Initialization

Figure 10.3 shows the service sequence diagram of a CSIFB initialization. Horizontal arrows show incoming and outcoming events while the two vertical lines show domain boundaries. The left part is the CSIFB domain while the right part depicts the underlying resource domain. Time increases downward. To summon and initialize a CSIFB, its INIT event input must be triggered with an INIT+ event (INIT event with QI input set to TRUE) along with the appropriate network parameters. Upon a successful initialization, the CSIFB emits a INITO+ event (INITO event with QO output set to TRUE). The INITO+ event comes as a result of the INIT+ event, so the former is placed below the latter. The vertical line indicates the relationship between the two events.

10.2.2.2 Unidirectional Data Exchange

PUBLISH FBs are used to send data in the network without expecting any acknowledgement, while SUBSCRIBE FBs receive information from the network without sending any confirmation. This allows multiple PUBLISH FBs to send to the same destination or multiple SUBSCRIBE FBs to listen to the same source. PUBLISH and SUBSCRIBE FBs are most commonly used to implement multicast technology and are usually paired in distributed IEC 61499 applications.

Figure 10.4 shows the service sequence diagram of a PUBLISH–SUBSCRIBE transaction. The initialization process of the PUBLISH FB is not related to the initialization process of the SUBSCRIBE FB. Both FBs must be initialized before any data transaction can occur. After both FBs have been initialized, to establish their connections to the underlying network, they are ready to handle as many data transactions as desired. Upon receiving a REQ+ event, the PUBLISH FB accesses the underlying network resource, which in turn triggers the SUBSCRIBE FB to emit an IND+ event along with the received data. The transmitted data are provided to the SD_1–SD_n data inputs of the PUBLISH FB (Figure 10.2) and become available at

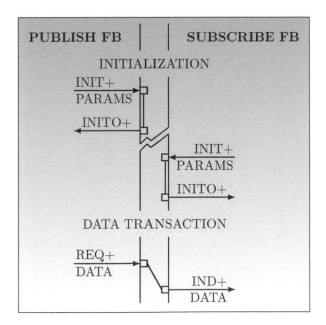

FIGURE 10.4
Service sequence diagram of a PUBLISH–SUBSCRIBE transaction.

the RD_1–RD_n data outputs of the SUBSCRIBE FB after the transaction is complete.

10.2.2.3   Bidirectional Data Exchange

CLIENT FBs are used to establish point-to-point connections. CLIENT FBs expect acknowledgment. On the other hand SERVER FBs accept connections from a source and reply according to the source's requests. CLIENT and SERVER FBs use acknowledgment and are most commonly used to implement the most deterministic master/slave technology, with CLIENT and SERVER FBs playing the roles of masters and slaves respectively. In distributed IEC 61499 applications, CLIENT and SERVER FBs are usually also paired.

Figure 10.5 shows the service sequence diagram of a CLIENT–SERVER transaction. Unlike the PUBLISH–SUBSCRIBE initialization, the CLIENT–SERVER pair most commonly has a certain initialization sequence. First the SERVER FB must be initialized establishing a server-like connection to the underlying network and then the CLIENT FB is initialized, establishing a connection to the SERVER FB through the underlying network resources. After both FBs have been initialized, they are ready to handle as many data

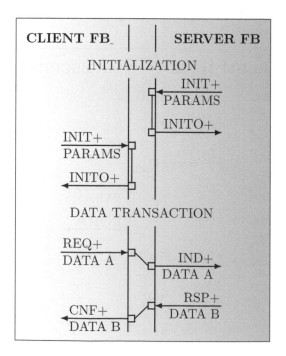

FIGURE 10.5
Service sequence diagram of a CLIENT–SERVER transaction.

transactions as desired. Upon receiving a REQ+ event, the CLIENT FB accesses the underlying network resource, sending a request to the SERVER FB, which in turn emits an IND+ event along with the received data. The transaction is not yet over. The SERVER FB has to respond with a confirmation indicating that all data was received or carrying requested data back to the CLIENT FB. This is realized by triggering the SERVER FB with a RSP+ event which sends the acknowledgment back to the CLIENT FB through the underlying network resources. Upon receiving the confirmation message, the CLIENT FB emits a CNF+ event.

Note that both the request and confirmation message can carry data across the network. The transmitted data attached to the request message are provided to the SD_1–SD_n data inputs of the CLIENT FB (Figure 10.2) and become available at the RD_1–RD_n data outputs of the SERVER FB. The transmitted data attached to the confirmation message are provided to the SD_1–SD_n data inputs of the SERVER FB and become available at the RD_1–RD_n data outputs of the CLIENT FB. Thus if a CLIENT FB requests data, these data will be available at its RD_1–RD_n data outputs when the transaction is complete.

10.2.2.4 CSIFB De-Initialization

To disable a connection established by a CSIFB to the underlying network, the CSIFB must be de-initialized. Figure 10.6 shows the service sequence diagram of a CSIFB termination. The CSIFB is triggered with an INIT− event (INIT event with QI input set to FALSE). Upon a successful termination the CSIFB releases all system resources and emits a INITO− event (INITO event with QO output set to FALSE). This is essential in some cases of CLIENT–SERVER connection where the CLIENT FB must disable its connection to the SERVER FB for another CLIENT FB to connect to it.

10.2.3 Compliance Profiles

The development of a CSIFB service depends on the software engineering environment in which the IEC 61499 blocks are developed. Normally the environ-

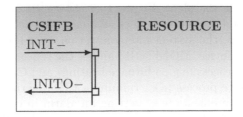

FIGURE 10.6
Service sequence diagram of a CSIFB termination.

ment will provide CSIFB templates and facilities for inserting the algorithms implementing the network protocol and specifying the service sequences. If the template is a program in Java or C++ or other language, a compilation of the program will generate the executable code of the CSIFB, ready to be used as programming construct through its graphical interface in any application program.

In order to remain adaptive to new programming and communication technology, IEC 61499 avoids explicitly defining how CSIFBs are used to implement a particular communication protocol. Instead the standard defines a mechanism to extend the standard and define user-specific implementations of communication protocols and other features. The mechanism is called compliance profile [17] and attempts to achieve compliance between different platforms and programming environments developed by different vendors.

A vendor implementing a feature left undefined by the standard must publish a compliance profile describing the details of his platform-specific solutions to a communication or other issue. Other vendors can consult this compliance profile to develop compliant solutions so that devices, platforms or programming environments with different internals can meet compatibility requirements and cooperate with each other. However, vendors often fail to follow these instructions and thus ignore the distributed nature of the IEC 61499 standard as explained in Section 10.7. On the other hand, the open source community carves its own way by publishing open compliance profiles offering solutions to some IEC 61499 issues.

In the next section, an example of a distributed application is given. This example is used in subsequent sections to describe how the CSIFBs that handle application communications are deployed in the application's FB network.

10.3   Example of Distributed Application

The event-driven nature of IEC 61499 applications allows their distribution to multiple devices. Each FB can be assigned to a certain resource lying in a certain device, according to the capabilities of the resource and the position of the device inside the plant. Figure 10.7 shows the system view of an application distributed on three devices. We can see each device connected to a network segment, the system interconnection, and the particular resource contained in each device. MONITOR handles the sensors and actuators monitoring a physical process (e.g., a furnace heating) and CONTROL handles the control of the above process (e.g., running a PID algorithm), while SAFETY handles safety requirements (e.g., setting an alarm in case of an emergency). Figure 10.8 shows the FB network of the distributed application. We can see each FB assigned to a certain system device.

MONITOR contains FBs SENSOR and ACTUATOR. FB SENSOR is

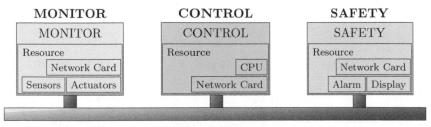

NETWORK SEGMENT

FIGURE 10.7

A distributed system of three devices. MONITOR monitors a physical process and contains a resource equipped with sensors and actuators. CONTROL handles the control of the physical process and contains a resource with a programmable CPU. SAFETY handles safety requirements and contains a resource with an alarm system. All resources contain network cards, so they can physically connect to the network.

connected to a physical sensor (e.g., a thermal sensor). When triggered with a REQ event, it reads the current sensor value and emits it along with a CNF event. FB ACTUATOR is connected to a physical actuator (e.g., a heating resistor). When triggered with a REQ event, it sets or changes the actuator's value (e.g., the power of the heating resistor) according to its VALUE data input. When triggered with an OFF event, the actuator is disabled and the FB emits a CNF event along with a log file containing the last values of the actuator for safety check reasons.

CONTROL contains a FB PID that controls the physical process monitored by FBs SENSOR and ACTUATOR. When triggered with a REQ event, it reads the given sensor value and runs some algorithm (e.g., a PID loop) to calculate the appropriate actuator value which in turn emits along with a CNF event.

The sensor value is also provided to the SAFETY device for safety check. There the CHECK FB, when triggered with a REQ event, checks whether the value is within accepted safety limits. If the value exceeds these limits or a value is not provided for a critical period, the FB emits an ERROR event. This event triggers the ALARM FB to enable a red alert and also triggers the ACTUATOR FB of the MONITOR device to disable the physical actuator. If the actuator is successfully disabled, the ACTUATOR FB emits a CNF event along with the log file as explained above. The CNF event triggers the ALARM FB of the SAFETY device to drop to a yellow alert and print the received log file to a monitor, thus completing the emergency management.

Dashed lines in Figure 10.8 denote transitions between devices. Communication between FBs in different devices is implemented by inserting IEC 61499 standard CSIFBs in these positions (see [25, pp. 64–67]). Eventually a FB in one device can trigger FBs in other devices.

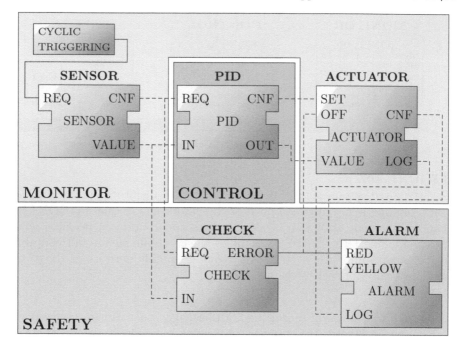

FIGURE 10.8
An application distributed on three devices. FBs SENSOR and ACTUATOR are assigned to MONITOR, FB PID is assigned to CONTROL, while FBs CHECK and ALARM are assigned to SAFETY. Dashed lines denote connections that go through devices' boundaries.

10.4   UDP/IP and TCP/IP Communication Protocols in IEC 61499

The publicly available *IEC 61499 Compliance Profile for Feasibility Demonstrations* maintained by Holobloc Inc. [10] attempts to map the unidirectional and bidirectional models of communication described in IEC 61499 to the User Datagram Protocol [22] over Internet Protocol [23] (UDP/IP) and Transmission Control Protocol [24] over Internet Protocol [23] (TCP/IP), respectively. The UDP/IP and TCP/IP implementations in IEC 61499 are presented in the following subsections.

10.4.1   User Datagram Protocol/Internet Protocol

A brief description of the UDP/IP implementation in IEC 61499 is given in [38]. UDP/IP is utilized for PUBLISH and SUBSCRIBE FBs. To communicate with each other, both PUBLISH and SUBSCRIBE FBs are given the same parameterized ID data input:

mulicastAddress:port

mulicastAddress sets the internet multicast address used by the PUB-
LISH/SUBSCRIBE FB to send and receive the transmitted data
port sets the internet port number used for the above multicast transmission

For example, the ID input of a PUBLISH and SUBSCRIBE FB communicating over the multicast IP address and port number 230.0.0.1:4445 could be
represented as "230.0.0.1:4445". The given Internet multicast address and port
number serve as a key binding the two CSIFBs together so that all data transmitted by the PUBLISH FB are received by the corresponding SUBSCRIBE
FB.

There can be multiple PUBLISH and SUBSCRIBE FBs with the same ID
data input. Any data sent using a particular ID by any PUBLISH FB will be
received by any SUBSCRIBE FB using the same ID. UDP/IP is an unreliable
protocol, considering industry requirements, because it lacks handshaking and
acknowledgment features, thus suffering from packet loss. If it is to be used
in an industrial application, the programmer must integrate application-level
solutions. However UDP/IP is a ready-to-implement protocol and is usually
preferred over TCP/IP in simple applications with low requirements for reliability and high requirements for timing.

10.4.2 Transmission Control Protocol/Internet Protocol

A brief description of the TCP/IP implementation in IEC 61499 is given
in [38]. TCP/IP is utilized for a CLIENT and SERVER pair. The SERVER
FB is given the following ID data input:

internetAddress:portNumber

internetAddress set to the IP address of the SERVER FB TCP/IP server
or "localhost" or "127.0.0.1" or the device's name in the network
portNumber sets the listening port of the SERVER FB TCP/IP server

For example a SERVER FB running in a device named m51568 with IP address "161.153.19.227" could have an ID input represented as "m51568:1499",
"161.153.19.227:1499", "localhost:1499" or "127.0.0.1:1499". The corresponding CLIENT FB has the following ID data input:

internetAddress:portNumber

internetAddress set to the IP address or name of the SERVER FB's device
portNumber set to the listening port number opened by the SERVER FB

A CLIENT FB accessing the SERVER FB of the above example could have an ID input represented as "m51568:1499" or "161.153.19.227:1499". Having this information, the CLIENT FB can start exchanging data by connecting to the SERVER FB.

It is obvious that CLIENT and SERVER FBs work in pairs, establishing a reliable connection with one another. This does not mean that the time of receipt of any message is guaranteed, but the message will be sent repeatedly until received. This adds to the reliability of the transaction, but offers no guarantee regarding the response time. The use of TCP/IP is limited because of this feature.

10.4.3   Application Example of UDP/IP and TCP/IP Implementation

The distributed application shown in Figure 10.8 will be used as an example of applying the UDP/IP and TCP/IP protocols in an IEC 61499 application. As explained previously, the example refers to an application that monitors, controls and secures a physical process. Three devices, MONITOR, CONTROL and SAFETY handle these functions respectively.

There are two network event and data flows in the system. One handles the communication between the MONITOR and the CONTROL devices, while the other handles the safety communication. To better monitor the physical procedure, the SENSOR FB must be triggered frequently to produce a vast number of values; therefore the simple UDP/IP protocol will be used to handle the MONITOR–CONTROL communication—saving network bandwidth. On the other hand the SAFETY device must be able to reliably shut down the ACTUATOR of the MONITOR when needed; moreover this is not a frequent transaction but occurs only in case of an error. Thus the more reliable TCP/IP protocol will be used to handle the SAFETY–MONITOR communication. CONTROL–SAFETY communication will also be handled by a UDP/IP connection.

The full application along with the inserted CSIFBs that handle the system's communication can be seen in Figure 10.9. The initialization process of the CSIFBs is omitted but a detailed presentation of the procedure can be found in Section 10.2.2. Note that each set of CSIFBs carries its own ID key. Also note that the multicast feature of the UDP/IP protocol is utilized to send the sensor data to both the CONTROL and SAFETY devices. The system transactions are as follows. First the SENSOR FB is triggered, calling the PUBLISH_A FB to publish the sensor data on the network. Both SUB-SCRIBE_A and SUBSCRIBE_A2 FBs can receive the sensor value. Using different ID keys makes the CSIFBs able to distinguish between the relevant and non-relevant UDP packets on the network. The SUBSCRIBE_B FB will not receive data published by the PUBLISH_A FB. When the SUB-SCRIBE_A FB receives the sensor value, it calls the PID FB to calculate the appropriate value for the ACTUATOR FB. When the PID FB has the value

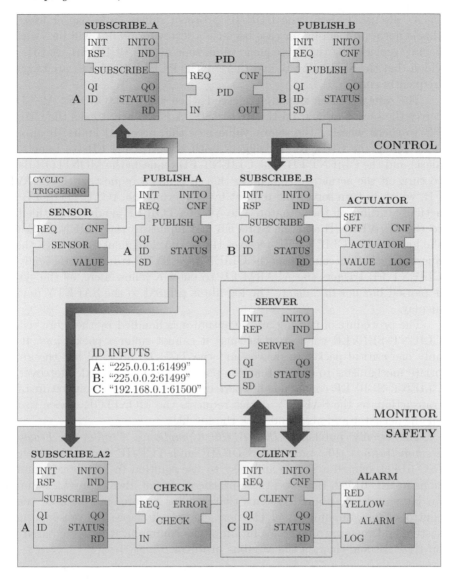

FIGURE 10.9
Deployment of CSIFBs in the application shown in Figure 10.8. PUBLISH and
SUBSCRIBE FBs use the UDP/IP protocol, while CLIENT and SERVER
FBs use the TCP/IP protocol.

ready, it calls the PUBLISH_B FB to publish the actuator value on the network. Only the SUBSCRIBE_B FB will receive the actuator value published
by the PUBLISH_B FB and in turn will call the ACTUATOR FB to set the
appropriate value to the physical actuator.

As discussed above, the UDP/IP protocol suffers from packet loss, but this is not a problem in our case because a lost sensor value will be—with no harm—replaced by the subsequent value sent by the PUBLISH_A FB, provided that the SENSOR FB, and thereby the PUBLISH_A FB, are triggered frequently enough.

The safety transactions are triggered when the SUBSCRIBE_A2 FB receives the sensor value from the PUBLISH_A FB. Then it calls the CHECK FB to check whether the sensor value is within accepted limits. If an error is detected, the CHECK FB triggers both the ALARM FB (to issue a red alert) and the CLIENT FB. The CLIENT FB requests the MONITOR device to turn off the actuator. To achieve this, it sends a request to the SERVER FB, which in turn serves the request by triggering the ACTUATOR FB and actually turning off the actuator. The ACTUATOR FB sends the log file back to the SERVER FB and the SERVER FB replies to the CLIENT FB with the log file. The CLIENT FB receives the acknowledgment along with the requested data, thus completing the CLIENT–SERVER transaction. Finally the CLIENT FB triggers the ALARM FB to enter a yellow alert, as the system is paused but not in danger. The log file is printed to the SAFETY device's display.

The procedure of turning off the actuator is handled by the more reliable CLIENT–SERVER connection because it cannot suffer a packet loss. It has only one critical packet to deliver and the TCP/IP protocol has the appropriate mechanisms to ensure that the packet will be delivered. Moreover the CLIENT–SERVER connection fits perfectly to the request–reply nature of the transaction, as the SAFETY device requests the MONITOR device to turn off the actuator and also send back the log file.

The openly published *IEC 61499 Compliance Profile for Feasibility Demonstrations* [10], providing UDP/IP and TCP/IP implementations in IEC 61499, offered a simple and easy to use solution to handle communications in IEC 61499. However, these protocols do not always provide solutions for large scale distributed systems that require a more sophisticated central control of the communication. The next chapter discusses and presents such a system and then introduces a solution of the communication issue based first on UDP/IP and then on the Modbus protocol.

10.5 Example of SCADA System

The majority of distributed control and automation systems of a certain size require a central control of all cooperating subsystems. A special supervisory control and data acquisition (SCADA) system is set above all subsystems to gather and analyze real-time data in order to monitor and control devices in different sub-systems. Typical operation of a SCADA system is to parameterize remote devices, collect data for HMI diagrams and handle the central communication system. IEC 61499 notations are appropriate to describe such a system. Figure 10.10 shows the system view of a simple automation system incorporating a SCADA unit. Each MONITOR_n device monitors a physical process (e.g., the movement of a conveyor belt) by using a sensor. Each CONTROL_n device has the ability to control the physical process monitored by the corresponding MONITOR_n device by running a control algorithm and providing the resulting data to an actuator contained in each CONTROL_n device. SCADA is used to collect all sensor data to a log file for HMI purposes. Figure 10.11 shows the actual FB network of the application and the assignment of each FB to the appropriate system device.

Each MONITOR_n device contains a FB of type SENSOR. Each SENSOR_n FB is connected to a physical sensor (e.g., a speedometer). When triggered with a REQ event, each SENSOR_n FB reads the current sensor value and emits it along with a CNF event.

Each CONTROL_n device contains a FB of type PID and a FB of type ACTUATOR that control the physical process monitored by the corresponding SENSOR_n FB. When triggered with a REQ event, each PID_n FB reads the given sensor value and runs an algorithm (e.g., a PID loop) to calculate the appropriate actuator value, sending it along with a CNF event to the corresponding ACTUATOR_n FB. Each ACTUATOR_n FB is connected to a physical actuator (e.g., a conveyor motor). When triggered with a REQ event, it sets the actuator's value (e.g., the electric power of the motor) to the value written in its VALUE data input.

Each sensor value produced by the relevant SENSOR_n FB is also provided to the corresponding LOGGER_n FB contained in the SCADA device to be appended in a log file.

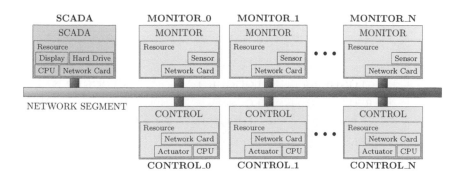

FIGURE 10.10
A distributed system of multiple control pairs supervised by a SCADA device. Each device is connected to a single network segment, thus achieving the system interconnection. Each device contains a specific resource providing the means to handle the device's operation.

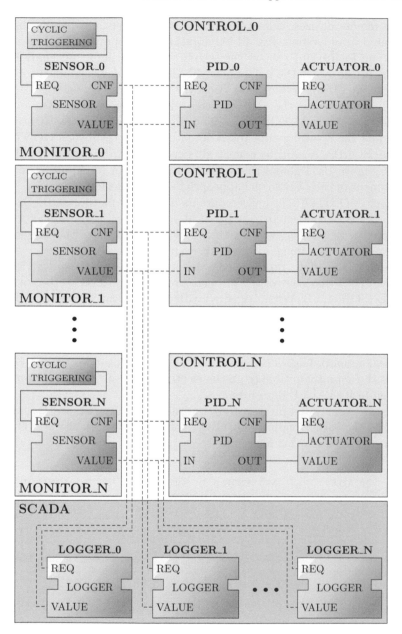

FIGURE 10.11

An application distributed on multiple devices along with a SCADA device. FBs SENSOR_n are assigned to the corresponding MONITOR_n devices. FBs PID_n and ACTUATOR_n are assigned to the corresponding CONTROL_n devices, while FBs LOGGER_n are all assigned to device SCADA. Dashed lines denote connections that go through devices' boundaries.

10.5.1 Applying UDP/IP Protocol on SCADA System Example

Applying the UDP/IP protocol in the SCADA system example of Figure 10.11 would be fairly easy. A PUBLISH FB is inserted after each SENSOR_n FB to broadcast the corresponding sensor value to the network, while a SUBSCRIBE FB is inserted before each PID_n FB to collect the corresponding sensor value from the network. Another SUBSCRIBE FB is inserted before each LOGGER_n FB to also collect the relevant sensor data from the network.

The full application along with the inserted CSIFBs that handle the system's communication can be seen in Figure 10.12. The system transactions are realized as follows: First each SENSOR_n FB is cyclically triggered, calling the corresponding PUBLISH_n FB to publish the corresponding sensor data on the network. Both the corresponding SUBSCRIBE_n and SUBSCRIBE_nb FBs can receive the sensor value. Using different ID keys makes the CSIFBs able to distinguish between the relevant and non-relevant UDP packets on the network. When each SUBSCRIBE_n FB receives the corresponding sensor value, it calls the corresponding PID_n FB to calculate the appropriate value for the corresponding ACTUATOR_n FB, which in turn sets the corresponding physical actuator accordingly, thus realizing the corresponding control loop. Each SUBSCRIBE_nb FB receives the relevant sensor data and calls the corresponding LOGGER_n FB to append the data to a joint log file.

To better monitor the physical procedure, the SENSOR_n FBs must be frequently triggered, producing a vast number of values. UDP/IP protocol uses a simple low bandwidth consuming messaging system, but sometimes it is not capable of handling this communication, as the network overflows with the broadcasted data especially when the number of control pairs increases significantly. Trying to solve this issue by lowering the frequency of the sensors' cyclic triggering would impair the control loop efficiency. A central monitoring system is required to exploit the available network bandwidth without overflowing the system.

In the next section, the commonly used SCADA applications Modbus communication protocol is presented. The application example described above is then used to show how CSIFBs are deployed in the FB network of Figure 10.11 to handle its supervising communication requirements using the Modbus protocol.

10.6 Modbus Communication Protocol in IEC 61499

Modbus is one of the oldest, simplest and most widely accepted communication protocols in industrial automation ([3, 12, 36], [4, p. 508-510]). Its

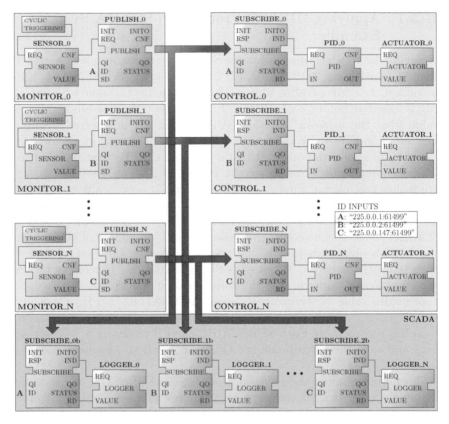

FIGURE 10.12
Deployment of CSIFBs in the application shown in Figure 10.11 to handle the communication using the UDP/IP protocol.

specifications were openly published in [28, 27, 29] and through its simplicity, robustness and efficiency it became a *de facto* standard.

Modbus protocol is based on a master/slave network. In each Modbus network only one master can exist, but several slaves can be connected. Each slave starts a server service and waits for requests. The master sends requests to read data or/and write data to the slave's internal memory. Only the master can trigger a transaction and the slaves send data to the network only when requested by the master. Thus the whole communication is handled by a single master device.

10.6.1 Modbus Protocol Implementation in IEC 61499

A simple IEC 61499 implementation of Modbus protocol is introduced in [38, p. 508-510]. The use of a CLIENT FB for the Modbus master and a SERVER

FB for the Modbus slave is suggested. But a new implementation favoring IEC 61499 abstract communication model is presented here. A more flexible and easy-to-use CLIENT–PUBLISH/SUBSCRIBE approach is utilized. The master is implemented using CLIENT FBs and the data exchanges in the slave device are handled by PUBLISH and SUBSCRIBE FBs.

Figure 10.13 shows the three different data transactions of the implementation. Figure 10.13(a) shows the publish operation of the slave device. As mentioned before, the slave device starts up a Modbus server service. Instead of sending the data on the network, a triggered PUBLISH FB writes the data to a special Modbus Data Table making them available to the Modbus Server Service without involving any network activity.

Figure 10.13(b) shows the read operation of the master device. When triggered properly, a CLIENT FB on the master device will send a read request on the Modbus server service on the slave device requesting data. Having the data available through the publish operation described above, the Modbus server service responds back to the master device sending the requested data, thus realizing the read operation.

Figure 10.13(c) shows the write operation of the master device. When triggered properly, a CLIENT FB on the master device will send a write request on the Modbus server service on the slave device. The Modbus server service responds by writing the data to the Modbus data table, while triggering any SUBSCRIBE FBs related to these data, and sends a confirmation back to the master device.

Note that a mixed read and write request can also be handled by the implementation by combining both operations. Also note that multiple PUBLISH and SUBSCRIBE FBs can be associated to a single Modbus Sever Service and function simultaneously on the slave device side, while multiple CLIENT FBs can exist and function simultaneously on the master device side.

10.6.2 Configuring CSIFBs to Use Modbus Protocol

Modbus comes with three variants, TCP, RTU and ASCII. The TCP variant runs on top of an Ethernet TCP/IP network, while the RTU and ASCII variants run on top of a serial line (RS232 or RS485). A notation is required to distinguish between CSIFBs using different Modbus variants or different protocols in general. In [8] it is proposed to use the same CSIFBs (see Figure 10.2) for all protocols and change the user-specific ID data input to set the use of the preferred protocol. This approach is followed here and extended to address the Modbus variant distinction issue. A string is passed to a CSIFB's ID data input to properly configure it for handling Modbus transactions according to a certain variant. To configure a CLIENT FB to handle Modbus master functionss the string has the following format:

modbus[variant:slaveAddress:(function:readAdds:sendAdds):(params)]

modbus sets the CLIENT FB to use the Modbus protocol

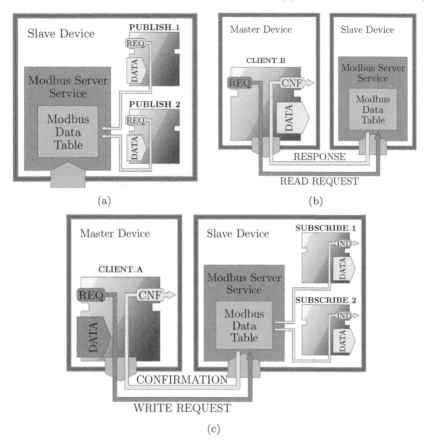

FIGURE 10.13
Three data transaction operations of the Modbus implementation: (a) A PUB-
LISH FB on the Modbus slave side writes some data to the Modbus data table
of the Modbus server service. (b) A CLIENT FB on the Modbus master side
sends a read request to the Modbus server service on the Modbus slave side.
(c) A CLIENT FB on the Modbus master side sends a write request to the
Modbus server service on the Modbus slave side.

variant sets the Modbus variant—valid values are tcp, rtu and ascii

slaveAddress sets the slave address to which the request is addressed—for
the TCP variant the value is the IP:port address of the slave device in
the TCP/IP network and for the RTU and ASCII variants the value is a
number from 1 to 247 corresponding to the slave serial address or 0 for
broadcast function mode

function sets the Modbus function code to be used—it is optional and used only for backward compatibility with devices not using IEC 61499

readAdds lists slave device data table addresses to be read—used only for backward compatibility

sendAdds lists slave device data table addresses to be overwritten—used only for backward compatibility

params sets protocol and variant specific parameters—it is optional; default values are used when omitted

PUBLISH/SUBSCRIBE FBs are configured to handle Modbus slave functions using the following ID data input:

modbus[variant:slaveAddress:(dataType:dataAdds):(params)]

modbus sets the PUBLISH/SUBSCRIBE FB to use the Modbus protocol

variant sets the Modbus variant—valid values are TCP, RTU and ASCII

slaveAddress sets the device's address—for the TCP variant the value is the IP:port address of the device in the TCP/IP network and for the RTU and ASCII variants the value is a number from 1 to 247 corresponding to the serial address of the device

dataType sets the relevant writing/listening Modbus data type—used only for backward compatibility

dataAdds sets the relevant writing/listening Modbus data addresses—used only for backward compatibility

params sets protocol and variant specific parameters—it is optional; default values are used when omitted

All optional and backward compatibility parameters have a significant role in configuring the Modbus communication network, but the details are omitted here for brevity.

10.6.3   Application Example of Modbus Implementation

The distributed application shown in Figure 10.11 will be used as an example of applying the Modbus protocol in an IEC 61499 application to set a supervising control mechanism of the system's communication. As explained previously, the example refers to a system consisting of multiple control pairs and a SCADA device.

The full application along with the inserted CSIFBs that handle the system's communication using the Modbus protocol can be seen in Figure 10.14. Note that the MONITOR_n–CONTROL_n pairs remain unaltered with only the ID inputs of the PUBLISH_n–SUBSCRIBE_n pairs changed as they are configured to use the Modbus protocol. On the other hand the SCADA device is significantly modified to take a supervising role in the communication scheme.

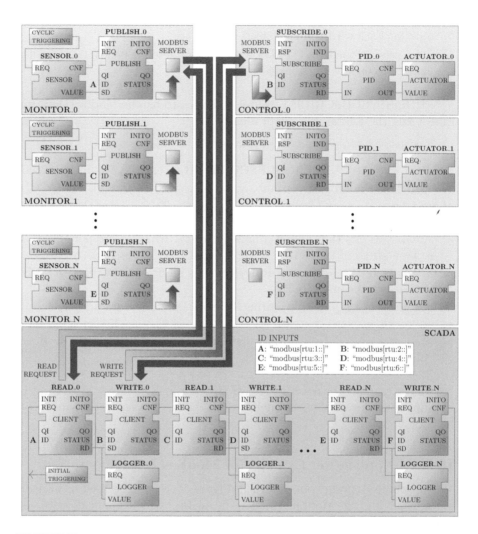

FIGURE 10.14

Deployment of CSIFBs in the application shown in Figure 10.11 for setting a supervising communication control using the Modbus protocol.

All SENSOR_n FBs are simultaneously and cyclically triggered, calling the corresponding PUBLISH_n FBs to write the corresponding sensor values on the corresponding internal Modbus data tables of the Modbus servers. Thus the MONITOR_n devices have no real access to the network. All transactions are carried out by the SCADA device as follows. First the READ_0 FB is triggered with an initial event—the small arrow in Figure 10.14 corresponds to a logical OR between the event lines coming from the INITIAL TRIGGERING box and the WRITE_N FB. The READ_0 FB acquires the sensor data from the MONITOR_0 device (previously stored to the Modbus server by the PUBLISH_0 FB). When the transaction is complete the sensor data is appended to the log file by the LOGGER_0 FB and is written to the Modbus server on the CONTROL_0 device by the WRITE_0 FB. When the writing is complete the PID_0 FB is triggered to calculate the appropriate value for the ACTUATOR_0 FB and the READ_1 FB is triggered to perform the next transaction related to the MONITOR_1–CONTROL_1 pair. Figure 10.14 shows only the transactions related to the first control pair. All subsequent transactions are performed in the same way until the MONITOR_n–CONTROL_n pair is served, thus completing the communication cycle and triggering the READ_0 FB to start a new communication cycle.

Having these communication cycles executed one after another exploits the full network bandwidth with no starvation or overflow issues, as the individual devices do not interfere with the communication system on their own initiative. In a similar manner, CSIFBs using the Modbus protocol can be configured to apply SCADA operations in large scale distributed systems containing multiple remote devices and subsystems.

The UDP/IP, TCP/IP and Modbus implementations in IEC 61499 have been widely accepted by the academic community, since they provide simple and ready-to-use solutions in the IEC 61499 framework. As researchers and vendors became more active in the use of the IEC 61499, they started providing different solutions and implementations to the IEC 61499 standard open communication issues.

10.7 Implementations of Other Communication Protocols in IEC 61499

The IEC 61499 standard was formed with the fieldbus IEC 61158 standard [19] and many of IEC 61158's concepts were taken into consideration during its development (see [25]). This is why communication and distribution became very important aspects of IEC 61499 standard. It was envisaged that applications would be distributed on multiple devices, while device replacement would be a simple vendor-independent task. Several fieldbus communication protocols have been implemented in IEC 61499, but as industry turned to Ethernet-based communications [30], IEC 61499 research followed. Now IEC 61499 development tools use a variety of IEC 61499 implementations of Fieldbus and Ethernet-based communication protocols, but only a few compliance profiles [17] have been published to the best of the authors' knowledge. UDP/IP and TCP/IP implementations have been published in [10] and used widely in (4DIAC [1] and FBDK [9] projects), while common industrial protocol (CIP) [32] implementation has been published in [39], but may be outdated. There was also an attempt to integrate communications of power industry standard IEC 61850 [15] into IEC 61499 as its communication strand [7].

10.7.1   Commercial IEC 61499 Tools

10.7.1.1   ISaGRAF

ISaGRAF [14] is an industrial development tool currently promoted by ICS Triplex ISaGRAF. Combining both IEC 61131 and IEC 61499 standards, but focusing on the former, ISaGRAF is less effective in creating distributed IEC 61499 applications. The evaluation platform of ISaGRAF comes along with three network segments, ETCP for TCP/IP, ISARSI for RS232 links and HSD (host system driver) for resource-to-resource data exchange on the same device. ISaGRAF provides no resource view of distributed applications and handles communication automatically. The user is not provided and does not have to use CSIFBs. After adding a particular network segment to the system's deployment view, ISaGRAF automatically adds the corresponding communication code for every event and data connection line that crosses the device boundaries (for a more detailed presentation see [37]). This is no derogation from the rules of IEC 61499, as the standard proposes that engineering tools will be able to automatically insert CSIFBs in distributed applications. The problem is that there is no publicly available compliance profile for ICS Triplex ISaGRAF implementation of these communications (a compliance profile is only mentioned in [13]), so no other vendor or academic researcher can develop compatible solutions or extensions.

10.7.1.2   nxtSTUDIO

nxtSTUDIO [31] is an industrial engineering tool owned by the Austrian Company nxtControl GmbH. It is mainly focused on IEC 61499 but also supports IEC 61131 standard's languages. It's demo version comes along with many network segment types such us Ethernet, ControlNet, DeviceNet and Profibus. nxtControl has created several CSIFB libraries to implement these communication protocols. In contrast to ISaGRAF, nxtSTUDIO offers a separate view of distributed applications on resources, but similarly provides an automatic way to insert the necessary CSIFBs that handle the communication. Although the nxtSTUDIO supports the *IEC 61499 Compliance Profile for Feasibility Demonstrations* [10], no published compliance profiles of all the other implementations exist, preventing other vendors or researchers from testing or extending the solutions offered by nxtControl.

10.7.2 Open IEC 61499 Tools

As an alternative, an open source development tool for IEC 61499 applications named 4DIAC has emerged in the last few years [1] (Framework for Distributed Industrial Automation and Control) supplementing Holobloc's FBDK [9] (Function Block Development Kit) as the first IEC 61499 research platform. 4DIAC was recently accepted as an official Eclipse project [5], while Holobloc has recently published the second version of FBDK [11].

Both projects are based on their own run time system, FORTE and FBRT for 4DIAC and FBDK, respectively, but can work in unison. FBRT and FORTE also share the same principles regarding the implementation of different communication protocols. The open source nature of both projects has drawn the attention of the research community, specifically in a number of publications regarding the implementation of communication in IEC 61499.

As already mentioned, in [8] it is proposed to use the same CSIFBs for all protocols and change the user-specific ID data input to set the use of the preferred protocol, giving an overall solution to CSIFB's distinction between protocols. Moreover as already discussed the *IEC 61499 Compliance Profile for Feasibility Demonstrations* [10] introduced the implementations of UDP and TCP protocols in IEC 61499, while a CIP protocol implementation was published in *IEC 61499 Compliance Profile for CIP based Communication Function Blocks* [39]. A description of the CIP implementation is also given in [40], where it is noted that its multimaster nature makes it most suitable for the distributed nature of IEC 61499, while it offers modes to support the cyclic, sequential and parallel execution models of IEC 61499.

10.8 Programming Example

The example presented in this section demonstrates how one could program a control application of a wearable knee joint exoskeleton used in rehabilitation tasks using the open source development tool 4DIAC [1].

The exoskeleton mechanism fits the wearer's leg to which it is fixed by means of straps. The mechanism is driven by an actuator and it is subjected to the human effort delivered by the muscles of the lower limb acting on the knee joint. The wearer is supposed to be in a sitting position performing flexion and extension movements of the knee joint without taking into account the ground contact. The torque delivered by the exoskeleton is controlled in order to track a desired trajectory determined with respect to the wearer's intentions or according to certain exercises. The wearable device is equipped with sensors that detect the wearer's applied torque as well as the angle and angular speed of the knee joint. The torque value is used to estimate the desired trajectory while the angular data are provided to a controller in another device to calculate the power of the assistance that needs to be provided by the exoskeleton mechanism for the knee to reach the desired angle at the desired angular speed. A third device is used as a panel to insert the data for the exercises to be applied to the knee joint or alternatively to select the automatic operation by which the exoskeleton follows the wearer's intentions. The 4DIAC system configuration view of the application is illustrated in Figure 10.15.

Each device of the network is given an IP address and is loaded with an EMB_RES resource capable of holding an IEC 61499 application. The devices are connected over an Internet network and the parts of the applications that reside on each device communicate with PUBLISH–SUBSCRIBE CSIFBs using the UDP/IP protocol.

The application view is shown in Figure 10.16. As proposed by the IEC 61499 standard, the application view does not include the CSIFBs used by the application to exchange data. The CSIFBs appear only in the separate resource view of the distributed application shown in Figures 10.17, 10.18 and 10.19. There the PUBLISH and SUBSCRIBE FBs that handle the communication can be seen, as well as the special START FB that initiates the application in each resource. This way the algorithmic part of the application is separated from the applied communication. Moreover, IEC 61499 suggests that CSIFBs for applying a certain communication protocol should

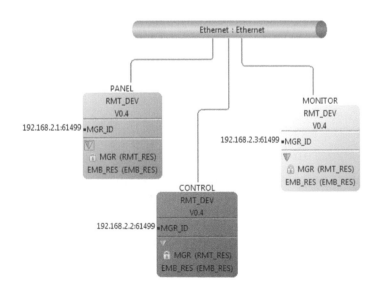

FIGURE 10.15
4DIAC system configuration of the distributed control application of the exoskeleton mechanism.

be automatically inserted, a feature expected to be available in 4DIAC's next versions.

The algorithmic and communication operation of the application is realized as follows. The MONITOR device activates the actuator that makes the exoskeleton mechanism apply the torque required to be added by the exoskeleton mechanism to the wearer's effort. It also reads the sensor measurements of the applied torque, the angle and the angular speed of the knee joint. The angular data is published to the CONTROL device which calculates the appropriate torque that the exoskeleton mechanism must apply. This torque value is published back to the MONITOR device. The wearer's applied torque along with the angular data is published to the PANEL device which estimates the wearer's desired angle and angular speed of the knee joint. The estimated values are published to the CONTROL device to be used in the calculation of the exoskeleton auxiliary torque. Note that the angular data is published both to the CONTROL and PANEL device using the UDP/IP multicast feature. The PANEL device is also used to select the manual mode of operation where the desired angular data are not estimated and are forced in order to apply certain exercises to the limb.

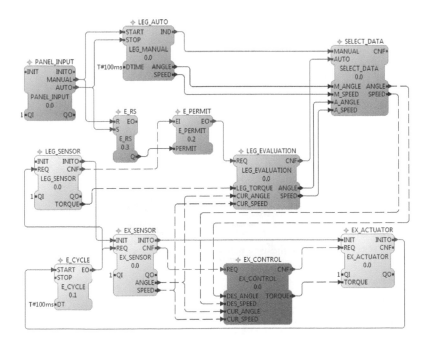

FIGURE 10.16
4DIAC Application view of the program.

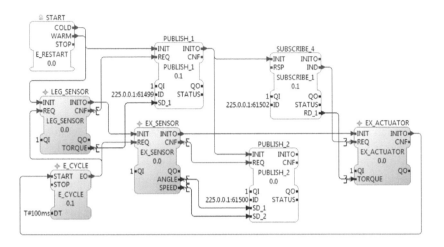

FIGURE 10.17
Partial application distributed to the MONITOR device.

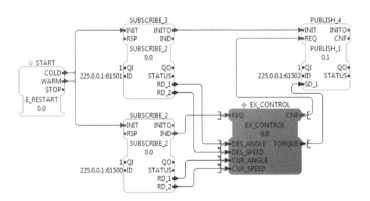

FIGURE 10.18
Partial application distributed to the CONTROL device.

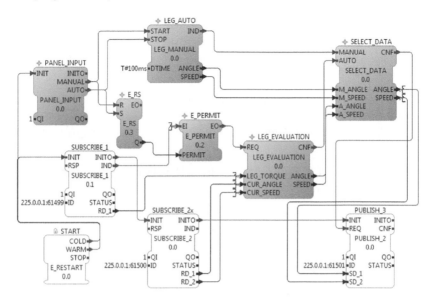

FIGURE 10.19
Partial application distributed to the PANEL device.

Bibliography

[1] 4DIAC Consortium. 4DIAC Framework for Distributed Industrial Automation and Control, Open Source Initiative. http://www.fordiac.org/14.0.html, 2014.

[2] CAN in Automation. CiA 404 CANopen: Device Profile for Measuring Devices and Closed Loop Controllers. http://www.can-cia.org/standardization/technical-documents/, 2002.

[3] M. De Sousa and P. Portugal. Modbus. In B. M. Wilamowski and J. D. Irwin, editors, *The Industrial Electronics Handbook: Industrial Communication Systems*. CRC Press, Boca Raton, 2011.

[4] B. Drury. *Control Techniques Drives and Controls Handbook*. Institution of Engineering and Technology, Stevenage, United Kingdom, second edition, 2009.

[5] Eclipse Foundation, Inc. 4DIAC Framework for Distributed Industrial Automation and Control. https://projects.eclipse.org/projects/iot.4diac, 2015.

[6] Fieldbus Foundation. Foundation Fieldbus Technical Overview, 2003.

[7] N. Higgins, V. Vyatkin, N. Nair, and K. Schwarz. Distributed power system automation with IEC 61850, IEC 61499, and intelligent control. *IEEE Transactions on Systems, Man, and Cybernetics, Part C: Applications and Reviews*, Vol. 41, No. 1, 81–92, 2011.

[8] M. Hofmann, M. Rooker, and A. Zoitl. Improved communication model for an IEC 61499 runtime environment. In *16th IEEE International Conference on Emerging Technologies & Factory Automation*, pages 1–7, 2011.

[9] Holobloc Inc. FBDK Function Block Development Kit. http://www.holobloc.com/doc/fbdk/, 2011.

[10] Holobloc, Inc. IEC 61499 Compliance Profile for Feasibility Demonstrations. http://www.holobloc.com/doc/ita/index.htm, 2013.

[11] Holobloc Inc. FBDK 2.3: Function Block Development Kit. http://www.holobloc.com/fbdk2/index.htm, 2015.

[12] P. Huitsing, R. Chandia, M. Papa, and S. Shenoi. Attack taxonomies for the modbus protocols. *International Journal of Critical Infrastructure Protection*, Vol. 1, , 37–44, 2008.

[13] ICS Triplex ISaGRAF Inc. ISaGRAF 5.1 assessment according to IEC 61499. http://www.isagraf.com/get/isagraf_5-1_1499-tuvreport.pdf, 2007.

[14] ICS Triplex ISaGRAF Inc. ISaGRAF - IEC 61131 and IEC 61499 software. http://www.isagraf.com/, 2014.

[15] International Electrotechnical Commission. Communication networks and systems for power utility automation. Part 6: Configuration description language for communication in electrical substations related to IEDs, 2009.

[16] International Electrotechnical Commission. Function blocks. Part 1: Architecture, 2012.

[17] International Electrotechnical Commission. Function blocks. Part 4: Rules for compliance profiles, 2013.

[18] International Electrotechnical Commission. Programmable controllers. Part 3: Programming languages, 2013.

[19] International Electrotechnical Commission. Industrial communication networks: Fieldbus specifications. Part 1: Overview and guidance for the IEC 61158 and IEC 61784 series, 2014.

[20] International Electrotechnical Commission. Industrial communication networks: Fieldbus specifications. Part 6-10: Application layer protocol specification: Type 10 elements, 2014.

[21] International Standards Organization. Open systems interconnection-basic reference model: The basic model, 1994.

[22] Internet Engineering Task Force. Request for Comments 768: User Datagram Protocol. https://www.ietf.org/rfc/rfc768.txt, 1980.

[23] Internet Engineering Task Force. Request for Comments 791: INTERNET PROTOCOL. https://www.ietf.org/rfc/rfc791.txt, 1981.

[24] Internet Engineering Task Force. Request for Comments 793: TRANSMISSION CONTROL PROTOCOL. https://www.ietf.org/rfc/rfc793.txt, 1981.

[25] R. Lewis. *Modeling Control Systems Using IEC 61499: Applying Function Blocks to Distributed Systems.* Control Engineering Series. Institution of Engineering and Technology, London, 2001.

[26] P. S. Marshall and J. S. Rinaldi. *Industrial Ethernet.* ISA, Research Triangle Park, NC, 2005.

[27] Modbus Organization, Inc. Modbus Messaging on TCP/IP Implementation Guide V1.0b. http://www.modbus.org/docs/modbus_messaging_implementation_guide_v1_0b.pdf, 2006.

[28] Modbus Organization, Inc. Modbus over serial line specification and implementation guide V1.02. http://www.modbus.org/docs/modbus_over_serial_line_v1_02.pdf, 2006.

[29] Modbus Organization, Inc. Modbus Application Protocol Specification V1.1b3. http://www.modbus.org/docs/modbus_application_protocol_v1_1b3.pdf, 2012.

[30] P. Neumann. Communication in industrial automation – What is going on? *Control Engineering Practice*, Vol. 15, No. 11, 1332–1347, 2007.

[31] nxtControl GmbH. nxtSTUDIO: Engineering tool for the planning of control, visualization, process integration, simulation, testing and documentation. http://www.nxtcontrol.com, 2014.

[32] Open DeviceNet Vendors Association, Inc. Volume One: Common Industrial Protocol (CIP) Specification. http://www.odva.org/default.aspx?tabid=79.

[33] Open DeviceNet Vendors Association, Inc. Device Net Technology Overview. https://www.odva.org/default.aspx?tabid=107, 2015.

[34] Rockwell Automation. Controlnet Network. http://www.ab.com/en/epub/catalogs/12762/2181376/214372/1809768/, 2015.

[35] T. Sauer. Fieldbus system fundamentals. In R. Zurawski, editor, *Industrial Communication Technology Handbook*. CRC Press, Boca Raton, 2015.

[36] J.-P. Thomesse. Fieldbus technology in industrial automation. *Proceedings of the IEEE*, Vol. 93, No. 6, 1073–1101, 2005.

[37] V. Vyatkin and J. Chouinard. On comparisons of the ISaGRAF implementation of IEC 61499 with FBDK and other implementations. In *6th IEEE International Conference on Industrial Informatics*, pages 289–294, 2008.

[38] V. Vyatkin, M. De Sousa, and A. Zoitl. Communication aspects of IEC 61499 architecture. In B. M. Wilamowski and J. D. Irwin, editors, *The Industrial Electronics Handbook: Industrial Communication Systems*. CRC Press, Boca Raton, 2011.

[39] F. Weehuizen and A. Zoitl. IEC 61499 Compliance Profile for CIP based Communication Function Blocks. http://www-ist.massey .ac.nz/functionblocks/cipcompliance.htm, 2006.

[40] F. Weehuizen, A. Zoitl, et al. Using the CIP protocol with IEC 61499 communication function blocks. In *5th IEEE International Conference on Industrial Informatics*, volume 1-3, pages 261–265, 2007.

11

Adapted Design Methodology to IEC 61499 for Distributed Control Applications of Machine Tools

Carlos Catalán

Universidad de Zaragoza

Alfonso Blesa

Universidad de Zaragoza

Félix Serna

Universidad de Zaragoza

José Manuel Colom

Universidad de Zaragoza

CONTENTS

11.1 Short Motivation

The design of the control for machine tools is a non-trivial and relevant domain in the development of industrial software. It is non-trivial because this software must consider many different technologies to realize an integrated control of the mechanical, electro-mechanical, electric, and electronic components. Moreover, there are functions not strictly related with control that must be integrated in the final application that must be implemented in a distributed environment. This complexity gives rise to restrictions such as attention to hard real-time constraints and integration with other information systems within the company that must be taken in account. All this leads to the need for methodologies that may be capable of dealing with the design from a multidisciplinary perspective and integrating multidisciplinary teams. Some of the methodologies of software engineering could be adapted to this context. But given the long tradition within the domain of machine tools and the economic importance of the domain, these methodologies should integrate, at least partially, the pre-existing modes of design as well as inherited components. This chapter describes such a methodology (or design guides) adapted to this application domain to help close the gap between the IEC 61499 standard and the industrial reality. It presents COSME, an execution platform designed to support this methodology, which takes into account the features of the machine tools problem in a IEC 61499 context. Finally, as a case study, the work briefly summarizes the characteristics of a flat glass cutting table and the aspects that affect its control software design process (e.g., cycle time, number of axes, and number of I/Os).

11.2 Introduction

The manufacturing industry must offer faster responses to market changes, while still being able to maintain low-costs. This has caused a change of the production strategies toward smaller adapted series to fulfill these changing demands. Flexible manufacturing systems (FMSs) have been technological answers to this scenario. They are based on the use of standardized general purpose machine tools (e.g., computer numerical controls), which can produce different parts, products and transport lines to move them.

Most control systems for machine tools are implemented using programmable logic controllers (PLC). PLCs are model of centralized and cyclic control; i.e., acquisition of the input variables, implementation of control algorithms and updating of output variables. IEC 61131-3 [11], widely accepted in industry, provides a standard for model-based PLC systems. This standard defines data types, function blocks (FB) and languages for control applications and is a major step forward in enabling portability of applications between different models and development platforms.

The new challenges of the industry demand a control system with new features; e.g., easily adaptable, reconfigurable (even dynamically) and reusable, and this standard does not comply with these new requirements successfully [28]. Outside of the automation domain, software engineering has been using technologies that facilitate software development. One such technology is the component orientation, which allows flexibility and reuse of software and has gradually been accepted in the field of control applications. Another significant development has been the advance in communications that has facilitated the shift from a centralized paradigm to a distributed one.

The previous requirements and challenges have been covered in the IEC 61499 standard [12]. IEC 61499 applications are basically distributed FB networks. For the definition of the model, the following must be taken into account: (a) alternative selection algorithms, (b) models of planning and communication, (c) response time, (d) implementation based on events, (e) software reuse and (f) agility [12]. This standard has not provided methodologies or guidelines to design FB networks and some works have proposed design patterns such as the well known proxy pattern [5] and the failure management one [20] as examples.

The IEC-61499 has been a breakthrough in the design of control systems because, among other features, it enables the design of distributed systems and has an execution model based on events. There is currently a debate about the advantages and disadvantages of this execution model. Concerns related to the predictability of execution time have not been addressed in the industry [26]. Several works [28, 35] indicate possible actions required to promote its adoption.

This chapter presents an FB model and a distributed control platform, named COSME intended to make distributed control software design based on IEC 61499 easier. These concepts also present easier implementation and good scalability. Both are adapted to machine tool control software in the context of agile manufacturing systems (AMS). The rest of the chapter is organized as follows. Section 11.3 shows relevant basics of the IEC 61499 standard. Section 11.4 describes the scope of the COSME platform and Section 11.5 the COSME design goals. Sections 11.6, 11.7 and 11.8 present the COSME FB model, platform architecture and design process, respectively. Finally, Sections 11.9 and 11.10 present an implementation, tests and conclusions.

11.3 Control Software for AMs: IEC 61499 Standard

The rules to control software based on PLCs have been established by the IEC 61131 standard [11] and are widely accepted and used in the industry. In order to meet the new manufacturing requirements, IEC 61499 has emerged [12, 28]. This standard defines an architecture and models that seek to facilitate the development of reconfigurable and distributable control systems. The most significant model is the FB, which extends the FB defined in the previous standard IEC 61131. An IEC 61499 application consists of a function block network (FBN). This FB model incorporates an event-driven execution, which is more efficient than the IEC 61131 scan execution. IEC 61499 function blocks execute control algorithms only when events indicate that there are new data to process, whereas the IEC 61131 function blocks execute control algorithms cyclically, even if there are no new data.

On the other hand, a distributed control system consists of multiple controllers interconnected by networks, without a central controller where communication and synchronization issues play a central role. IEC 61499 models and architecture were defined from the beginning to develop such systems. FBNs can be deployed across different resources (i.e., runtimes), which are hosted on devices (i.e., controllers). Event and data links between controllers are abstracted by communication service interface function blocks (CSIFBs) with a publish/subscribe and client/server models.

The goal of reconfigurable software can be achieved using the component software paradigm. IEC 61499 function blocks can be considered software components [14]. Reconfiguration is possible through changes in FBNs by creation, deletion, replacement and interconnection of FBs on the fly and deployment and installation of new FB types. The standard provides a management model of resources, devices and function blocks to support reconfiguration issues.

Unlike IEC 61131, industry has not adopted the new standard yet. Some early studies as [25] and [9] have analyzed possible reasons, mainly immaturity of the standard. [28] indicates the actions taken to promote its adoption (next appearance of a new version, first industrial experiences, etc.). There are several tools and platforms for IEC 61499. FBDK and its runtime FBRT [10] are implemented in Java and cannot meet real time constraints, but have been the references for the standard in many academic works. An important open source initiative is 4DIAC [35], with a tool based on the Eclipse framework and runtimes for PC and ARM. Other proposals may be seen in [28]; this work also indicates that it is hard to develop applications with these academic tools. However, it is important to have tools and platforms that may have practical use by control engineers outside research laboratories. Finally, two commercial proposals have emerged from the manufacturers of automation equipment: a) ISaGRAF, based on a previous framework for IEC 61131 in [29] can be seen as comparable to FBDK and FBRT), and b) NxtStudio [17]. The first industrial experiences are being developed with them.

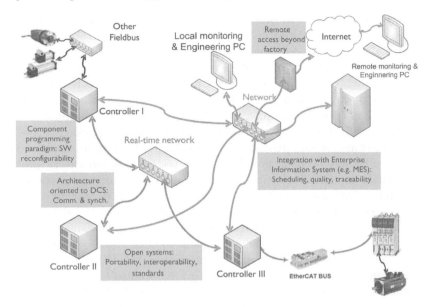

FIGURE 11.1
Intercontroller communication types scenario.

11.4 Communicating Machine Tools with IEC 61499: Scope of COSME Platform

This work is focused on machine tools, that are important elements of any manufacturing system. From a general view, a machine tool performs operations such as cutting, drilling, milling, etc. These operations require a certain number of movements of pieces, machining tools and finished items. In order to perform these tasks, control software should consider: a) precise positioning, b) loading and unloading of pieces and finished items, and c) operating modes and failure management.

Communication aspects are key in machine tools. That is, they must be integrated in the different levels of factory communication networks. In order to perform this integration, control software should consider communication: a) with other elements of the same factory workflow, and b) with external applications (e.g., supervisory control, data acquisition and production management). Since machining operations may change, in an agile manufacturing context, another aspect to considerer is the software reconfigurability. Figure 11.1 shows the scenario considered in this chapter for distributed control for machine tools.

An important feature of these machine tools is failure management, which should consider both their own failures and those caused by other machines in the same workflow. In order to establish a response actions policy (i.e., failure management) could be considered the several operation modes such as normal operation, degraded operation and failure operation. Also, it is necessary incorporate features of preventive and predictive maintenance.

Some works on the use of IEC 61499 for machine tools [31] present a framework and methodology for adaptive assembly operations using FBs for dynamic shop operations. Hagge and Wagner [8] use FBs to implement open modular architecture control (OMAC) state models for packaging machines. An architecture for an open and distributable CNC system based on STEP-NC and FBs has been proposed [16] and tested in a laboratory prototype with FBDK and FBRT. Xuemei [32] considered a reconfigurable control for a PLC and a CNC but presented no experimental results except for a brief case study. None of these proposals has been tested in real-life applications

A real machine tool may have dozens of axes and independent movements, so its control based on IEC 61499 may result in large and reconfigurable FBNs with hundreds of FBs. In a first approximation, the FBN design may start by dividing the control in basic parts (e.g., axis control). Those parts appear repeatedly and can be implemented with one sub-FBN. Then control may be implemented by composing these sub-FBNs and adding of some business logic. FBN scalability is an important goal for maintaining good performance and shortening design times. That is, these issues should not depend on the number of axes or movements. The authors identified the control parts of a specific machine tool and proposed design guidelines [21].

These parts are sufficiently general to be applicable to any machine tool. In summary they are: a) Control: precise positioning of axes, b) automation: movement sequences, c) failure management, d) business logic: specific functionality and operating modes, e) interface process: sensors and actuators via field bus, and f) external communication: gateway functionality with the external applications.

Regarding distributed control, an FBN of composed sub-FBNs of control parts can be deployed (i.e., distributed) from one to dozens of controllers, depending on the distribution granularity. That is, one controller can control a single axis or a complete machine. Again, from a first approximation this distributed FBN may establish links between controllers using CSIFBs.

11.5   COSME Platform Design Goals

11.5.1   Facilitate Application Design

From a component paradigm programming view, FB types should be designed carefully to allow reuse in different contexts (i.e., interaction with other FBs). An FBN is more than a monolithic design split into parts that are implemented as FBs; it must be designed using software engineering methodologies. IEC 61499 does not establish methodologies or guidelines that describe FB types or define FB interactions. Software based on model-driven methodologies related to IEC 61499 have been summarized [28]. Other authors focused on design patterns [19, 20, 22] and guidelines [3] for the standard.

A handicap is that control engineers are not familiar with IEC 61499 paradigms (i.e., event-driven and software components), or with software engineering methodologies. Therefore these subjects must be included in control engineering curricula (e.g., see [25] and [36]).

The COSME approach is to provide to control engineers with an adapted FB model preset to specific kinds of application, and sufficiently versatile. This adaptation attempts to predefine events and algorithms to facilitate application design and shorten development times.

11.5.2 Compliance with Principles of Distributed Systems

From the view of the standard, a distributed control application can be developed in two steps [30]: 1) the FBN is designed without considering distributional issues and 2) the FBN is split in dependent parts, which are later deployed on resources and coupled by CSIFBs. This process seeks to increase reusability of FBs and FBN parts. Accordingly, distribution issues (e.g., number or type of controllers) and FBN topology have no influence in the design. Design of distributed FBNs is not easy to implement [1, 24, 27]. For example, it is not possible to design without considering communication delays and synchronization between FBs [30]. In order to meet these issues, designers must add modifications to the initial FBN. Distributed control based on IEC 61499 can be addressed taking into account the generally accepted design principles of distributed systems [15]: transparency, reliability, performance, scalability, security, flexibility and heterogeneity. At this point we will focus on the first five, because the last two are intrinsic to the standard.

Transparency: This principle is considered hiding underlying distributed systems. Transparency is not met by IEC 61499, because designers must make the FBN modifications mentioned above, and then insert CSIFBs in the communication links. A solution is proposed [30] by which CSIFBs are inserted automatically by a development tool when the mapping between FBs and resources has already been established. Another proposal [7] replaces CSIFBs by a middleware layer. To meet this goal, the COSME approach predefines the communication links, which facilitates automatic insertion of CSIFBs and avoids further modifications in the FBN. Previously, communication types were be identified and characterized from specific application domains [2]; this is similar to communication cases presented by [23]. Table 11.1 shows the identified types, which are characterized by several features: latency, execution mode, reliability and amount of transferred data. These types are incorporated into the COSME FB model, increasing the distribution abstraction level (i.e.,

TABLE 11.1
Communication Machine Tools Types

Communication Type	Typical Latency	Execution Model	Reliability	Data Amount	Example
Application control	100 ms	Event-driven	Yes	Bytes	Operation modes Failure management
Process control	1 ms	Cyclic-driven	Yes	Bytes	Axis precise positioning by PID loop
Process command	100 ms	Event-driven	Yes	Bytes	Sequence of motion commands: "go XY"
Process synchronization	100 ms	Event-driven	Yes	Bytes	High semantic commands: "available piece"
Human-machine interface	100-500 ms	Cyclic and Event-driven	No/Yes	KBytes	Supervision/Alarms
Application data management	500 ms	Cyclic-driven	Mostly	KBytes	Preventive maintenance
Production data management	Few seconds	Cyclic-driven	Yes	MBytes	Communication with MES

transparency design principle) for distributed control software based on IEC 61499.

Reliability: Reliability in FMSs is considered especially from a hardware view, since traditional control applications are relatively simple and therefore, the software is also simple. PLCs are designed to maintain a high degree of reliability and availability for long periods in harsh industrial environments. Although reliability should increase through the use of already tested FBs, agile manufacturing systems (AMS) requirements increase software complexity, making safety-critical systems design more difficult to achieve. In particular, IEC 61499 distributed programming and event-driven features make it difficult to ensure the deterministic behavior of applications and therefore their reliability, [1]. One way to ensure reliability is using verification and validation techniques, but they are difficult to apply in this context [28].

The adapted COSME FB model with predefined events and algorithms leads to an execution model (see Figure 11.6) that facilitates determination of system behavior in design time, contributing to reliability. In addition, this model avoids designing from scratch, which also contributes to this goal.

Performance: Although, from the view of control engineers event-driven execution is more efficient than scan execution, its implementation adds overhead that decreases performance [28]. This affects worst case execution time

(WCET), which is important for determining compliance with real-time constraints. WCET calculation in IEC 61499 FBNs is complicated because of its event-driven execution (strongly dependent on its runtime implementation), and the reconfigurability feature. Some works have addressed this issue. For example, [13] performs static analysis based on model checking, and [7] calculates WCET to show the performance of a specific framework. A complete study of real time execution and WCET calculus in IEC 61499 systems has been presented by [34]. This work lists established guidelines that limit flexibility in FBN design, in order to improve WCET and performance.

The COSME execution model establishes a simple and regular FBN topology that seeks to obtain good performance and easy WCET calculation.

Scalability: It is desirable that a distributed system has a performance at least equivalent to a centralized system and is scalable. The scalability of IEC 61499 applications is affected by two factors: a) the number of controllers and b) the size of the sub-FBN on each controller. There are works that have studied IEC 61499 applications with large numbers of controllers [4, 33], but we know of no works about large sub-FBNs per controller.

In general, scalability depends on the communication aspects and the parallelism exploitation of the underlying platform; for large sub-FBNs the second issue is the most important. Parallelism is related to the concurrent execution of FBs and it is implicit in its event-driven model. In practice, scalable implementation of FBN is difficult to achieve because real applications (e.g., machine tools) may require a large number of FBs running concurrently [9] and consequently demand the use of many platform resources. These resources are related mainly to execution contexts (e.g., threads), that were analyzed for different IEC 61499 elements [24]. This work proposes the event chain concept to achieve an efficient FBN execution, but has not considered the scalability issue.

To achieve this goal, the COSME execution model needs a limited number of threads, independent of the size and complexity of the FBNs.

Security: Current manufacturing systems are integrated in communication networks which go beyond factory applications. If security is very important in any communication network; in manufacturing systems it is mandatory, since these systems can impact people, equipment (i.e., safety) and reserved information. The security aspects to be considered are integrity, confidentiality, non-reputability, availability and authentication. IEC 61499 does not take into account any of these, because they are *implementation dependent*, and abstracted by the CSIFBs. Nevertheless, some works have considered this issue under the standard view [6, 18].

The predefined communication links allow COSME to establish specific security mechanisms. For example, real time communications between controllers supported by a dedicated wired local network may not require data encryption; non-real time communication between controllers and external applications supported by a non-dedicated network necessarily requires data encryption.

11.5.3 Incorporation of Value-Added Features

As mentioned in Section 11.4, these applications, beside of control functionalities have extra-functional requirements such as preventive and predictive maintenance. The COSME platform incorporates services to meet them. In the first case, the adapted model can include continuous usage accounting (time or number of cycles of machine elements). In the second one, it can include FBs that infer failures (e.g., by rules) of monitoring machine elements [19, 22].

Control applications should take integration issues at all production management levels, for example, manufacturing execution systems (MESs). These systems manage mandatory functions such as production scheduling, quality control and product traceability. COSME includes features that allow control engineers to manage those functions.

Table 11.2 compares the COSME and IEC 61499 approaches. It considers the application domain, component-based paradigm, architecture for supporting distributed control and value-added features.

Table 11.2 summarizes the main COSME approach characteristics vs. the IEC 61499 general approach.

11.6   COSME FB Model

IEC 61499 indicates that FBs should accommodate multiple algorithms selectable by events or conditions. The COSME FB model (see Figure 11.2) is based on predefined algorithms taking into account common situations in control applications, such as initialization, normal operation, failure operation, etc. In Table 11.3 a more detailed description is shown.

The input and output events are predetermined, but the input and output data are not, and they depend on the functionality of FB type. The algorithms generate an event when they have finished. This allows composing FBNs to establish a daisy-chain topology, that connects events of the same type of algorithm. The extension to distributed control systems of this model can be simply done by incorporating CSIFBs in the inter-controller chains. These CSIFBs should be publish-and-subscribe types, because they are appropriate for the propagation of events along the controllers.

As indicated above, the COSME model aims to incorporate the communication types of machine tools. To do this (see Figure 11.2), the types are mapped with the predefined chains as follows:

COLD_INIT, WARM_INIT, FINALIZE and FAILURE RECOVERY chains correspond to the control application communication type. Only events are propagated; therefore CSFIBs without data are needed.

FAILURE also corresponds to the control application communication type,

TABLE 11.2
Comparison of COSME and IEC 61499 approaches.

	IEC 61499 approach	COSME approach
Application domain	General purpose automation systems	Specific purpose, machine tools for construction materials (e.g., glass, wood,...)
Component-based programming paradigm		
FB Model	Event-driven. Designers define event and data I/O, and ECC.	Event driven. Predefined events and algorithms (normal and failure operation, initialization, etc.) adapted to application domain
Execution models	FB invocation: sequential or cyclic; execution contexts: FB, resource or event chain; Non-preemptive multi-threaded resource; parallel	Daisy chain (based on event chain model)
Distributed FBN	Designers must deal with details (e.g., CSIFB, FBN modifications caused by synchronization issues)	Distributed daisy chain by means of predefined CSIFB. FBN shoud not be modified by synchronization issues
FBN design	Model-driven methodologies with little IEC 61449 support	FB adapted model (precooked). Designers only define data I/O and predefined algorithms code in FB adapted model, then establish daisy chains
Architecture oriented to support distributed control		
Transparency	Insert CSIFB manually or using middleware	COSME developtment tool inserts CSIF automatically based on predefined communication types.
Reliability	Difficult to ensure deterministic behavior	COSME FB model facilitates determination of system behavior in design time.
Performance and scalability	Strongly dependent of FBNs complexity, execution models and its platform implementation.	COSME execution model and its implementation needs limited number of resources independent of size and complexity of FBNs. Good: 200 FB network on real application
Security	Platform dependent	Predefined communication types allow specific security mechanisms: (1) process synchronization. (2) application and production data management
Value-added features	—	COSME platform allows features such as preventive and predictive maintenance or integration with enterprise system information

but it has associated information about failure condition; in this case CSIFB with data is needed.

NORMAL_RT and EXT_EVENT chains correspond to the process control communication type. Usually, they are not distributed (i.e., intra-controller) due to realtime constraints with small cycle times in the first case and immediate response needs in the second one. Otherwise, the communication delays should be calculated to determine that these constraints are met and data must include a timestamp.

NORMAL_NRT and BACKGROUND chains correspond to the process command and synchronization, HMI and management data communication types. Unlike the previous chains, they have no hard real time constraints. They have associated information and CSFIBs with data are used. These chains could be distributed (i.e., inter-controller) and cycle time could be so long that communication delays can be ignored.

11.7   COSME Platform Architecture

COSME is the layer between control applications and the underlying system (i.e., hardware, operating system and communications). A typical COSME environment is shown in Figure 11.3. The platform follows the event chain (EC) concept [24] for execution. Implementation is done by associating these ECs to different priority threads according to a model named *Event path per*

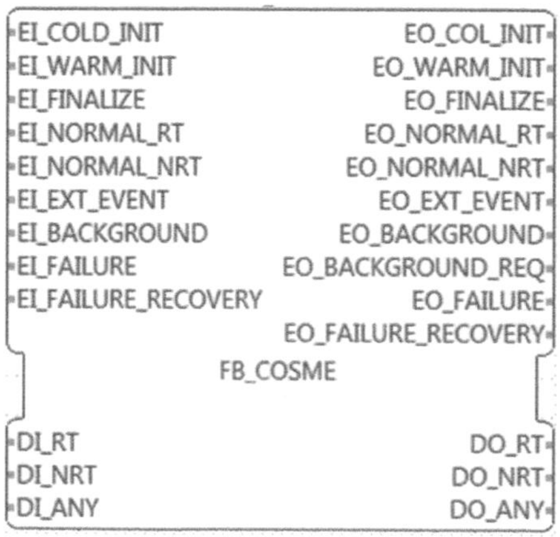

EI_COLD_INIT	EO_COL_INIT
EI_WARM_INIT	EO_WARM_INIT
EI_FINALIZE	EO_FINALIZE
EI_NORMAL_RT	EO_NORMAL_RT
EI_NORMAL_NRT	EO_NORMAL_NRT
EI_EXT_EVENT	EO_EXT_EVENT
EI_BACKGROUND	EO_BACKGROUND
EI_FAILURE	EO_BACKGROUND_REQ
EI_FAILURE_RECOVERY	EO_FAILURE
	EO_FAILURE_RECOVERY
FB_COSME	
DI_RT	DO_RT
DI_NRT	DO_NRT
DI_ANY	DO_ANY

FIGURE 11.2
The COSME FB model. The inputs and outputs events are predetermined; data inputs and outputs are not predetermined and depend on FB functionality.

TABLE 11.3

FB COSME Algorithms

Name	Description
Cold Init	It establishes the initial state of FB. It is executed aperiodically with the highest priority.
Warm Init	It performs resume actions after a scheduled stop. It is executed aperiodically with the highest priority.
Finalize	It performs finalization actions. It is executed aperiodically with the highest priority. A typical action can be deactivating actuators.
Normal RT	It performs actions with realtime constraints. It is executed periodically with high priority. A typical action can be a PID loop. If an abnormal condition occurs, it can generate warnings of operation degradation or failure to other FBs.
Normal NRT	It performs actions with non-realtime constraints. It is periodically executed with less priority than normal RT. A typical action can be a supervision or HMI communication. Like normal RT, it can generate an event to notify of abnormal operation or request a background action.
Background	It performs actions to background. It is sporadically executed with the lowest priority. A typical action can be calculation of CNC trajectories.
Failure	It performs failure mode actions. It is executed periodically with high priority. Typical actions can be setting safe mode and logging failure data.
Failure recovery	It performs recovery actions from failure mode. Typical actions can be rearms.
External Event	It handles external events. It is executed aperiodically with the highest priority. A typical action is handle data from the I/O field bus.

task, event path-based priority, [7]. Control algorithms must be executed in a short time because they block the system to the ECs with a lower priority. Due to the limited number of ECs and threads, the implementation is easily scalable. Thus, this model is especially suitable for applications with a large number of FBs and it also facilitates determining the system behavior in design time.

The COSME elements that perform execution are the execution server, name server and application loader. The loader, loads FB types, creates FBs and makes intra- and inter-controller connections, then orders the execution boot. The platform has a persistent storage element to perform warm startup of the applications.

Although the current COSME version is focused on application design and distribution aspects, it allows simple reconfiguration of the FBNs in execution time by means of commands (instantiation, reconnection, etc.). In a new COSME version, these commands will be sent by an external reconfiguration manager application, that will receive data from control and from MES to implement dynamic reconfiguration. The COSME elements that perform this function are the component server, name server and gateway.

Inter-controller communication links are based on publish and subscribe mode. The messages contain message number, publisher and subscriber names, timestamp and data (when necessary). Inter-controller communication uses a wired dedicated network and a switched Ethernet device. These devices guarantee deterministic delays without special hardware or software. An authentication policy is used to guarantee security.

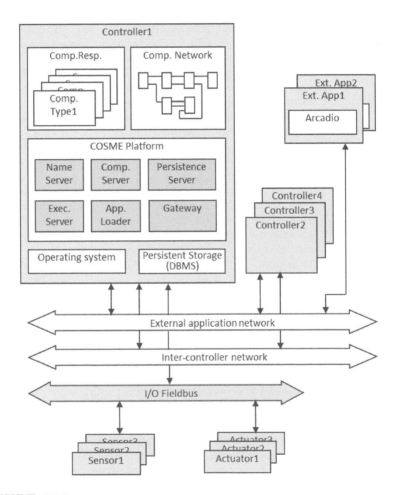

FIGURE 11.3

A typical COSME environment. There are four controllers and two external applications: HMI and management production. Each COSME platform has services (name, application, persistence and component server, loader and gateway), FB repositories and FB network.

Communication with external applications is performed with client and server or publish and subscribe messaging models. In both cases, messages have the same format as in the inter-controller case. The Arcadio library [21] performs communication in external applications, hiding from designers the implementation details. Here, non-dedicated networks are used; therefore for security reasons, connections use authentication and encryption. During booting, the application loader establishes all communication links and they are managed later by the gateway element. Finally, process interface communication is performed by a field bus.

11.8 COSME Design Process

Briefly, the process consists of mapping each application control part identified with one FB or sub-FBN and then composes them together. A sub-FBN can be composed by several FB types. For example, an axis precise positioning has one PID control loop, one setpoint generator and one gearbox. Each FB type should have the following: a) FB interface by means of output and input data; b) FB state by means of internal variables and c) FB behavior by setting actions in the predefined algorithms.

To make FBNs, the COSME composition rules are: a) connecting output and input data between the interrelated sub-FBNs; b) establishing daisy chains with input and output events considering data dependencies (i.e., it should ensure that FB data producers are executed before consumer ones); c) establishing CSIFBs in the distributed daisy chains corresponding to the communication types previously indicated. To facilitate the design process, COSME incorporates the Domiciano programming environment [2]. Domiciano is an IDE (integrated development environment) that allows development of control applications for the COSME platform. Domiciano has been designed to take into account the characteristics of COSME, such as establishing the daisy chains from the FB execution order given by the designer. Then, it inserts CSIFBs when necessary.

A simplified application example is shown in Figure 11.4. This application controls a loader, which provides glass sheets to a cutting glass table. Both machines perform movement sequences and business logic (AUTOMATION), I/O conditioning (SENSORS and ACTUATORS) and process interfaces (FIELD_BUS). The table also has a two-axes precise positioning tool (PID), data preparation for monitoring and control (SUPERVISION), communication with HMI/SCADA (FB_HMI) and background calculations of CNC trajectories. (TRAJ_CALC). Daisy chains are distributed of CSIFBs, except NORMAL_RT due to its small cycle time; it is used for movement sequences and precise positioning. NORMAL_NRT is used for process commands (e.g., CNC commands), process synchronization (e.g., transfer protocol

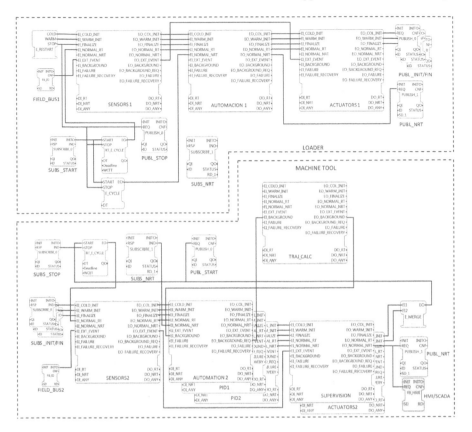

FIGURE 11.4

Simplified COSME application. To facilitate the representation, data connections and some event connections are hidden and some FBs connected in daisy chain are overlapped. In order to deploy COLD_INIT, WARM_INIT, FINALIZE AND NRT_NORMAL chains, CSIFBs (PUBL and SUBS FBs) are included on the FBN automatically by the COSME platform.

of sheets) and supervision. Finally, although not shown in Figure 11.4, a hierarchical failure management has been used. This is based on a design pattern presented by the authors in [20].

In conclusion, the COSME daisy-chain is an intermediate model between events-based and the scan-based. Designers must write response code to situations, like event-driven, but the situations are predefined. COSME has been designed with machine tool control software in mind, although it could be used in other applications. The rules of composition are simple and can be applied by engineers with a classical education in control engineering.

TABLE 11.4

Communication Latencies of Controllers

E_CYCLE	Max.	Min.	Avg.	Std. Avg.	Median
50	10.19	1.99	2.58	0.65	2.40
100	10.49	2.01	2.46	0.57	2.30
300	9.57	2.15	2.19	0.51	2.16

Note: All measurements are in milliseconds.

11.9 Implementation Issues

The current implementation of COSME runs on a Linux PC with real time application interface (RTAI) and Java SE6. RTAI has been used to implement the component server, name server and execution server. The other elements have been partially implemented with Java and RTAI. COSME provides the Arcadio communication library to third party external applications, so that they can connect to COSME to utilize applications related with user interface, debugging, logging tasks, remote monitoring, MES integration, etc. Arcadio and Domiciano are fully implemented in Java. COSME has been used by control engineers of the machine tools manufacturer TUROMAS-TECNOCAT Group to design the control software of a new cutting glass line [21].

The cutting table has nine axes, which have a cutting accuracy lower than one millimeter and up to 22 m/s^2 acceleration. This line has loaders and stores glass sheet and finished items. There are two external applications: HMI and production management, both implemented in Java. A simplified description of this test application has been given in Section 11.8. The real application tested has been included for the purpose of giving some performance measures of the implementation of COSME operation in a real environment, the effectiveness of the approach and the interest of using COSME versus more generic platforms.

The cutting glass machine tool can have up to 35 FB types, 200 instances and 70 automatic sequences. Application hardware is composed of two B&R Automation PC 620 PC controllers with an Intel Celeron 1.1 GHz CPU , each with two 100 Mb network interfaces and one PCI-7841Adlink CANopen master card. Also, there are two X20 B&R remote I/Os, two s0700 SMC valve islands, three Sdrive Power Stages standard motor drives and two CISCO SLM200 Gigabit smart switches.

The NORMAL_RT periodic chain in charge of the precise positioning and the movement sequences has a 2 ms RT_E_CYCLE to achieve the cutting accuracy required. To establish a cycle time of NORMAL_NRT, communication latencies between controllers have been measured with different E_CYCLE times (see Table 11.4). Results depend on Java performance, but they have similar values for this configuration. To meet requirements, a 50 ms E_CYCLE has been chosen.

The main factors of performance and scalability to be considered are number of axes and number of movement sequences. These factors greatly increase the complexity and size of the FBN. The number of axes is the key factor because more complex sub-FBNs are required to implement it. Table 11.5 shows maximum and minimum execution times and scalability performances for three axis numbers. Note that the results are close to linear behavior.

Table 11.6 shows NORMAL_RT execution times of relevant application parts.

11.10   Conclusions

This chapter shows the difficulties of designing distributed control applications of AMSs based on IEC 61499. The proposed solution characterizes the distributed control of specific domains, such as machine tools. This solution can successfully address the design of applications that must meet real time requirements or concurrency (with many axes). Technological changes should not alter the application architecture (for example, change the fieldbus serving I/O) and finally, this approach allows adaptation to the changes imposed by the market (i.e., easy modification from functional and regulatory restrictions).

This allows us to define an adapted FB model and an execution model. This approach is based on a methodology (which abstracts the system as a set of components and a network that interconnects) and well-defined semantics (supported in its execution cycle) in a communication model and the definition of the behavior component. This proposal is accompanied by an engineering

TABLE 11.5

Axis Control Execution Times

Number of Axes	Max. (μs.)	Min. (μs.)
3	646.35	276.59
6	1285.60	601.78
9	1670.72	821.54

TABLE 11.6

Execution Times of Application Parts

Function	Max. (μs.)	Min. (μs.)
Axis PID control	8.72	3.69
Loading sequence	45.63	8.68
Drive CANopen(3 axes)	179.48	104.35
I/O conditioning	0.90	0.33

tool that supports the methodology with predefined semantics and adaptable and/or expandable to technological changes.

To verify its effectiveness, the COSME platform has been implemented. The results obtained in real and complex applications show good performance and scalability. In addition, these models reduce the complexity of the design, thus facilitating the work of control engineers.

Bibliography

[1] R. Brennan. Toward real-time distributed intelligent control: A survey of research themes and applications. *IEEE Transactions on Systems, Man, and Cybernetics. Part C: Applications and Reviews*, Vol. 37, No. 6, Sept. 2007.

[2] C. Catalán, F. Serna, A. Blesa, J. Rams, and J. Colom. Communication types for manufacturing systems: A proposal to distributed control system based on IEC 61499. In *Automation Science and Engineering, IEEE Conference on*, pages 767–772, Aug. 2011.

[3] C. Catalán, F. Serna, T. Civera, A. Blesa, and J. Rams. Control design for machine tools using Domiciano, an IDE based on software components. In *Emerging Technologies Factory Automation, IEEE 16th Conference on*, pages 1–4, Sept. 2011.

[4] J. Chouinard, J. Lavallée, J.-F. Laliberté, N. Landreaud, K. Thramboulidis, P. Bettez-Poirier, F. Desy, F. Darveau, N. Gendron, and T. C.-D. An IEC 61499 configuration with 70 controllers: Challenges, benefits and a discussion on technical decisions. In *IEEE Conference on Emerging Technologies and Factory Automation*, Sept. 2007.

[5] J. H. Christensen. Design patterns for systems engineering with IEC 61499. *Verteilte Automatisierung - Modelle und Methoden für Entwurf, Verifikation, Engineering und Instrumentierung*, 2000.

[6] I. M. Delamer and J. L. Martínez. Information security for reconfigurable manufacturing systems using networked embedded controllers. In A. D. M. E. Pereira, editor, *Information Control Problems in Manufacturing 2006*, pages 129–134. Elsevier, Oxford, 2006.

[7] G. Doukas and K. Thramboulidis. A real-time-linux-based framework for model-driven engineering in control and automation. *IEEE Transactions on Industrial Electronics*, Vol. 58, No. 3, 914–924, March 2011.

[8] N. Hagge and B. Wagner. Implementation alternatives for the OMAC state machines using IEC 61499. In *Emerging Technologies and Factory Automation, IEEE International Conference on*, pages 215–220, Sept 2008.

[9] K. Hall, R. Staron, and A. Zoitl. Challenges to industry adoption of the IEC 61499 standard on event-based function blocks. In *IEEE Conference on Industrial Informatics*, 2007.

[10] Holobloc. http://www.holobloc.com/doc/fbdk/index.htm, 2012.

[11] International Electrotechnical Commission. *IEC 61131-3 Edition 2.0: Programmable Controllers. Part 3: Programming Languages.*, 2003.

[12] International Electrotechnical Commission. *IEC 61499-1, Function Blocks. Part 1: Architecture*, 2.0 edition, 2005.

[13] M. Kuo, L. Yoong, S. Andalam, and P. Roop. Determining the worst-case reaction time of IEC 61499 function blocks. In *IEEE Conference on Industrial Informatics*, July 2010.

[14] R. Lewis. *Modelling Control Systems Using IEC 61499: Applying Function Blocks to Distributed Systems*. Number 59. Institution of Engineering and Technology, 2001.

[15] M. Liu. *Distributing Computing: Principles and Applications*. Pearson-/Addison Wesley, 2004.

[16] M. Minhat, V. Vyatkin, X. Xu, S. Wong, and Z. Al-Bayaa. A novel open CNC architecture based on STEP-NC data model and IEC 61499 function blocks. *Robotics and Computer-Integrated Manufacturing*, Vol. 25, No. 3, 560–569, June 2009.

[17] NXTControl. http://www.nxtcontrol.com, 2012.

[18] C. Schwab, M. Tangermann, and A. Lüder. Integration of scalable safety and security actions in IEC 61499 control applications. In *IFAC Conference on Field Bus Systems and Their Applications.* Nov. 2005.

[19] F. Serna, C. Catalán, A. Blesa, J. Colom, and J. Rams. Predictive maintenance surveyor design pattern for machine tools control software applications. In *Emerging Technologies Factory Automation, IEEE 16th Conference on*, pages 1–7, Sept. 2011.

[20] F. Serna, C. Catalán, A. Blesa, and J. Rams. Design patterns for failure management in IEC 61499 function blocks. In *Emerging Technologies and Factory Automation, IEEE Conference on*, pages 1–7, Sept. 2010.

[21] F. Serna, C. Catalán, A. Blesa, J. Rams, and J. Colom. Control software design for a cutting glass machine tool based on the COSME platform. case study. In *Automation Science and Engineering, IEEE Conference on*, pages 501–506, Aug. 2011.

[22] F. Serna, C. Catalán, A. Blesa, J. Rams, and J. Colom. Preventive maintenance manager design pattern for component based machine tools. In *Industrial Informatics, 9th IEEE International Conference on*, pages 615–620, July 2011.

[23] S. Sierla, J. Peltola, and K. Koskinen. Real-time middleware for the requirements of distributed process control. In *Industrial Informatics, 3rd IEEE International Conference on*, pages 1–6. Aug. 2005.

[24] T. Strasser, A. Zoitl, J. Christensen, and C. Sünder. Design and execution issues in IEC 61499 distributed automation and control systems. *Systems, Man, and Cybernetics, Part C: Applications and Reviews, IEEE Transactions on*, Vol. 41, No. 1, 41–51, Jan 2011.

[25] K. Thramboulidis. IEC 61499 in factory automation. In K. Elleithy, T. Sobh, A. Mahmood, M. Iskander, and M. Karim, editors, *Advances in Computer, Information, and Systems Sciences, and Engineering*, pages 115–124. Springer, 2006.

[26] K. Thramboulidis. IEC 61499: Back to the well proven practice of IEC 61131? In *Emerging Technologies & Factory Automation, IEEE 17th Conference on*, pages 1–8, 2012.

[27] V. Vyatkin. The IEC 61499 standard and its semantics. *Industrial Electronics Magazine, IEEE*, Vol. 3, No. 4, 40–48, Dec 2009.

[28] V. Vyatkin. IEC 61499 as enabler of distributed and intelligent automation: State-of-the-art review. *Industrial Informatics, IEEE Transactions on*, Vol. 7, No. 4, 768–781, Nov 2011.

[29] V. Vyatkin and J. Chouinard. On comparisons of the ISaGRAF implementation of IEC 61499 with FBDK and other implementations. In *Industrial Informatics, 6th IEEE International Conference on*, pages 289–294, July 2008.

[30] W. Wagner, J. Bohl, and G. Frey. An IEC 61499 interpretation and implementation focused on usability. In *IEEE Conference on Emerging Technologies and Factory Automation,*, sept. 2008.

[31] L. Wang, S. Keshavarzmanesh, and H. Feng. Design of adaptive function blocks for dynamic assembly planning and control. *Journal of Manufacturing Systems*, Vol. 27, , 45–51, 2008.

[32] H. Xuemei. Distributed and reconfigurable control of lower level machines in automatic production line. In *8th World Congress on Intelligent Control and Automation*, pages 2339–2344, July 2010.

[33] J. Yan and V. Vyatkin. Distributed execution and cyber-physical design of baggage handling automation with iec 61499. In *Industrial Informatics, 9th IEEE International Conference on*, pages 573–578, July 2011.

[34] A. Zoitl. *Real-Time Execution for IEC 61499*. O3NEIDA and Instrumentation Society of America), 2009.

[35] A. Zoitl, T. Strasser, and A. Valentini. Open source initiatives as basis for the establishment of new technologies in industrial automation: 4DIAC case study. In *Industrial Electronics, IEEE International Symposium on*, pages 3817–3819, July 2010.

[36] A. Zoitl and V. Vyatkin. Face to face: IEC 61499 architecture for distributed automation: The glass half full view. *IEEE Industrial Electronics Magazine*, Vol. 3, No. 4, 7–23, 2009.

Part III

Industrial Application Examples

12

Flexible and Reusable Industrial Control Application

Gernot Kollegger

nxtControl GmbH

Arnold Kopitar

nxtControl GmbH

CONTENTS

12.1 Introduction

The concept of Industry 4.0 widely discussed in Europe is raising new challenges for industrial automation. Production will become more flexible, adaptive, integrated and even cross company borders. More components are becoming intelligent and are integrated in wide communication networks [1]. Automation is crucial to deliver the requested flexibility, autonomy and data consistency. A solution for the complexity in the engineering is as well crucial to enable the implantation of these modern but complex systems.

In automation of such complex systems, there is a need for a flexible and easy distribution of intelligence or control logic to several or many controllers. This is where the IEC 61499 demonstrates its strength. Beside that, the separation of hardware from application software is a big plus for flexibility [2]. A topology-independent and hardware-independent engineering becomes possible. The engineering tool needs to support the engineer to meet the challenges of distributed intelligence, easy adaptation of processes and data consistency from field level to cloud level.

Object-oriented engineering is the key to solve the problem. As visualisation such as human–machine interface (HMI) and/or supervisory control and data acquisition (SCADA) is more or less requested in all projects, it has to be integrated along with the connection of I/Os from software application to the physical hardware. Just imagine an engineer who needs to program all the communication paths between all the distributed controllers. Integrated documentation, test, simulation and commissioning support are other features helping engineers to master the complexity in engineering.

A challenge of increasing importance is the connection to IT systems. More accurate data is requested from the field level to be provided to superordinate management systems. On the other hand, autonomous and self-adapting control units make manufacturing more efficient and flexible. Managing and making available huge amounts of process data via Internet technology is becoming an important skill.

This chapter will show how object-oriented and hardware-independent engineering is designed to support solutions for the latest challenges and devise reuseable software objects to keep engineering costs reasonable despite increasing challenges.

12.2 Expectations to IEC 61499-Based Automation Solutions

What are the expected benefits when using IEC 61499-based applications?

To answer this question, we need to step back into the past. Let's reflect the state of the art for programming automation projects. Which possibilities do we have to create an automation program?

- Program a dedicated firmware from scratch for a piece of hardware

- Use IEC61131 standardized programming environment if it is supported by the hardware manufacturer

- Use the proprietary programming environment from a specific hardware manufacturer

All of these solutions are based on a bottom-up approach. This means, a projects starts with a selected piece of hardware and the application software relates to this hardware. Software can only be reused on exactly the same type of hardware. Hardware-independent parts of the software may be reused on the same hardware family. In few cases, it is possible to reuse the same software on different hardware platforms, from different hardware manufacturers. The IEC 61131 standard helped a lot to establish a common programming language across a wide range of automation devices. However, the resulting software is not hardware-independent and the automation programs need to be adapted to the different underlying hardware.

The software tools from hardware manufacturers differ a lot and there is no interchangeable format between the tools. Even if they are based on the IEC 61131 standard, the produced code is often not reusable in another programming environment or for other hardware platforms without rewriting at least parts of the code.

By supporting configurability and portability, the IEC 61499 helps to overcome the limitations mentioned above [3]. Another problem domain is the aliasing effect of communication links between cyclic-driven automation devices. The precarious quality of data, the resulting delay of at least one cycle and the need of a programmed hand shake mechanism to ensure quality, leads to very rare use of this solution as a distributed system. Even if a distributed system would fit better to the industrial requirements, the use of a centralized system is often preferred. Powerful centralized automation devices combined with field bus systems and decentralized I/O environment are used.

Starting from this situation, what are the expectations for a new IEC 61499 standard?

To move the existing code base to a new standard or to start with a new standard in automation, beside the existing capabilities, there is the need to focus on topics beyond the state of the art.

A system integrator is interested in efficient engineering and reliable applications. Both aspects are very much supported by the standard and nxtControl's automation software.

For efficient engineering reuse of already established solutions, integration of several automation tasks in one single engineering tool and a comprehensive set of features supporting the engineer are key factors. Distribution of control is increasing the complexity in engineering, even more if visualization via HMI or SCADA is part of the project. Engineering tools need to help master complexity in such applications.

An important step toward efficiency is the object-oriented approach in engineering. The encapsulation of several automation aspects (control logic, visual representation, I/O connectivity, test, simulation and documentation) of real devices and functions in software objects is essential. This integration represented in the engineering tool is the major factor promoting efficiency in engineering and tasks like HMI and SCADA are very quickly and easily established. Again the engineering tool will create the communication paths between controllers and visualization clients. The second big point is the hardware-independent nature of these software objects. Thanks to IEC 61499, this hardware independence is easily achieved. The third point is the easy way of distributing control logic to several or even many controllers by simple drag-and-drop. There is no need to program the cross communication between controllers. This is done by the engineering tool taking into account IEC 61499 methods.

The reliability of an application is very much supported by features and concepts mentioned above. Seamless integration of different engineering tasks in the engineering tool are eliminating interfaces between the different steps of the automation pyramid. Automatically established communication paths between distributed controllers and between controllers and several visualization clients are reducing the risk of faults. Finally the ability to test and simulate the application on the office desk before startup ensures reliable applications. Comprehensive watch and debug features offer efficient failure detection during commissioning. An automatically updated documentation of the application ensures reliability over its whole life-cycle.

12.3 Requirements to IEC 61499-Based Applications

12.3.1 Automation Application Requirements

An automation project must fulfill several requirements:

- A visualization system must be included.

- The visualization must include user management, monitoring and control elements, alarm management, journalling, trending and historical data access.

- Multiple visualization clients need to be connected in parallel to the different devices.

- Applications must consist of easily reusable components.

- Application software components have to be connectible to different hardware I/O systems

- Application must be freely distributable to a changing number of automation devices

- The distributed application must have a self-healing program structure.

- Parts of the program must be programmed as sequence control logic.

Some of the requirements are only reasonably feasible based on the concepts of IEC 61499. The event-oriented execution approach is a prerequisite to run distributed and integrated applications.

12.3.2 Real-World Components as Software Components

A prerequisite to reach the expected capabilities is the transformation of real world components into software components. These components are built as function blocks and they have to integrate all the properties, inputs and outputs as interface and configuration abilities to reflect the real component in all expected states of operation. This includes the need for operating and monitoring the state of the real world object. The connectivity to the visualization system has to be an integrative aspect within the software component. Nevertheless, the software components must not depend on the hardware and physical I/O structure on which the software is finally executed.

Hardware-Independent Software Components

To overcome the limitation of using software components representing real world components on dedicated hardware, the function blocks need to implement a physical I/O connectivity interface layer. All signals need to be transformed into a common range, not related to hardware. The software component needs to be flexible within its structure.

Reusable Software Components

With the different options of parameters, the software components should be adjustable to reflect different kinds of the same types of real world objects without the need to create a new software component if just some parameters differ. As an example by adjusting parameters, a software component should be able to reflect different expansion stages of valves with or without limit switches or varying the time to open or close.

A major work is to create and follow a concept for a library of software components which allows us to create more complex software components by combining basic software components. This includes the definition of naming conventions for function blocks and variables, signal range definitions, operation modes, a concept for initializing the program, update frequency of states and signals, time behaviour and others. Similar work has to be done for the visualization part to define a concept for the user interface and a common usability involving different kinds of objects including colour schemes, design guidelines, operation flows and others.

12.4   Power of Attributes

Attributes are defined within the IEC 61499 standard. They are used to declare metadata on different levels of a function block. The information is used, for example, to specialize functionality of function blocks, identify variables for additional use, define configuration information for function blocks where they are used to provide a simple and intuitive work flow for the user while programming an automation application. In fact, attributes are used to extend the functionality and increase the productivity without breaking the rules and definitions of the standard.

12.5   CAT: Composite Automation Type

The CAT is a specialized function block, introduced by nxtControl. From the IEC 61499 standard view, it implements a composite function block. nxtStudio uses this type to define multiple aspects of a software component and combines them into a single type. Basically a CAT does not limit the number of aspects it combines. Currently, control logic (IEC 61499 function block network), visualization (with symbols and face plates), hardware connection

(*SYMLINK* concept) and documentation (standardized HTML and docbook format) are included within a CAT.

To integrate the visualization part, a HMI SIFB is added automatically to a CAT definition which represents the data and event interface between the IEC 61499 runtime and the HMI runtime. On top of this interface, symbols and face plates are created with the capabilities of .Net GUI applications. CATs can be used in a hierarchical way, in a CAT-in-CAT structure.

12.5.1   Nested Software Components

Complex software components such as motors, control loops and dosing lines combine or integrate their functionality in a hierarchieal manner. They are built by the use of more basic function blocks into a hierarchical structure of objects in objects.

To create easily usable software components, different types of function blocks are combined and connected together in function block networks. A function block network can also contain the most *Basic function block*. The inner implementation of a basic FB consists of an execution control chart (ECC) (see Figure 12.1) which includes transitions and the states with actions and calls of algorithms defined in the FB. The ECC also defines the events fired in the different states of the basic FB.

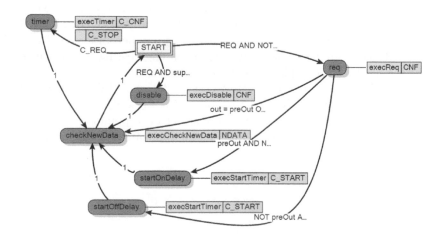

FIGURE 12.1
Basic FB event control chart.

12.5.2   Connecting Hardware-Less Applications to I/Os

To reach the highest grade of reusability of function blocks, it is necessary to have an interface to connect variables through all hierarchies of nested function blocks, without the need to expose this interface on each border of a nested hierarchy. Otherwise this would increase the number of variables to a huge number on the topmost function block.

Nevertheless, an automation system needs physical I/Os. The connection to I/Os is related to the hardware. nxtControl has developed a hardware abstraction layer concept to configure the hardware-related I/O signals as a bottom-up configuration, providing them to the automation application. Figure 12.3(a), shows hardware I/O signals. They can be local inputs directly on a device or a configured field bus system such as EtherCAT, ModBUS or others. Each piece of hardware has the possibility to be represented as a CAT which allows monitoring, manual setting (for commissioning) and configuration within the visualization system (Figure 12.3(b)).

Symlink Service Interface function blocks

To access physical I/Os, a mechanism is used to easily bind the software application to these physical signals. Complex components combine function blocks in a hierarchical way. Relevant I/O signals may be needed on different nesting levels. This design requires a virtual, symbolic access method to connect the application to the hardware signals within an automation project. nxtControl has implemented a symbolic link concept. The family of *SYMLINK* FBs are generic service function blocks. This means, the number and type of variables on the interface can be dynamically adjusted at engineering time per instance, to fit the needs of the application (Figure 12.2).

Instances of function blocks, using a service interface FB of the *SYMLINK* family, will populate all NAME variables with ${PATH} in their values, to the symbolic links list, where they can connect data points by name to each other (Figure 12.3(a)). To assign an exposed application hardware point to a

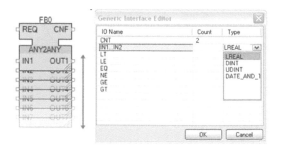

FIGURE 12.2
Generic FB interface and configuration.

physical I/O, drag and drop the signal from the symbolic link list onto the physical I/O.

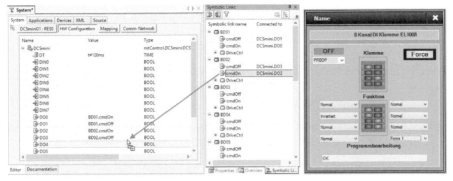

(a) Configuration View   (b) Monitoring View

FIGURE 12.3
Hardware abstraction configuation and monitoring.

12.6 Process Control Application

The design consists of an automation projects for a belt conveyor line, multiple dosing control lines, a reactor unit with temperature, pressure and flow regulations and pump stations with tank units and level control. The complex components are built on top of library elements designed for the process control units.

12.7 Concept and Components of Process Control Libraries

The library consists of a dedicated process control library and a library of basic components. We will look into some of the components to give an overview of the concept and the structures of these libraries.

12.7.1 Important Basic Components

Basic function blocks

modeBase is used to calculate and reflect operation mode states.

pt1EC is used as a PT-1 filter for analogue values.

pidSC provides the PID algorithm for control loops.

Composite function blocks

plcStart indicates the start of the PLC, provides a sequential startup procedure and prevents an event burst when starting a resource.

TIMEMONITOR is used to monitor time range, for example, to monitor the transition times of states for motors .

CATs

MODE is used as a sub-CAT in many other CATs. It provides the user interface for monitoring and controlling operation modes of motors, valves, analogue values, analogue controllers and other components.

BINGETVAL, BINSETVAL are very simple and basic CATs for monitoring and controlling binary values. They are used for simple tasks such as monitoring a single bit or setting a parameter value and have no further control capabilities.

ANAGETVAL, ANASETVAL, ANASETVALEXT are very simple and basic CATs for monitoring and controlling analogue values. They are used for simple tasks such as monitoring or setting analogue values having no further control capabilities.

MATHCURVE can calculate an output value with a mathematical expression with one parameter.

12.7.2 Generic Operation Modes and MODE CAT

A common concept and easy understanding of operation modes are important for a library to gain acceptance of users. An operation mode defines how a component works and operates throughout an application. The process library defines different operation modes for monitoring component states and controlling components. Some of the operation modes can be selected by the operator; others provide information and cannot be changed by the operator.

Informative Modes

LOOV local override The software functions of a component are overridden by signals in the real world components. This may be the case if there is a physical switch on the electrical cabinet to activate a motor. The software control has no effect. Available feedback signals can be used to track the real state in the software component.

FMAN forced manual A control signal forces the component to a certain state. For example, if a valve is closed, a motor must not be allowed to start and is forced manually to state off.

EMAN external manual The component's setpoint states are manually controlled by physical signals and cannot be changed by the operator with the visualization system.

Operator Selectable Modes

MAN manual The component's output states are set by the operator with the visualization system. No control algorithms are executed and the signal limitations and interlock functions are active.

AUTO automatic The component's output states are calculated by the component algorithm. The setpoint for the calculated output is provided by the operator with the visualization system.

CAS cascade The component's output states are calculated by the component algorithm. The setpoint for the calculated output is provided by a software signal from another component function block.

12.7.3 Advanced Components

Basic function blocks

stateCheckBin calculates the current state of a binary controlled component such as a motor or a valve. Up to eight input signals can be used to reflect a maximum of five different states of a binary component.

stateControlBin calculates the output signals of a binary controlled component such as a motor or valve. Up to eight output signals can be configured for a maximum of five different states of a binary component.

binDev1SEC calculates the requested setpoint of a binary controlled component such as a motor or valve. It includes the operation mode, permitted states, interlocking and local override signals in the calculation.

stateDevAnaEC calculates an analogue output signal. It includes the operation mode, interlocking, output limits and output change rate settings.

CATs

DSCBINDEV1S uses function blocks for binary components for operation mode monitoring control, for alarming and transition monitoring, and combines them to a CAT for controlling and monitoring binary devices with one active state. Such devices can be motors (off/on) and valves(close/open).

DSCBITCTRL simply controls a binary output signal without state transition monitoring. It uses the function blocks for monitoring binary signals and for operation mode monitoring and control.

BITMON monitors binary input signals. It uses the function blocks for monitoring binary signals, alarming and for operation mode monitoring and control.

DSCANAMON, DSCANACTRLPID, DSCANALIMITMON, DSCANASP and **DSCANAPULSCOUNT** monitor and control analogue-based values. They use function blocks for analogue signal calculation, operation mode monitoring and control, and alarming and combine them to build CATs for controlling and monitoring analogue process values.

12.7.4　Inner Component Structure

The structure and the implementation of the control logic will be explained based on the CATs DSCANAMON and DSCBITCTRL. Other components of the library follow the same concept and implementation rules.

Analogue Monitor CAT DSCANAMON

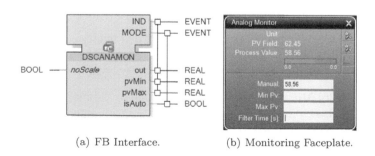

(a) FB Interface.　　　　　　　　(b) Monitoring Faceplate.

FIGURE 12.4
CAT DSCANAMON.

Figure 12.4(a) shows the outer interface and data and event association for this CAT's function block as well as the HMI faceplate for monitoring and controlling. The purpose of this CAT is to read analogue values from a physical I/O source, and scale the source value to the correct range to reflect the physically measured value. Table 12.1 gives an overview of the interface variables and events of the CAT function block.

The inner logic of the CAT uses the plcStart function block to get sequentially initialized during the startup phase of the runtime device. To read the process value a *SYMLINK* service function block is used and configured with the *${PATH}fvPct* as name parameter. The ioScale function block is used to scale the 0 to 100% normalized process value to the configured denormalized process value range, for example 0 to 10 bar. The range values are taken from the DSCANAMON_HMI service function block, which is the communication interface between the HMI runtime and the IEC 61499 runtime. The calculated process value is then forwarded to a PT1 filter function block. The PT1 controls a cyclic timer, to start or stop the cyclic calculation of the *PT1.out* value. To reduce the events in a system, PT1 calculation is stopped as soon as the deviation from the calculated *PT1.out* compared to the process value *PT1.pv* is <0.01%. The MODE CAT is used as a sub-CAT within the DSCANAMON to provide common operation modes. For this CAT, MAN and AUTO are allowed. Depending on the selected mode, the calculated and filtered process value *PT1.out* or the manual value from the HMI interface function block *IThis.manPV* is transferred to the *out* variable of the CAT.

TABLE 12.1
Interface variables of DSCANAMON

Direction	Name	Type	Description
Input	noScale	BOOL	Parameter; no scaling needed since raw value is already process value in physical units
Output	IND	EVENT	Event is fired on any change of calculated analogue monitor value
Output	MODE	EVENT	Event is fired on any change of operating mode
Output	out	REAL	Calculates analogue value used by other function blocks in funciton block network
Output	pvMin, pvMax	REAL	Configures range of analogue values can be used by other function blocks, e.g., to configure operation range of PID control loop automatically
Output	isAuto	BOOL	One indicates that function block operates in automatic mode and out value is derived from physical process value

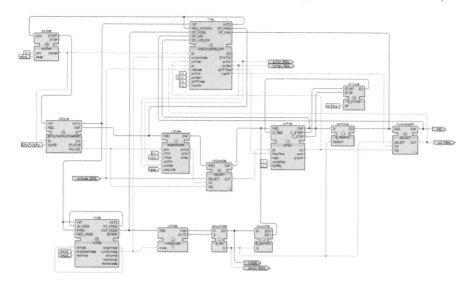

FIGURE 12.5
CAT DSCANAMON inner control logic.

HMI Service Interface function block within CAT

The HMI interface function block of a CAT is the communication entry point between the visualization client and the IEC runtime. It is not only used for visualization purposes but also to configure persistent user parameters and additional meta information for the IEC 61499 runtime system and the engineering environment. The communication to the HMI follows strictly the event-oriented execution model. This can be seen by the event inputs (see Figure 12.6(a)) for mode change (OP_MODE), process value monitoring (OP_UPD_PVF), calculated value monitoring (OP_UPD) and also by the output events for setting parameters (OP_PARA) and operating manual values (OP_MAN). Variables on the interface can be defined as persistent; this will allow to set default values (if no persistent value exists) with the *%input-name%>* variables and to load persistent stored values for the variables when starting up the device. When an operator changes values through the HMI, these values are stored and reloaded on boot up of the device.

The engineer can define attributes for each variable (see Figure 12.6(b)). In this example, archiving is activated for the *pv* variable. The values of an attribute defined on type level can be overridden on every single instance of the DSCANAMON CAT. This allows storage of selective data and helps to calculate the needed database storage and communication bandwidth within an automation system.

Faceplates (Figure 12.4(b)) and symbols (Figure 12.9(a)) for the HMI are

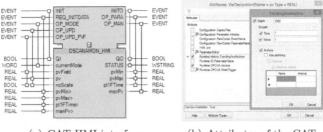

(a) CAT HMI interface.　　　　(b) Attributes of the CAT.

FIGURE 12.6
CAT DSCANAMON HMI interface.

designed on top of this interface and can be used to create process views for monitoring and controlling.

Binary Command Control CAT DSCBITCTRL

Figure 12.7(a) shows the outer interface and data/event association for this CAT function block as well as the HMI faceplate for monitoring and controlling. The purpose of this CAT is to control binary signals. Table 12.2 gives an overview of the interface variable and events of the CAT function block.

The inner logic of the CAT, as shown in Figure 12.8 follows a similar structure as the DSCANAMON described earlier. It uses the plcStart function block to get sequentially initialized during the startup phase of the runtime device. The *sp* input is used as setpoint in automatic mode. The binMonBasicFB is used to calculate the control value. Delay times can be parametrized for calculating the output value. A preliminary output value is calculated without any delay time constraints. The MODE CAT is used as a sub-CAT within DSCBITCTRL to provide common operation modes. For this CAT, the MAN and AUTO modes are allowed. Depending on the selected mode, the setpoint

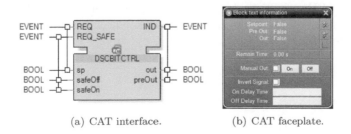

(a) CAT interface.　　　　(b) CAT faceplate.

FIGURE 12.7
Application and HMI view of the DSCBITCTRL CAT.

TABLE 12.2

Interface Variables of DSCANAMON

Direction	Name	Type	Description
Input	REQ	EVENT	If fired, call normal control algorithm
Input	REQ_SAFE	EVENT	If fired, call algorithm for safety operations
Input	sp	BOOL	Setpoint used in automatic mode
Input	safeOff	BOOL	Forces CAT to operation mode FMAN and switches off bit control
Input	safeOn	BOOL	Forces CAT to operation mode FMAN and switches on bit control
Output	IND	EVENT	Event is fired on any change of calculated bit control output values
Output	out	BOOL	Calculats output value (after any configured delay times) for bit control CAT
Output	preOut	BOOL	Calculates preliminary output value (configured delay times are not considered) for bit control CAT

for the output calculation is taken from the *sp* variable or from the manual value from the HMI interface function block *IThis.manPV*. The HMI function block is used as usual for communicating with the visualization system and configuring persistent values and define the attributes for archiving and other functionalities.

This CAT and most of the others from the library do not contain *SYMLINK* service function blocks for accessing hardware values. The CATs are designed in a generalized way and are prepared to be used as sub-CATs in more specialized CATs. For example, a valve can have one or two physical outputs, can have physical inputs such as limit switches or can implement a logic for external manual control and so on. But the base control logic is a DSCBINDEV1S CAT used as nested component in the specialized CAT. The dedicated implementation uses *SYMLINK* service function blocks to read from and write to physical process values.

12.8 Belt Conveyor Lines Application

The belt conveyor lines vary in terms of size and number of segments. Each segment contains a motor to drive the belt. Several segments are grouped to a zone. If the belts of a zone start, a warning horn and light operate prior to

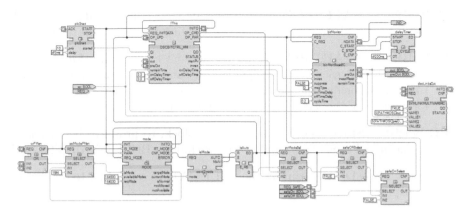

FIGURE 12.8
CAT DSCBITCTRL inner control logic.

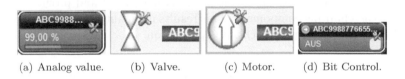

(a) Analog value.   (b) Valve.   (c) Motor.   (d) Bit Control.

FIGURE 12.9
CAT symbol examples.

each start. A zone consists of at least one safety signal to stop belts in case of emergency.

A belt-driven transport line needs to be started in sequence. This means, the last belt has to be started first, and after the belt is up and running the preceding belt can start, and so on. If a belt stops for some reason, all preceding belts need to be stopped immediately.

Since the distance of a belt transport can be large, distributed electrical cabinets to power the belt motors are needed. Instead of using a centralized PLC with a distributed I/O system, each cabinet should contain a small PLC to control and monitor the motors assigned to the cabinet. Therefore, the control application needs to support the distributed approach. Furthermore, the solution should be easily extendible and modifiable. On top of these requirements, the following software components should be implemented:

CAT BELT_CONTROL controls and monitors a belt line and has start, stop and status indications.

CAT BELT_DRIVE is a belt motor with dedicated physical I/O signals. Internally it uses the CAT DSCBINDEV1S as a sub-CAT to control and monitor the motor.

CAT BELT_SAFESTOP reacts to a safe stop request. Additionally, the software state is tracked and forced according to a safe stop request.

CAT BELT_STARTSIG is used to represent a signal unit for a belt line zone. Prior to each start, a signal horn sounds for a certain amount of time. Additionally, the signal lights are used to indicate belt zone status for workers in the danger area of the belt line.

FB beltStartState uses a basic function block, to produce the signals for the horn and signal lights and is used within the BELT_STARTSIG CAT.

Adapter IBeltDrives connect the BELT CATs to each other to build a belt transport line.

The BELT CATs are designed so that they can be easily used and connected together with adapters. The adapter interface consists of all data needed between the BELT CATs to build the line's control logic. Each component holds the complete control logic to manage itself and provide requested data to neighboring components. This results in an automatically programmed and therefore unrestricted scalable application.

12.8.1 IBeltDrive Adapter Definition and Usage

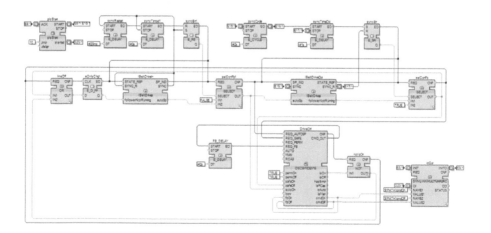

FIGURE 12.10
CAT BELT_DRIVE inner control logic.

A major solution is the use of an adapter to connect individual components together. One can imagine the adapter connection as a kind of virtual bus system between software components. Signals are transported and forwarded from one component to the next. Each component reads and/or modifies data or produces and actively sends signals on the bus. Figure 12.10 shows implementation detail to how the adapter interface is used in the BELT_DRIVE component.

The CAT contains a plcStart FB to sequentially initialize the CAT during the startup phase of the runtime device. The adapter consist of setpoint and status information and synchronization events. The synchronization events are very important, since they are used to build a distributed application which will be mapped to multiple devices. As the status of devices in a network and the health of the network cannot be guaranteed, a kind of keep-alive and synchronization mechanism was needed. If the connection of related devices is lost, the belt motors need to change state into the safe position to prevent damage on the belt transport line.

Through the use of components of the library, the remaining parts of the logic are straightforward and consist of the *SYMLINK* for the control signals; feedback is internally generated and the DSCBINDEV1S CAT is used to control the motor.

12.8.2   CAT Instances of Belt Transport Line

By use of the implemented software components, a belt transport line is easily built. There is the need to instantiate at least one item of each CAT. The BELT_CONTROL should be the first, followed by one or more instances of the type BELT_DRIVE. The last two instances have to be of type BELT_SIGSTART followed by the type BELT_SAFESTOP. However, additionall instances of these type can be placed anywhere in the line to build security zones and safety zones. A instance of type BELT_SIGSTART or BELT_SAFESTOP is always responsible for all previous instances of type BELT_DRIVE. Figure 12.11 should clarify the structure of a belt line.

The instances of this application are mapped to three control units and cooperate as a distributed application. Thanks to the event-driven execution model, the needed CPU load and network bandwidth are very low. The HMI application (Figure 12.16(b)) allows the operator to control and monitor the belt transport line.

12.8.3   Mapping of Application

The belt line control application is mapped to three devices (see Figure 12.11, Control Units 1 to 3). The communication network between the devices is created automatically by the engineering tool. The whole application will be analyzed when the mapping is accomplished. If a connection of two function blocks mapped to different devices is identified, additional logic must be added to the application for the communication task.

Communication takes place using Ethernet UDP protocol between the devices. The engineering tool automatically adds send and receive function blocks for data and event connections to the application and configures them

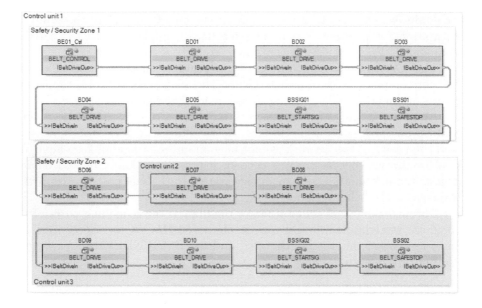

FIGURE 12.11
Belt line application.

according the links of the distributed function block network. The configuration ensures that a data and event transfer takes place only if the interface is unchanged on both sides. Signatures for event and data wires are part of the communication protocol. If the signatures do not match, the message will be discarded to prevent unreliable or inconsistent data and event transfer between devices. The event-oriented approach of IEC 61499 and the implementation from nxtControl gives the capability to emit events and data through a distributed environment without aliasing effects and in a reliable and fast way.

For the belt line control program, it was an easy task to distribute the application to several devices. However, to ensure a reliable environment, some time-out monitoring logic was implemented into the function blocks. This logic permanently checks the link between components and if a time-out is excessive, the belt drives switch off into a safe state to prevent damage. As soon as the link between software components can be verified again, the application re-enters into the automatic operation as defined for the belt line logic control.

12.9   Distributed Sequence Control Approach

A common task of automation programs is the sequential execution of commands for valves, drives, regulators and other components. The IEC 61131 standard offers the SFC language for such tasks. This may be used for single monolithic function blocks, but there is no way to distribute a SFC. To use the capabilities from SFC in a distributed approach, a concept was needed to separate the aspects of a SFC into function blocks to be used as partial networks and therefore distributable on top of multiple devices.

The goal was to create a distributable sequential control (DSC) logic which can be compared partially to the SFC implementation. It should be possible to formalize sequential control algorithms which span multiple devices to build a sequential control program. It should also be possible to combine partially created DSCs to a complete DSC program by interconnecting the DSC parts with adapters. The DSC concept must be implemented so that the sequential program initializes itself during startup to determine the structure of the whole DSC implementation.

12.9.1   DSC Function Blocks

The following FBs are mainly used to create a DSC program:

CAT DSC_HEAD is the main control function block for the DSC application.

CAT DSC_STEP represents a step in the DSC program. It also contains the direct transaction evaluation that must follow a step.

CAT DSC_TRANSACTION represents an alternative transaction for a DSC. It is used to run an alternative branch of a DSC program.

CAT DSC_ACTION_ENTER represents an action to execute. The FB is executed when the step is activated. Multiple actions can be added to one step.

Composite cpDscParallelAlloc is used to crate a parallel branch in a DSC program.

Composite cpITransJumpBackCmd creates an entry point for a jump back operation (equivalent to a loop structure).

Adapter IDscChain connects the DSC steps to DSC flow control FBs.

Adapter IDscParallelAllocChain connects parallel branch FBs.

Adapter IDscParallelMergeChain connects parallel branch merge FBs.

Adapter IAction connects action FBs to the associated step.

Adapter ITransaction connects alternative transaction FBs to a step.

Adapter ITransJumpBack creates a jump-back operation.

Adapter interfaces are heavily used to build the sequence structure and to communicate within the DSC. The FBs use a SIFB to determine whether adapter interfaces are connected on the FB interface. Depending on the result, the logic builds the control sequence and message flow through the DSC program.

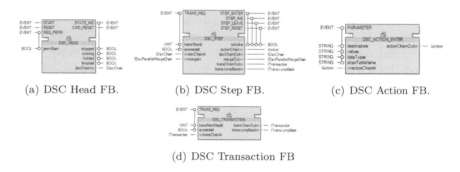

(a) DSC Head FB.   (b) DSC Step FB.   (c) DSC Action FB.

(d) DSC Transaction FB

FIGURE 12.12
DSC CATs function block interface.

The *DSC_HEAD* FB is the first function block in a DSC program and is used to initialize the sequence structure and control the operational state of the sequence. Initializing means, counting the number of function blocks and verifying a valid structure of the sequence. Several alternative symbols and faceplates are used to monitor and control the sequence operational states. The FB interface has event and data inputs to control the states from other FBs or to prevent the sequence from starting. Subsequent FBs are connected to the *dscChain* adapter output

A DSC step is defined by *DSC_STEP* FB. It has the *dscChain* adapter input and output interface to connect other steps or flow control composites. Flow control composites also may use *mergeIn*, *mergeOut*, *transChainOut* and *transJumpBack adapters*. Most important for issuing commands of steps is the *actionChain* adapter by which the *DSC_ACTION* FBs are connected and executed by a step.

There might be other *DSC_ACTION* FBs in the future; for now only the *DSC_ACTION_ENTER* FB is available. This FB will be called and executed when the connected step is activated. Through *actionChainIn* and *ActionChainOut*, multiple *DSC_ACTION_ENTER* FBs can be connected to a step FB. The position within the action chain is also the order, in which the action FBs will be executed. Beside the adapter, the *SYMLINK* FBs are other major

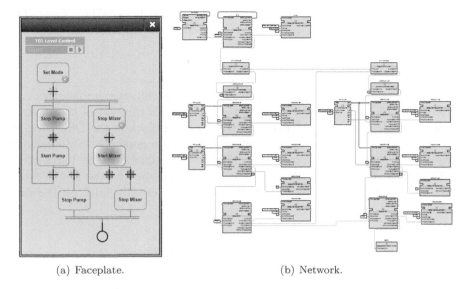

(a) Faceplate.　　　　　　　　　　(b) Network.

FIGURE 12.13
DSC sequence, HMI and logic view.

components of DSC. Objects are prepared with an internal interface (basically the same idea as for hardware connections) to issue commands into and read back states from DSC-enabled components. For example, the *DSCBINDEV1S* consists of a DSC interface for *setMode, rcasOn, currentMode, hasError, isOff*, and *isOn*. This allows the DSC sequence program to write control commands and to read back status information for transaction definitions.

The interface variables *destinations, values, dataTypes* specify the action commands to be sent. All variables are comma-separated string lists. Each element matches a command. The number of elements is the count of commands sent by this FB. All three variables must have the same number of elements. The variable *destinations* is the list of addresses where commands should be sent. For example, *TP01.DriveCtrl.setMode;T01_M01.DriveCtrl.setMode* will send change mode commands to *TP01.DriveCtrl* and *T01_M01.DriveCtrl*. The *values* variable lists values that should be sent to the element identified by the destination variable. The value is specified as a string and needs to be converted to the target format. To find the correct format, the *dataTypes* variable defines numbered data types responsible to the value which will be converted accordingly and sent to the target afterward.

It is also possible to create a completely reusable DSC sequence, but this needs an interactive instance variable replacement for *destinations*. To accomplish this, an additional *aliasTableName* variable is available. When using this approach, one can use place holders of the form *%VALVE%.setMode* in the *destinations* definition. The place holder will be resolved by querying a key/value pair table referenced by the name given in *aliasTableName*. This advanced usage of a DSC program is not used in the described project.

12.9.2   Mapping of DSC Application

The interconnection between the FBs of the DSC application is made by event, data and adapter connections. Based on this, a distribution of the program can be made easily by just mapping the function block of the DSC program to one or more devices. Cross communications are generated automatically by the nxtStudio engineering tool. There is a restriction that parts of the DSC application with *SYMLINKs* need to be mapped to the device where the *SYMLINK* target resides. *SYMLINK* references only work on the same device.

12.9.3   Visualization of DSC Applications

As usual in nxtStudio, the CATs for DSC sequences also have implementations as symbols and faceplates. To create a graphical representation of a DSC program, we need only to add the symbols to a canvas and draw connecting lines to represent the sequence program flow. If the DSC program is encapsulated in a new CAT with the DSC FBs as sub-CATs, a faceplate or symbol can be created for this CAT in the same way. In this case the sequence can be monitored and controlled by symbols and faceplates (Figure 12.13(a)).

12.9.4   DSC Applications in This Project

In this project, DSC sequences are used for controlling the lime supply to the lime reservoir before it reaches the production line. Another DSC sequence controls the dosing process for lime, water and chemicals, the heating and cooling process, and the drain sequence from the reactor. A managing DSC sequence starts, stops and controls the sequences as a simplified batch process.

12.10   Dosing and Reactor Application

Even if a process control program utilizes individual logic and cannot be reused in different projects, parts of the process are built upon the same concepts and work. These parts may be reusable by duplicating them for similar work. In this project, the dosing sub-process is used multiple times to produce the end product. Therefore it is logical to create a custom CAT that can be used multiple times for the individual ingredients. The object-oriented function block approach provides the basic technologies for this work.

Dosing Unit Application

The dosing application CAT *DOS_UNIT* is a standardized component which includes the sensors and actuators as sub-CATs. A separate DSC control sequence for each dosing instance was created. The *DOS_UNIT* CAT consists of a flow meter sensor and limit switches, an impulse counting quantity meter and a valve. The impulse counting quantity meter can also provide control logic for the dosing process since it provides setpoint and pre-setpoint control. The IEC 61499 logic of the sensors and actors includes all the logic and control algorithms needed for the dosing work. Beside the automatic mode, each component has a manual mode enabling operators to use each single element manually and in remote auto mode, which is used by the distributed sequence control to use the components in the DSC application. Figure 12.16(b) shows the visualization part of the application and faceplates for the counting quantity meter and a DSC control chart.

A DSC program and a recipe definition calculates doses for each individual ingredient and transfers it to the reactor. A sequence program is used in the reactor for controlling the temperature and the dosing sequences to add the correct amount and control time processing of ingredients in the reactor.

12.11   Hardware Configuration and Monitoring Application

Two time-consuming tasks within a project are the definition the configuration of the controller hardware and the I/O subsystem. Usually the definition is done in an early stage of the project. Within a distributed project utilizing IEC 61499 and nxtControl, this task need not be done as a first stage. However, the concept of a distributed project requires a separation of the technological system application and the hardware definition. The *SYMLINK* concept enables the engineer to connect the hardware to the system application for interoperating with the real world.

The I/O definition is a bottom-up concept. For this reason, the hardware configuration is added to the resource of a device. For the supported hardware devices for example, bus coupler and I/O clamps, a CAT is available for each element in dedicated hardware libraries.

This project is built up on a Profibus I/O sub-system. The library for Profibus was added to the project which made the supported modules available to be configured and added to the program. Figure 12.14 shows the hardware configuration view of the engineering system. The I/O variants consist of common digital and analogue input and output modules and also special modules for motor starter and weight sensors. The hardware is based on a modular I/O system with a Profibus coupler and clamps for a variety of signals. Each

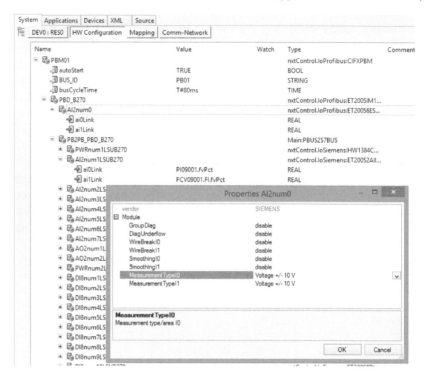

FIGURE 12.14
DSC hardware configuration view.

coupler is configured by adjusting the parameters per clamp according to the added modules. The resulting configuration file is used in the startup phase of the controller by the Profibus master to interact with the Profibus I/O subsystem.

After the hardware modules are added to the project, a graphical view can be created by adding canvases for the hardware view and placing the symbols of the I/O modules on the canvas. The hardware CATs represent the physical I/O point and help the system integrator to easily set up the hardware environment and support the technicians in the commissioning phase of a project.

To achieve that, the hardware CATs consist of symbols and faceplates. The faceplates are used to monitor signals, input clamps, and control signals of the output clamps. Furthermore, it is possible to interactively set parameters to manipulate the connected I/O signals, for example, invert digital inputs. In the commissioning phase of a project, the hardware CAT can be used to quickly check the I/O connections of the real world by monitoring the digital and analogue input signals and simulate application signals for digital and analogue output signals. For the hardware check of a project, only the visualization application (Figure 12.15) for the hardware is needed. The technicians can simply use the visualization view for their work.

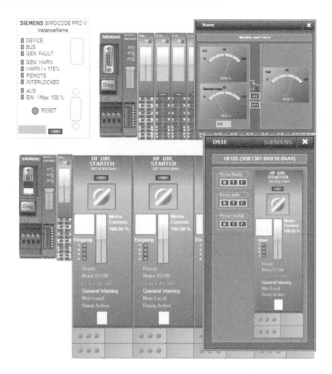

FIGURE 12.15
Monitoring view of DCS hardware components.

12.12 Conclusion

Industrial automation is changing. More flexible solutions and seamless integration with ITsystems, as discussed in Industry 4.0 and in the Internet-of-Things, will exert their impacts. More intelligent devices will be distributed in automation systems and the engineering of such systems will become more complex. The project described above is just one example to which IEC 61499 can be adapted for specific customer needs and easily designed for the engineer. The object-oriented engineering approach is well suited for mastering the increasing complexity. Finally the conclusion is that IEC 61499 can be used for all applications where IEC 61131 is used today, but offers many more possibilities. This is true for distributed control systems where the advantages are obvious.

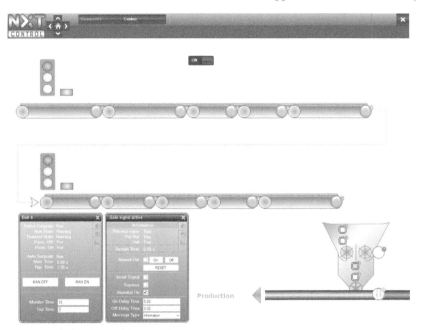

(a) Belt line and faceplates for motor control and binary signal monitoring.

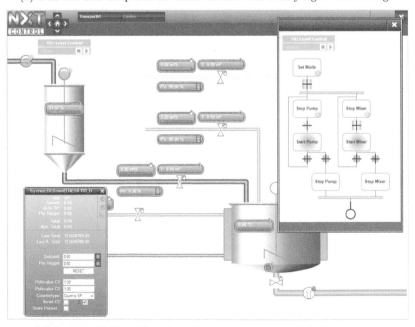

(b) Reactor HMI and faceplates for a counter and detail of a sequence.

FIGURE 12.16
Project visualization.

Bibliography

[1] Forschungsunion und Deutsche Akademie der Technikwissenschaften. *Umsetzungsempfehlungen für das Zukunftsprojekt Industrie 4.0, Abschlussbericht des Arbeitskreises Industrie 4.0.* Forschungsunion und Deutsche Akademie der Technikwissenschaften, 2013.

[2] B. Vogel-Heuser, C. Diedrich, A. Fay, S. Jeschke, S. Kowalewski, M. Wollschlaeger, and P. Göhner. Challenges for software engineering in automation. *Journal of Software Engineering and Applications*, pages 440–451, 2014.

[3] V. Vyatkin. *IEC 61499 as Enabler of Distributed and Intelligent Automation: State-of-the-Art Review.* IEEE, 2011.

13

Building Automation Simply Done

Gernot Kollegger

nxtControl GmbH

Arnold Kopitar

nxtControl GmbH

CONTENTS

13.1 Introduction

It is estimated that 40% of globally generated energy is consumed in buildings [2]. The European Community is setting the target to reduce CO_2 exhaust by 20% until the year 2020 [1]. New regulations on the energy efficiency of buildings will help to achieve this target. Beside structural measures, building automation is the key to energy-efficient buildings. The seamless integration and communication of different technologies is the challenge. The complexity of building automation is very often underestimated.

Several technologies from multiple vendors may be implemented in a building. For the system integrator interation is a complex task, because engineering efforts must compensates for the lack of interoperability. Sustainable building automation solutions should enable the efficient usage of energy and efficient engineering to master the complexity.

Projects for building automation are awarded through tendering. Comparison of products is easy but comparison of systems is not. That is why innovative building automation solutions are not implemented often enough. The most economical solution based on product comparison is selected, without taking into account that the solution may increase costs of engineering. For better building automation, cost transparency must become more common or building automation must become simpler. A way to make building automation simpler is to hide complexity.

The following chapter will show how complexity can be reduced for the system integrator by means of object-oriented engineering based on IEC 61499 concepts to promote a simpler and faster building automation engineering.

13.2   Building Control Application Requirements

Applications get more and more complex in when several technologies need to be integrated into one project. A building application is by its nature a distributed system. Therefore a distributed control environment such as nxtControl and IEC 61499 best fits the needs of such an application.

In the former chapter we described how to use the benefits of distributed control systems in the process control environment. A building control system has specific requirements to be addressed on different levels of a building. There are specific requirements for the engineering process, for operating and interacting with the users, and controlling the individual levels of a building.

In the heating, ventilation and air conditioning (HVAC) domain, the requirements are similar to those of process control applications with another granularity of distribution. A main ventilation unit, for example, must be controlled and monitored in a common process environment. However, air flow to various building areas is often decentralized and isolated from the main ventilation system even though the main task remains temperature and air quality control.

A further level of control is required in occupied areas where lighting and shading, termerature, air quality, and security are critical. People live and work in these areas and automation systems must support their individual needs and perform their tasks without notice. In many situations, both types of technologies work side by side.

For light control, the digital addressable lighting interface (DALI) system is a common technology. Shading motors are controlled by standard motor interfaces (SMI). Switches and room control units are managed by the standardized OSI-based network communications protocol for intelligent buildings (KNX) or by directly wired elements.

This mixture of systems was not a problem in the past even though the systems worked independently. Now, however, new automation control system concepts are needed. Sub-systems must be integrated to work together. Data flow between systems is critical to achieve consistent control of building at all levels.

Driven by these requirements, new automation control systems concepts are needed. Sub-systems need to work together. Data flow between the systems is a must to achieve common control of a building through all levels.

13.2.1 Hide Complexity with Adapter Interfaces

Component-oriented systems encapsulate algorithms in functions or function blocks with interfaces to supply components with parameters. If an engineer uses such a function within other functions as sub-component these parameters must be defined. When systems become complex and several levels of nesting are applied to a component, the interfaces often grow too complex to be handled by a single function block. Furthermore this becomes even harder if the goal is to integrate sub-systems into a common automation layer.

Adapter interfaces, sockets and plugs

The IEC 61499 standard includes the definition of adapter interfaces. nxtControl's concepts for automation applications for building automation make heavy use of them.

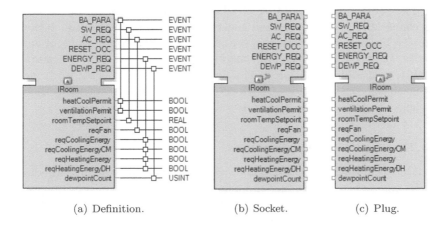

(a) Definition. (b) Socket. (c) Plug.

FIGURE 13.1
Adapter interface views.

Adapter interfaces allow bidirectional communication between function blocks. They can be used on the left (input side) as sockets or on the right (output side) as plugs of a function block. Figure 13.1(a) shows the definition, Figure 13.1(b) the representation within a function block when used as a socket and Figure 13.1(c) when used as a plug. The interface of an adapter consists of event and data inputs and outputs and their associations defined by the *WITH Construct*. Communication between source and destination adapters ensures a consistent data transfer in the granularity of the event and data association dataset construct. Even if the communication leaves the borders of a device, the data consistency is given in a distributed application. The datasets are transferred immediately to the destination adapter and the events are scheduled for execution. This reduces the delay of event execution to a minimum.

Hiding complexity

Adapters help to reduce the complexity for technicians using and combining function blocks in their projects. However, the creation of a library is more complex, but to reflect to the needs of the market to integrate a wide range of functionality and systems into one common set of automation objects. The creation of libraries is an expert task and should be quality checked, verified and version controlled.

Interfaces built from events and data variables are used in most all interacting function blocks. Adapters reduce the interface to one single connection point. Not all signals need to be collapsed into one adapter, but domains of variables can be built to connect certain types of function blocks to each other by a single line.

The source and destination of adapters must be compatible so they can be connected. This requirement makes the world of integration technicians simpler because it reduces the possiblity of faulty connections of function blocks; they are limited to implementing correct interfaces. For example, an adapter controlling lighting cannot be connected incorrectly to an adapter designed to control shades. Obviously, the software components should be designed to perform only the expected function.

13.2.2   Scalable Applications

Building applications at the level of room control can be highly standardized in the functions of controlling and monitoring. However, the advantage of high level standardization is reduced by the inconsistence of different technologies for controlling a room. To build scalable applications, adapters are used to interconnect these different technologies in a most flexible way and release the basic room control applications from hardware dependencies. Hardware-related operations have to be separated into specialized components and connected via adapters to the basic control application. The interface in form of adapters can master the mixture of technologies and join them into an efficient application.

For demand-driven applications, scalability is an important factor for the structure of the program. In an energy saving building, demand-driven provision of heating and cooling is a key property. However, the programming of such a system should not disturb the easy setup of a building application. Moreover, it must be a complete application and a component of the basic software components.

13.3 Control Application

13.3.1 Description of Work

The work consists of multiple ventilation control applications, heating and cooling supply systems, a geothermal heating and cooling system and room automation control for HVAC and shading. The building to be controlled by this application contain three floors with over hundred rooms. The rooms are standard offices, conference rooms and laboratories.

The heating and cooling supply systems should be controlled with demand-driven calculated setpoints. Whenever possible, geothermal energy should be used for heating and cooling.

More advanced control and monitoring are needed for the laboratories. As there chemicals may be used in certain laboratories, the pressure needs to be controlled in these rooms. Furthermore, exhauster control is needed in all areas where researchers work with chemicals and in the chemical storage rooms.

13.3.2 Building Control and Monitoring Application

To build the control application, four controllers were used. One controller per floor for room automation, provided heating, cooling and shading. Advanced control algorithms were added for special laboratory rooms. A fourth controller handled heating and cooling supply systems.

13.3.2.1 Room Control Applications

Room temperature control

The building required several levels of room control complexity. The room controls were programmed as CATs and included all needed parts to automate rooms. Each room had a dedicated control device connected by a serial Modbus to an automation device for the floor. Temperature controllers were installed in each room. To reach the desired temperature several actuators were used. Each room had an HMI to control and monitor its automation (Figure 13.20). Examples of actuators are:

Radiator regulated by a pulse controlled valve.

Air flow box to provide volume-controlled air flow from the HVAC system.

Heating coil located in the ventilation shaft just before the regulated inlet air flow box.

Heating and cooling ceilings can be used for either function, depending on the outside temperature.

Cooling aggregates are used in some laboratories to provide additional cooling.

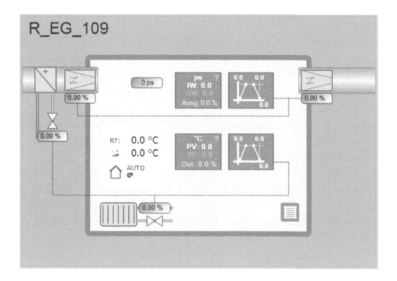

FIGURE 13.2
Advanced room control for different sources of heating shading.

Several types of rooms are used in the project and require different combinations of actor types. The Modbus-connected room control unit is used to inform the users about the status of the room control. Users can also adjust the room temperature setpoint in a predefined range.

Rooms with heating and cooling ceilings have additional dew point sensors and window contacts to prevent condensation when the system is in cooling mode. These sensors are connected to existing input channels on the room control device and the signals are available on the Modbus I/O subsystem.

If the system is in cooling mode and a window is opened, an event forces the ceiling valve to close. An informative alarm is generated and the room control device status is updated. Additionally the room user is informed about the situation. The room is not controlled by the system as long as the window

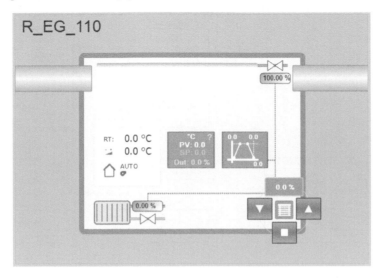

FIGURE 13.3
Basic heating and shading controls.

stays open. When the window is closed, the system switches back to automatic mode and tries to regulate the temperature according to the given setpoint.

Demand-driven energy request

Each room calculates its energy demand dependent on the setpoint and deviations from it. To connect the energy request to the HVAC supply system, adapters for energy supply request are implemented in each room as sockets and plugs. The rooms are interconnected to each other through this adapter. Within a room control CAT, the signals of other rooms are read from the socket and compared to the calculated requests of the current room. If there is a need to adjust the values, the new values are transferred to the plug.

This calculation is only done when a setpoint changes in the system or the measured temperature changes more than the configured deviation. The design of the CAT and the way the adapters are implemented reduces the events to a minimum for every change.

Each floor of the building can be individually adjusted in terms of HVAC control mode and base setpoints for the room temperature controllers. The operation modes *Off, protection, economy, precomfort* and *comfort* can be time based. Configurable schedulers are used to activate the modes, one for daily setup and another for special days. The setpoint CAT implements the same adapter as the room control CATs and is used as the first CAT in the program structure of a floor.

Figure 13.4 shows a simplified structure of the system to calculate demand-

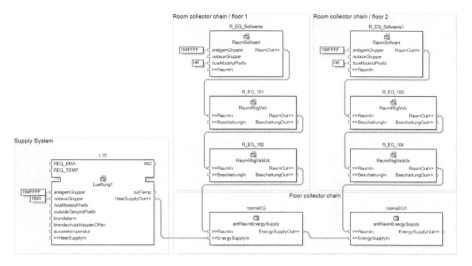

FIGURE 13.4
Demand-driven setpoint calculation.

driven setpoints. The setpoint and all room segment CATs are connected with the *IRaum* adapter and at the end to *anfRaumEnergySupply* composite for each floor. The composites are connected with an *IEnergySupply* adapter. The energy supply systems are connected to this adapter and the setpoints collected from the rooms are transferred to the ventilation systems, district head supply, cooling machine and geothermal subsystem.

Shading control system

Beside the temperature control, each room contains logic for controlling the shades. The shades have different levels of control:

Locked by the Weather station means that depending on the wind speed, the shades may be opened and locked automatically to prevent damage.

Automatic shading programs allow time-based positioning of the shades based on the position of the sun. Each shade can be assigned to a group related to the cardinal direction. For each group, an individual control program is created

Direct control by the user allows manual control of shades with wired switches.

Direct control on HMI room level allows for maintenance or control of the shades of a room from the HMI.

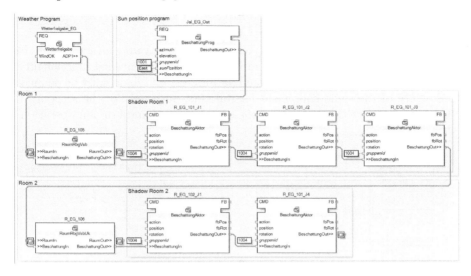

FIGURE 13.5
Shade control program structure.

Direct control and automatic control have the same priority and are able to override each other. The weather station overrides all other commands for safety reasons.

Each shade control is encapsulated in a CAT of the same *BeschattungAktor* type and implements the adapter *ISunblindCtrl*. Multiple shades in a room are connected and linked to the room. Rooms also implement the *ISunblindCtrl* and are connected to receive the control commands from the weather or group automation programs. Figure 13.5 shows a reduced structure for shade control.

13.3.2.2 Scalability of Room Segments

Based on the CATs for room automation and shading control it is easy to work with different numbers of shades per room, rooms per floor and floors in a building. To add shades to a room, simply connect them with the *ISunblind-Ctrl* adapter. To connect rooms to a floor, install the *anfRaumEnergySupply* composite and the *IRaum* adapter. Finally, with the *IEnergySupply* adapter, the floor energy request FBs are connected to the supply system.

13.3.3 HVAC and Supply Control Applications

To provide energy of correct quantity and quality for the building, several HVAC units are implemented. The different supply units are developed as CATs for reusability and minimized commissioning work for the on-site integration.

HVAC Systems

The project required a standard ventilation system and an exhauster. Figure 13.6 shows a ventilation system CAT. Most of the needed functionality is already included but some additional parameters and logic must be defined to allow use of the same equipment for different purpose. For the setpoint, the CAT implements the *IEnergySupply* adapter for demand-driven operation of the system.

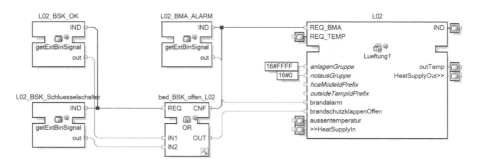

FIGURE 13.6
Logic network of a ventilation system.

To operate the ventilation units, a unit mode is globally calculated, based on the outside temperature. The *SYMLINK* concept is used to switch between the ventilation modes for *winter, summer* and *transition time.* Also room types with heating and cooling ceilings use the mode information to switch between heating and cooling operations.

The ventilation and exhaust types are built with standard CATs for valves, motors, analogue and digital sensors, regulators and simple generic function blocks. They are connected with standard event and data connections of IEC 61499. The CAT concept with its nested structure allows creation of a complete ventilation system as a single CAT. The integrated visualization and hardware abstraction makes it easy to add the CAT to a project. The remaining work for the system integrator is to drag and drop the CAT-type and create an instance in the IEC 61499 editor and in the HMI (Figure 13.7) editor and to connect the hardware I/O points to the bottom up created hardware of a controller.

Heating and cooling energy supply systems

Several heating and cooling systems wre used within the project. Geothermal energy was combined with district heating and cooling components. Heat and cold buffer storage systems were installed to optimize the efficiency of the geothermal unit. The automation program was required to control the geothermal unit and use it as much as possible to reduce the need to generate energy from the district system.

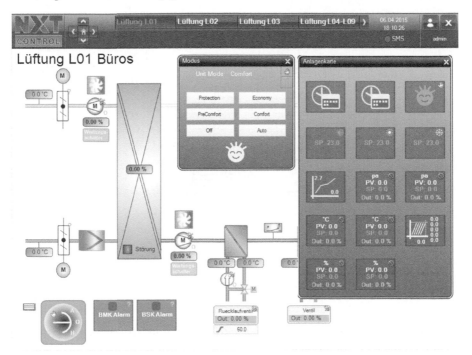

FIGURE 13.7
Ventilation system HMI.

This combination required control of complex interactions between the various units. A single system would destroy the reusability of the program code. Engineers found it necessary to design domains that could be encapsulated into components capable of interacting with each and also operate as stand-alone systems. This made the wiring of single components easy. The components are as follows:

- Cooling buffer control

- Geothermal buffer control

- Free cooling control

- Cooling machine

- Geothermal control

To ensure simpler integration in the automation structure, the components are connected (Figure 13.8) with adapters for buffer control *IPuffer* and the *IEnergySupply* adapter to set the demand-driven setpoints from the rooms. Some logic (*FreecoolingWT*) for the free cooling task was added and also the

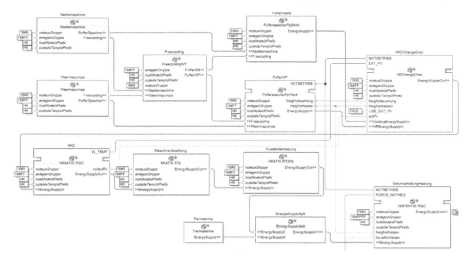

FIGURE 13.8
Heating and cooling supply systems.

adapters *IFreecooling and IFreecoolingK* were added to the geothermal and cooling machine components.

The components use basic CATs for sensors and actors to create the needed functionality. Data and event connections are used for the signal flow. *SYM-LINK* function blocks are added to provide the access points for the hardware connectivity. The parameters to setup and configure the system also require basic CATs. The CATs have symbols and faceplates which are used on the canvases of the HMI. In the commissioning phase, the parameters are entered through the HMI. The values of the parameters are stored on the controller and automatically restored every time a device is restarted.

Customized software components

Along with source supply components, there are control elements specific to this project. The elements implement the *IEnergySupply* adapter and use the demand-driven calculated setpoints for decision and control algorithms. The adapters provide a common interface to implement customized program functions but keep them reusable and integratable.

13.3.3.1   Historical Data and Alarm Archiving

Two important tasks are alarm archiving and trending. The configuration for the alarm and archiving can be a time-consuming task.

The structure of IEC 61499 covering nesting and attribute capabilities supports the engineer in this task. nxtStudio, provides additional support to easily add and manipulate attributes to software components, even if they are nested within a project. This is called attribute over-parametrizing. Basically, the creator of library elements can define the attributes for an object at type level. This includes, in the case of archiving, the attributes to enable the online archive capability and control the depth of the online buffer and parameters for time and value filtering. Similar attributes are available for alarming (alarm class and text).

Another notable attribute is the *HMIAlias*. The path of a software component is used to clearly identify a component. However, this path is not easy recognizable in the domains of alarming and archiving. The HMI alias gives the option for human identification of a component. The HMI aliases of nested elements are automatically concatenated to provide a legible name for an object. This helps identify elements in the alarm list view or in the archive list view.

Re-parametrizing of attributes

Attributes are basically defined at type level. If a function block is directly used on application level, the defined attributes may be re-parametrized. This is important for individualizing alarm text, changing the alarm class, modifying the depth of the online trend buffer, and altering other settings. nxtStudio provides a powerful editor for manipulating attributes with filters, copy and paste functions and selectively collected attribute lists.

13.3.4 Hardware Connectivity within Project

Room control and HVAC equipment in a building project impose the challenge of integrating a wide range of hardware sub-systems wiht different protocols, fieldbus systems, and signal ranges. The technologies required to generate the needed information include:

- EtherCAT for slice clamp-based digital and analog signals and shading control modules

- Serial Modbus for room control devices

- Serial Modbus for weather station

- Serial Modbus for emergency light control

Some of the hardware signals are directly available on the controller. Most are clamp-based signals and the Modbus connected room control devices. Other signals, for example, the values from the weather station, are connected to one controller and distributed through the application to all other controllers.

The hardware CATs are responsible for normalizing the different signal sources to a common value range which is compatible to all software components with the *SYMLINK* concept. Within the hardware configurator, the

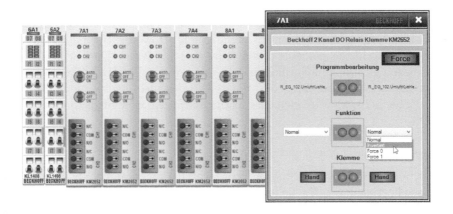

FIGURE 13.9
Hardware HMI.

parameters for sub-system connectivity are defined. They are used to generate the configuration file, which is downloaded to the controller during the deployment phase and used to configure the respective drivers.

The hardware clamps used in this project have also the capability to manually control the output of a channel by switches located on the clamps (Figure 13.9). The status of the manual switches is read back and used by the control elements to indicate the locally overridden mode directly, for example, in the ventilation system HMI control canvas. These switches are commonly used only if there is some misconfiguration in the software or to override the signal of the automatic program.

13.4   Conclusion

Distributed control has become a reality in building automation, especially with the widespread use of room controls. Several technologies must interact within a building to archieve the energy savings required for economical operation. Building automation is a complex undertaking. Advances in building automation (smart buildings) can achieve measurable reductions of carbon dioxide emissions. The project described in this chapter clearly shows that IEC 61499 is well suited for such applications.

The ease of distributing control logic to several control devices and the object-oriented approach are key for a simple building automation solution. The building automation architectures for floors, aisle, rooms and segments are represented perfectly. Furthermore, the software objects and library created during the project are reusable. Automation has cut the engineering efforts in succeeding projects dramatically. Ready-to-use libraries and the IEC 61499 are facilitating simple building automation.

Bibliography

[1] European Parlement. RICHTLINIE 2012/27/EU DES EUROPÄISCHEN PARLAMENTS UND DES RATES vom 25. Oktober 2012 zur Energieeffizienz, zur Änderung der Richtlinien 2009/125/EG und 2010/30/EU und zur Aufhebung der Richtlinien 2004/8/EG und 2006/32/EG. 2012.

[2] VDI Verein Deutscher Ingenieure e.V. Positionspapier Klimaschutz und Energiepolitik: Kriterien fÃijr die Bewertung der Nachhaltigkeit von GebÃđuden. 2011.

14

Control Software for Cutting Glass Machine Tool Built Using COSME Platform: Case Study

Félix Serna

Universidad de Zaragoza

Carlos Catalán

Universidad de Zaragoza

Alfonso Blesa

Universidad de Zaragoza

José Manuel Colom

Universidad de Zaragoza

Josep Maria Rams

TUROMAS Group

CONTENTS

14.1 Introduction

This chapter presents a case study showing the design of a glass cutting machine tool. This design is the first technological demonstrator of the COSME platform, its guidelines and tools. It represents an approach to simplify the design of control software in the manufacturing context, by shortening time to market and aggregating value-added features. The presentation covers the life-cycle phases of the software application: (1) construction of the model based on components from patterns related to the functions of control, and patterns related to preventive maintenance; (2) implementation and deployment of the model using function blocks adapted to the standard IEC 61499; (3) monitoring and follow-up to the execution of the application. Finally, we present experimental results obtained from the implementation on machine tools operating in a real industrial scenario, which together with the experience gained in the design of an application of this type allows us to make an assessment of the standard in a typical industry in this sector.

The current manufacturing industry must offer faster responses to market changes, while maintaining low costs. Flexible manufacturing systems (FMSs) are technological answers to this scenario. Basically, they are based on standardized general purpose machine tools which can produce different parts and products, and transport lines to move them. These elements are organized on the factory floor to work in a coordinated fashion. Control of FMSs has been traditionally implemented in a centralized manner, with manufacturing cells and islands controlled by programmable logic controllers (PLCs). Usually, PLCs perform some type of weak communication between them to synchronize manufacturing processes. The rules for PLC-based systems have been established by the IEC 61131 standard [13], widely accepted and used in the industry. Moreover, CNCs (computer numerical controls), can be standalone devices communicating with PLCs via buses (e.g., Profibus, CANopen) or integrated as a PLC function.

The global economic development is causing life cycles of the products to be further shortened and customers to demand different characteristics. To adapt to this new scenario and meet changing demands, the manufacturing systems must be able to rapidly develop small series (even up to a unit). These new requirements have led to a new generation of manufacturing systems based on technologies and forms of industrial production that go beyond flexible manufacturing, identifing a new paradigm: agile manufacturing (AM). The new AM requirements have been analyzed by several authors (e.g., see [12, 16, 18]), and they are characterized by the use of reconfigurable hardware and software architectures that do not stop the production process, minimize bugs and their impacts (e.g., through systems of preventive and predictive maintenance, and fault tolerance) and fully integrate manufacturing into corporate information systems via Internet technologies (e-manufacturing).

The industry for handling glass is involved in this technological change, and as a result, manufacturers of machinery for this industry should take into account certain considerations in the development of their products. Our case study concern TUROMAS company [26], which designs and manufactures machinery for handling, machining and storage of flat glass. TUROMAS wants its new manufacturing systems meet requirements, reduce its development costs and improve its post-sale service based on Internet technologies (e.g., e-assistance).

Compliance with the above requirements makes the control software development for new manufacturing systems much more complex software for the current FMS. Logically, this increased complexity carries increased development costs. As indicated [25], these costs can reach up to 80% of the total costs (currently 40 to 50%). This increased complexity has required adaptation of methodologies and paradigms previously used by other software engineering application domains [10]. This scenario has emerged the IEC 61499 standard [14] for developing control applications.

In this context, we present a case study on the control of a machining glass line developed through technological cooperation between TUROMAS and our research group. This cooperation involved the development of the COSME (control systems and modeling environment) platform, to develop the new control of the TUROMAS machinery. COSME includes a framework along with development, debugging and supervision tools to develop machine tool control. COSME is based on a slimmed-down approach to the IEC 61499 standard, but it still proves powerful enough for this kind of application and shortens development time.

14.2   IEC 61499-Based Design versus Application Domain

The IEC 61499 standard defines an architecture and models seeking to facilitate the development of reconfigurable and distributable control systems. The most significant model is the function block (FB), which extends the FB defined by the previous standard IEC 61131. An IEC 61499-based application consists in a function block network (FBN). This FB model incorporates an event-driven execution, which is more efficient than the IEC 61131 scan execution. IEC 61499 function blocks execute control algorithms only when events indicate that there are new data to process, whereas the IEC 61131 function blocks execute control algorithms cyclically, even without new data.

The goal of reconfigurable software can be achieved using the component software paradigm due to an IEC 61499 FB considered analogous to a software component [17]. Reconfiguration is possible through changes in FBNs by means of creation, deletion, replacement and interconnection of FBs, as well as deployment and installation of new types of FB, "on the fly" (i.e., without stopping execution). To this end, the standard provides a management model to support reconfiguration issues. For a survey of reconfiguration of industrial automation with IEC 61499 see [2].

A distributed control system consists of multiple controllers interconnected by networks, without a central controller and where communication and synchronization issues play a central role. IEC 61499 models and architecture are used in a natural way on such systems and FBNs can be deployed across different resources (i.e., runtimes), which are hosted on devices (i.e., controllers). Event and data links between controllers are abstracted by communication service interface function blocks (CSIFBs) with publish/subscribe and client/server models.

Unlike IEC 61131, industry has not adopted this new standard yet. Some early studies [9] and [31] analyze possible reasons, mainly immaturity of the standard. Vyatkin indicates [28] actions taken to promote its adoption (next appearance of a new version, presentation of the first real experiences, etc.).

There are several initiatives to offer tools and platforms. FBDK and its runtime FBRT [11] implemented in Java cannot meet real time constraints but have been the references for the standard in many academic works. An important open source initiative is 4DIAC [1], with a tool based on the Eclipse framework and runtimes for PC and ARM. Other proposals may be seen in [27] and this work also indicates that it is hard to develop applications with these academic tools (e.g., due to a lack of debugging tools). However, it is important that the academic world develop tools and platforms that may have practical use by control engineers outside research laboratories. Finally, two commercial proposals have emerged from the manufacturers of automation equipment: a) ISaGRAF [15] based on a previous framework for IEC 61131 and b) NxtStudio [20].

From the view of development, the distributed FBN design is not easy. That is, it is not just a monolithic design split into parts implemented as FBs and distributed on controllers. Distributed systems should be designed carefully to allow reuse in different contexts (i.e., interaction with other FBs) using software engineering methodologies. IEC 61499 does not establish methodologies, or guidelines, that indicate how to identify FB types or define FB interactions [27]. An approach is to use design patterns of FBNs, with the first proposals presented in [6], or more recently proposals presented in [5, 22, 23].

Control engineers may not be familiar with software engineering methodologies and IEC 61499 paradigms (i.e., event-driven and software components). A survey of industrial control systems [7] identifies the characteristics and analyzes the different ways of implementing automation systems by the industrial partners participating in the European MEDEIA project. The results show that control applications are still handcraft works. The translation from the design specification to implementation is usually left to the experience and competence of engineers. These implementations are based on classical PLC in languages that are more or less compliant with IEC 61131 or on industrial

PC in language C. The COSME approach to these shortcomings is by means of IEC 61499 adapted models ("precooked") to specific applications but versatile enough to facilitate the distributed FBN design and avoid handcraft work. These models are supported by a platform and tools described in later sections.

This work is focused on machine tools as important elements of any manufacturing system. From a general view, a machine tool performs operations that define work such as cutting, drilling, milling, etc. These operations are composed of a certain amount of movements of pieces, machining tools and finished items and control software should consider: a) precise positioning of pieces and machining tools, b) loading and unloading of pieces and finished items. As a support to development of machine tools for new manufacturing systems, different initiatives of international consortiums address the use of open systems in the open architecture control (OAC), for example, OSACA in Europe, OMAC in United States and JOP in Japan. (see [21] for overviews of these and other initiatives).

This open architecture control must be based on a platform concept that encapsulates the specific characteristics of hardware, operating system and communication to facilitate easy porting of an application and its interoperability in heterogeneous distributed environments. Along with the platform is the application software subdivided in modules within independent units that contain both structural and functional information (e.g., IEC 61499 function blocks).

Communication aspects are key in the machine tools considered. They must be integrated in the different levels of factory communication networks. To perform this integration, control software should consider communication: a) with other controllers in the same machine tool (i.e., distributed control), b) with other elements of the same factory workflow (i.e., manufacturing cell or line), and c) with external applications (e.g., supervisory control and data acquisition, management production).

Another important feature of these machine tools is failure management to handle all failures in the same workflow. A response actions policy should consider the following operation modes:

1. Normal operation, i.e., the machine operates under normal conditions.

2. Degraded operation, related to performance of some component below nominal values (e.g., speed or acceleration of axes would be limited in case of overheating).

3. Failure operation which should be detected in parts of machines and in other workflow machines.

In addition, preventive and predictive maintenance features as a means to avoid unscheduled stops due to failures should be considered.

A real machine tool may have dozens of positioning axes and automatic sequences, so a control based on IEC 61499 require large FBNs with hundreds of FBs. In a first approximation, the FBN design may be done by dividing the control into basic parts (e.g., axis control) that appear repeatedly and can be implemented with one sub-FBN. Then, control may be implemented by composing these sub-FBNs plus the addition of some business logic. With this perspective in mind, FBN scalability is an important goal to maintain good performance and shorten design times. These issues should not depend on the number of axes or sequences. From the distributed control view, the resulting FBN of composed sub-FBNs of control parts can be deployed (i.e., distributed) from one to dozens of controllers, depending on the distribution granularity and links between controllers and CSIFBs.

The following example may illustrate the point. A coarse granularity could mean having a controller for each machine in the cell or production line system, while a fine grain would have a controller for each axis or automatic sequence. In the latter case, the controllers may be simpler (since fewer tasks would attend) than the former. The designer must decide the number of controllers by establishing a compromise between communication delays (which penalize fine grain models) and the computational resources required to perform the tasks assigned to each controller (which penalize coarse grain models). Reconfigurability can also determine the number of controllers in the sense that the reconfigurable parts of the machine tool may require independent control (mechatronic approach).

Moreover, due to characteristics of manufacturing systems (e.g., reconfigurability, fault tolerance) there must be master controllers that direct (or orchestrate) other controllers. The operation of the distributed application can be coordinated (or choreographed) establishing a business logic implemented via messages between the controllers. The semantic (or meaning) of these messages is related to the granularity of the distribution: the lower the granularity, the higher the semantic level of messages due to the increased complexity of the functionality implemented on each controller. In this sense, an approach to the design of distributed control and other types of architecture is the multilayer design. A proposal that follows this approach for IEC 61499-based applications is presented in [29]. It establishes three layers: interface process (access to sensors and actuators), operations (sequences and positions) and application (business logic). Another approach is to identify and characterize the types of communication based on the application domain [4]. The characterization includes: a) typical latency, b) communication model periodic or event, c) reliability and d) bandwidth.

There are several works related to hardware platforms for machine tools on agile manufacturing systems. A framework and methodology for adaptive assembly operations using FBs is presented in [8], where function blocks allow dynamic job shop operations. In [19], FBs are used to implement open modular architecture control (OMAC) state models for packaging machines. The architecture for an open and distributable CNC system based on STEP-NC and FBs has been proposed in [30] and tested in a laboratory prototype with FBDK and FBRT. A reconfigurable control for a PLC and CNC controller is considered [30], but experimental results are not presented except for a brief description of a case study. These proposals have in common that they have not been tested in real applications or considered the aspects mentioned.

14.3 Glass Machining Modelling

A machining line for laminar and/or monolithic glass (see Figure 14.1) is a well-defined product. Its functionality is determined by the type of material to be handled: monolithic glass sheets or laminated glass sheets (i.e., two glass sheets with an intermediate butyral layer). Main requirements concern: a) types of machining operations: warming, cutting, breaking, etc., b) types and measures of glass sheets, c) production rates estimated, d) types of movements of glass sheets and parts: loading, unloading, squaring, cutting, etc, and e) number of axes based on the previous feature.

This kind of machine is constrained by different market considerations that must be met (characteristics of the markets to be targeted as local safety normative, expected price, product family segmentation, etc.). Besides machining elements these lines involve items such as loaders and unloaders, storage and

FIGURE 14.1
Laminated cutting glass machine. (Photo courtesy of TUROMAS.)

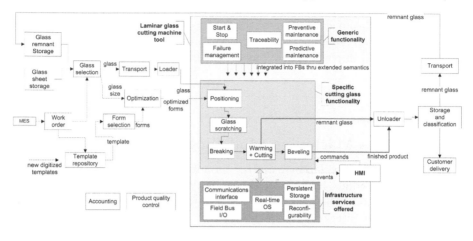

FIGURE 14.2

Cutting glass machine based on its environment.

transport facilities, Other issues must also be taken into account: human–machine interface (HMI), integration with external MES (manufacturing execution system) and ERP (enterprise resource planning) systems and other non-functional ones that add value to the product. Mechanical, electronic and system designs (e.g., movement types, actuator types, number of axes) establish software requirements that can be organized in specific and generic functionalities detailed below.

In Figure 14.2, elements belonging to the machine tool are shown in the dashed box and organized as mentioned above.

14.3.1 Specific Functionality

Most of functional elements present in the tasks mentioned above are technology-dependent. As an example, laminar glass has a butyral layer that must be warmed prior to the glass cutting. That warming can be achieved by different means, depending on the technology available, for example, infrared lamps, laser beams, etc. In this case the functionality (warming butyral) must be accommodated in the machine tools, no matter how the process performed.

This functionality can be addressed by PLC and CNC functionalities. PLC is related to movement automation operation modes. CNC deals with both tools and parts fine positioning. The elements related to PLC functionality are the following:

1. Loader. It handles the glass sheets to be cut. It needs some pre-established sequences corresponding with sheet movements from storage to the cut table (e.g., turning over, displacement, etc.). Certain "glue logic" is also needed in order to orchestrate the sequences (i.e., finite state machines).

2. Machining. The same scheme mentioned above applies. The pre-established sequences correspond with glass sheet and tools movements (e.g., glass squaring, tool pre-positioning). Again, glue logic will coordinate the different sequences. There are two machining processes. The first involves two-dimensional scratching (complex forms can be drawn), while the latter only permits straight line scratching, followed by butyral warming and glass breaking.

3. Unloader is similar to the loader, but the glass items are removed and transported to intermediate storage, before the next manufacturing steps.

CNC functionality corresponds to precise positioning. Different elements form an abstraction hierarchy:

1. Axis. Allows precise one-dimensional positioning. Axes are composed of a PID (proportional integral derivative) control loop, a gearbox (software element which scales and performs unit conversions, e.g., converting encoder pulses to meters) and a set point generator.

2. Axes group encapsulates two or three axes, allowing for two- or three-dimensional positioning (e.g., circle form).

3. Trajectory calculator receives the desired forms to be cut, generates intermediate trajectory points and calculates velocities and accelerations for each intermediate point.

Related to process interface (sensors and actuators), a field bus is used (e.g., CANopen). Servo motors are commanded through CANopen servo motor drives. Electro valves are commanded using a CANopen valve island. Finally, the different digital inputs and outputs are handled using CANopen digital I/O modules.

14.3.2   Generic Functionality

The generic functional elements are of paramount importance. They appear in machine tools, not used in the the glass market. Choosing a common approach to these topics will make interoperability between machines easier. This functionality can be grouped into the following parts:

1. Operation modes include cold and warm start, scheduled and unscheduled stop, operational and others, following guidelines like GEMMA (Guide d'Etude des Modes de Marches et d'Arrts).

2. Failure management should be established to minimize unscheduled stops due to faults.

3. Preventive maintenance analyzes timing and number of cycles of machine elements to determine times of replacement.

4. Local HMI. A local user interface is needed to select machine operation modes, determine specifics forms to be cut, generate maintenance messages, and allow maintenance staff to order certain actions.

5. Control application data management. Certain data produced during execution of the control application will need to be managed. In particular, information regarding preventive maintenance or failure management must be persistently stored and retrieved.

6. Production data management. Machine tools receive working orders specifying what forms must be cut, size, glass sheet type, number of items and customer identifier. Whenever the different working orders have been committed, external applications (e.g., MES and ERP systems) will have to be notified.

7. Communications. Machines need to communicate with other applications such as SCADA, MES and ERP systems, and custom-built solutions.

14.3.3   FB Design and Design Patterns

From the IEC 61199 view, all the previous functionality must be implemented using function blocks (FBs). FB design is a very important part of the design process. FBs should be designed by qualified control engineers. Moreover, the design of the FB must consider all specific and generic aspects listed above. FBs grouped into the following categories must be identified:

1. FBs related to interface with process. A machine tool controls to process through physical devices (i.e., sensors and actuators). Each device must be modelled by a component. These elements are accessed via a fieldbus (e.g., CANopen, AS-I, etc.) that must be modelled by one or more components too. Therefore, two levels of abstraction are established in relation to the interface with the process. In order to facilitate reconfigurability there may be a third level intermediate between them, a logical connector component. All these interface components should be reusable for any machine type.

2. FBs related to control and automation. The first corresponds with the precise positioning of machine axis (e.g., sheet and tools movements) and are based on PID control. Normally, these components should be reusable. Automation components correspond with sequences of machine operations (e.g., loading and unloading phases). These components are less reusable than the previous ones because they depend on a specific logic.

3. FBs related to generic functionality (e.g. failure management). Although the component includes failure management, it is necessary to consider specific components that facilitate it. The first is the failure zone component that divides control of the system into denominated zones that consist of several components. Each failure zone receives notifications of faults

of its components and the notifications are propagated to other zones so that possible response actions can be formulated. Each failure zone should establish its own fault propagation policy. The mission of the failure supervisor component is to detect faults arising from several components. Faults are propagated to the relevant failure zone. finally, the preventive and predictive maintenance supervisor component manages related issues. The components may be parametrized by rules to make them reusable.

The next step is to establish the design patterns. These are sub-network components and may be grouped into the same categories as the components:

1. Design patterns related to interface with process. The pattern of the fieldbus (e.g., CANopen) is composed of a specific hardware driver, a generic master driver and several types of specific devices (e.g., digital and analog I/O, servo motor driver, electrovalves).

2. Design patterns related to control and automation. The control pattern (e.g., axis PID control) is composed of a PID controller, a setpoint generator and gearboxes. The latter adapts the setpoint and feedback signal. To establish patterns for automation, the components (i.e., business logic) may be divided into several parts, each corresponding to a design pattern, for example, lading and machining. This allows the patterns to be reusable in other machines.

3. Design patterns related to to generic functionality:

    (a) Failure management: The failure management design pattern has several components (e.g., interface, control, automation, failure supervisor). The proposal based on a design pattern has been presented in a previous work [23]. This proposal identifies, characterizes and classifies fault conditions and their appropriate response actions. A topological criterion is used to group faults coming from machine elements into failure zones, allowing the creation of a hierarchical tree. When a failure occurs, the affected element notifies its failure zone, that can propagate that event to the rest of elements and possibly also to other failure zones. When an element receives such a notification, it determines its appropriate behaviour. All this prevents a single failure from making the entire machine non-operational, unless it is unavoidable. Five failure zones have been established in this design, corresponding to four physical parts: loader, unloader, monolithic machining and laminated machining, plus a general failure zone acting as a root in the hierarchical failure management. Emergency switches are handled by an emergencies element, so that when an operator presses any switch, the general failure zone will be directly notified.

    (b) Preventive maintenance: Again, a design pattern presented by the authors [24] applies to two elements of the glass cutting machine:

venting fans and the moving parts for the scoring wheel positioning. The former has one associated maintenance action: fan alignment checking and its usage magnitude is working time. The latter has two associated maintenance actions: lubrication and cable substitution. Their associated usage magnitude is distance travelled by the scoring wheel.

With the types of FBs identified and the design patterns established, detailed design is an incremental and iterative process that generates a set of components and reusable design patterns.

The next is designing the specific machine control after determining the required features by considering: a) types of machining operations such as warming, cutting and breaking; b) types and measurements of glass sheets; c) estimated production rates; d) movements of glass sheets and supplies, e.g., loading, unloading, squaring and cutting; e) number of axes controlled based on movements listed in d); and f) characterization and classification of faults for formulating response actions.

Faults are classified by potential for damage. A *warning* means that an incident has occurred but functionality remains complete. *Degraded* performance means a fault occurred and impacted functionality. *Failure* indicates that a fault occurred and the system cannot perform to its assigned functionality level.

The design of specific machine control must consider the following steps:

1. Design of business logic. As indicated previously, machine automation can be split into functional parts. In this step, the parts already designed previously must achieve the specific functionality of the machine. Design of the business logic could be considered from the view of distributed systems. Each controller can handle one or more of these functional parts. These parts should be loosely coupled, with links at a high semantic level for synchronizing the operating processes. For example, the loader can tell the machining tool that a new sheet is available and specify its type. Another facet of distributed control is axis control. The links with this part should be positioning commands (e.g., go position XY, go zero). Moreover, each part also manages the start and stop modes from HMI or SCADA commands. The links between functional parts and their process interfaces are based on field bus design patterns described.

2. Design of failure management and of predictive maintenance. Although the failure management design patterns establish zones managed by a zone failure component, it is necessary to glue them together for the specific machine. This means interconnecting zones so that a zone may be notified of the failures in others and propagate the failure to its components. In this way, a hierarchical failure management can be established. Failure response actions are performed by components and should be considered when designing them. Another issue in this step is to determine the rules,

for monitoring components, for predictive maintenance. These rules should ensure that possible faults are inferred and notified to zones.

3. Design of communication aspects with HMI and SCADA applications. Control software must have appropriate local and remote HMIs for the operation of the processes under control. Usually, the HMI application is executed independently of the control software in multitasking environment. In the approach presented, the platform allows applications to have limited access to components. Thus, in this step, access is determined to implement the communication with the HMI. This access is also possible for remote interfaces.

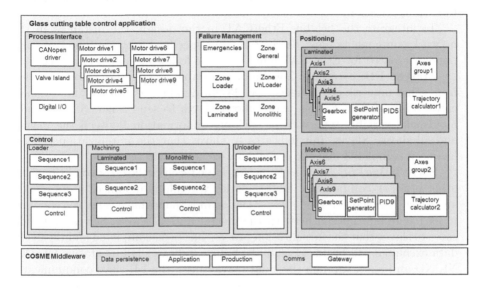

FIGURE 14.3
High level model of glass cutting table. It will be implemented with interconnected networks of FBs (FBNs).

The final design is a network of components composed of different loosely coupled parts, where business logic, failure management and HMI communication are distributed on multiple controllers. This facilitates reconfigurability, because changes in parts require less effort and/or time due to lower dependence among them. Also, generic features of machines are implemented as aspects in the component model.

In Figure 14.3, the architectural elements are hierarchically grouped according to their functionality types.

14.3.4　Detailed Design of FBN: EtherCAT Network

EtherCAT bus (ethernet for control automation technology) is a high performance communication bus based on Ethernet. The main objective of EtherCAT is the ability to use certain layers of Ethernet in automation and control applications requiring small data update times (also called cycle times), with low jitter values (for synchronization purposes) and low cost hardware. It is suitable to be used on machine-tools application domain.

The proposed use case is based on a PC control that attends a drive and an I/O module. A network of specific FBs implementing the EtherCAT fieldbus must be designed. This solution encapsulates the complexity associated with master and slaves on FBs. The designer should focus on functionality specific to each application.

To integrate the fieldbus EtherCAT in the design, the instances in the network components must be created from specific FB types. The two components created for EtherCAT must have the following inputs and outputs:

Master FB: *a) Inputs*: Zone failure variable receives the error condition from other network components (see [23] for details). *b) Outputs*: failure indicates failure status. Enable_write allows slaves to send a new SDO. Hands: handshake variable type used to exchange information between the master and all slaves. Slaves_init indicates to slaves that the initialization process has been completed and the cyclic data exchange will begin.

Slave FB: *a) Inputs*: Zone failure variable receives the error condition from other network components. Enable_write variable is modified by its master and allows the slave to send a new SDO. Hands: handshake variable type used to exchange information between the master and all slaves. Slaves_init indicates to slaves that the initialization process has been completed and the cyclic data exchange will begin. *b) Outputs*: failure indicates failure status. DI, DO, NI, NO and axes: through these data structures, the FB slave communicates with the numerical computation. The sense is bidirectional; the FB slave modifies these variables with the value of the PDO and SDO and communicates them to the numerical computation. Otherwise, numerical computation FB defines these values and communicates them to the FB slave to be transmitted using SDOs or PDOs to bus devices.

Figure 14.4 shows a simplified schema. The FBs described are highlighted (MST_ETHERCAT for master FB, and ETHERCAT_SLAVE FBs for motor drive and I/O module slaves), and the main data connections to other FB of the application are shown. There are digital and analog I/O, gearbox (to change signal scale), control numeric and PID calculations, and so on. FB data I/O related with background state, external event, cold and warm initiation are hidden for simplicity.

Note that event connections are not shown. As explained further, due to the execution model of COSME (based on event chains), these parts of the implementation are predefined by COSME.

14.4 Implementation

14.4.1 COSME Platform

IEC 61499 standard indicates that FBs should accommodate multiple algorithms selectable by events or conditions. The COSME FB model [3] is based on predefined algorithms taking into account common situations in control applications such as initialization, normal operation and failure operation. The input and output events are predetermined, but the input and output data are not, and they depend on the functionality of FB type. The algorithms generate an event when they have finished. This allows composing FBNs to establish a daisy-chain topology, where events of the same type of algorithms are connected. The extension to distributed FBNs can be simply done by in-

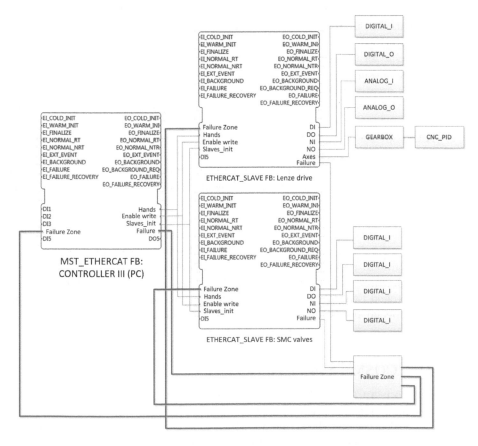

FIGURE 14.4
Template for EtherCAT FB network scheme.

corporating CSIFBs in the inter-controller chains. These CSIFBs should be of publish and subscribe type, because they are appropriate for the propagation of events along the controllers.

The COSME daisy-chain is an intermediate model between the event-based and the scan-based. Designers must write response code to situations, like event-driven, but here situations are predefined. COSME has been designed for machine tool control software, although it could be used in other applications. The rules of composition are simple and can be applied by engineers educated in control engineering.

COSME platform is the layer between control applications and the underlying system (i.e., hardware, operating system and communications). Related to execution, the platform follows the event chain (EC) concept. Implementation is done by associating these ECs to different priority threads, according to a model named "event path per task, event path-based priority". Control algorithms must be executed in a short time because they block the system to the ECs with lower priorities. Due to the limited number of ECs and threads, the implementation is easily scalable. Thus, this model is especially suitable for applications with a large number of FBs, and it also facilitates determining the system behavior in design time. The COSME elements that perform execution are the execution server, name server and application loader. The loader loads FB types, creates FBs and makes intra and inter-controller connections and it orders the execution boot. The platform has a persistent storage element ready to perform warm startup of the applications.

Although the current COSME version is focused on application design and distribution aspects, it allows simple reconfiguration of the FBNs in execution time by means of commands (instantiation, reconnection, etc.). In a new COSME versions these commands will be sent by an external reconfiguration manager application, which will receive data from control and MES to implement dynamic reconfiguration. The COSME elements that perform this function are component server, name server and gateway. Inter-controller communication links are based on publish and subscribe mode. The messages contain number, publisher and subscriber names, timestamp and data (when necessary). Inter-controller communication uses a wired dedicated network and a switched Ethernet device. These devices guarantee deterministic delays without special hardware or software. An authentication policy guarantees security.

14.4.2   COSME IDE

Domiciano is a COSME IDE tool built to facilitate control software development. The Domiciano features are determined by the IEC 61499 component model (FB) used, and the scope for which it was created (i.e., development of control applications for machine tools). Its goal is handle to the final application domains. The knowledge of those domain characteristics allows Domi-

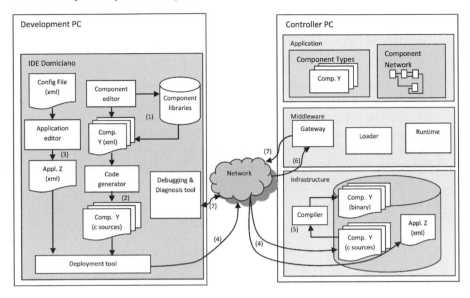

FIGURE 14.5
Development workflow showing components and application generation, application deployment, debugging and control operation.

ciano to hide some aspects to designers. So that, they do not need to deal with issues related with the standard [9]. Domiciano allows the following tasks:

1. Creating and editing components based on the model used.

2. Managing repositories of components (i.e., libraries).

3. Editing applications by creating component instances, interconnecting them, and using predefined templates of networks.

4. Generating the native code of applications by different code generators.

5. Applications deployment from a remote development platform.

6. Debugging applications by watching internal variables.

7. Applications management by start and stop actions, loading of applications and component types.

Domiciano provides editors, code generators, deployment and debugging tools. Figure 14.5 shows the workflow to develop an application which, in essence, can be summarized as follows:

1. Create components using the component editor and adding them to the components library.

2. Code generator creates the needed source files for the selected platform from the corresponding component file definition. Local compilation avoids syntactical errors.

3. A new application is created, instantiating the needed components. These instances are parameterized and a connections network is established using the application editor. When the number of instances and connections is large, Domiciano organizes them using views. Each view shows only a part of the network. This way, designers can focus on the different aspects of the application. Finally, the application editor generates an XML file that includes a components network representation.

4. The native code files and the application file are deployed (locally or remotely) into the target by the deployment tool.

5. The target compiler is invoked to generate the binary files.

6. The application is built and is started by the loader before it has loaded the component types.

7. The application can be monitored for debugging. Domiciano is communicated with middleware by a gateway.

The Domiciano elements needed to perform its functionality are the following:

Component Type Editor: Domiciano creates new component types, which are saved in a specific format in XML files. To do that, designers must designate the components internal variables data input and data outputs and define their types (Figure 14.6). The IDE provides a set of predefined types, but it also allows definition of new customized types. Component internal variables can be declared persistent, which means that their values are periodically stored during execution and just before the application is stopped, using a specific middleware service. They are also restored when the application starts. After, defining internal variables and component inputs and outputs, designers must write the stage functions source code. This source code can be written in any of the available languages installed in Domiciano by means of plug-in. Also, it is possible have functions written in different languages. Currently, there is a plug-in available for C language and an IEC 61131-3 FBD language plug-in is in progress. The first one provides an integrated C editor. To structure this stage functions C code, designers can define user functions which are called from the previous ones. The IEC 61331-3 FBD plug-in will provide a graphical editor and a library of standard functions (numerical, Boolean, selection, comparison, timers, counters, etc.). The execution of FBD graphical code is performed by a middleware service. The component type editor also can define IEC 61131-3 SFC (also denominated grafcets) by a graphical editor. A component can have multiple SFCs. This SFC editor is a tool for designing and presenting charts, providing a toolbox to insert initial steps, steps, transitions, convergences and divergences. The SCF editor verifies whether the chart is well formed. The execution of SFC is also performed by a middleware service. Component types are grouped as specific and generic. The first are those that are only used in particular applications, for example, to define the application business logic, and they are associated to an application. The second ones are reusable in the applications, for example, a PID control. These compose the components libraries.

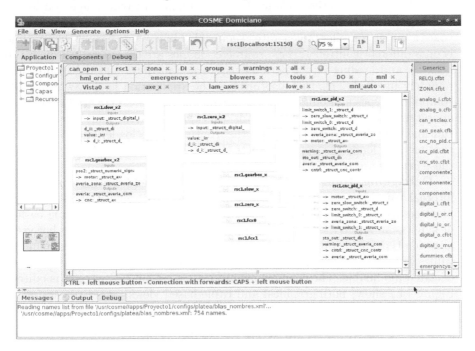

FIGURE 14.6
Component and FB network editor.

Application Editor: Domiciano includes a graphic editor for the components network. To make editing easier, views and layers can be defined. A view is a graphic representation of one or several layers and a layer is a logical structure that contains a subnetwork. Component and connections can be shown in one or more layers. This structure allows separation of the different aspects of the application and facilitates browsing the network. Designers must select the components to create the instances and then connect the instances input and output data. The editor only allows connection of inputs and outputs between compatible types. Designers can customize instance initialization by assigning specific values to internal variables. Instance internal variables have default values defined when the components were created. Applications are saved in a specific format in XML files. An XML configuration file allows Domiciano to load predefined sub-network templates (e.g., for failure management tasks) and component types when a new application is created. The XML configuration file indicates the available code generators and deployment communication profiles; it also saves Domiciano workspace (i.e., desktop state).

Code Generator: This element automatically generates source native code files for a specific runtime. It can generate code for any runtime for which a plug-in is installed. The available runtime plug-in in Domiciano is RTAI. It generates native code for Linux systems with real time extensions.

Deployment Tool: This element deploys native source code files and invokes a target compiler to generate executable binary code files. If the runtime is on a different controller, these source code files are transferred through the network and loading into a permanent storage medium and the compiler is remotely invoked. This tool also invokes the application loader, which loads component types and builds and starts the application. There are two loading operating modes. The first one loads the application without stopping the runtime and the second one stops the runtime. The loader then starts the runtime and loads the application. If the runtime is on a different machine, these actions are triggered remotely.

Debugging Tool: This element allows reading and writing components internal variables, inputs and outputs to debug the application. It communicates with the middleware via the gateway. To make the debugging easier, designers can watch the evolution graphically. This tool provides another visualization type, using tables with the current values. Also, the tool can highlight values outside a predefined range. Designers can change values for variables declared modifiable. The component property is established when it is edited. Figure 14.7 shows both graphical and table-based views.

14.4.3   Communication with External Applications

Communication with external applications is performed with client-server or publish-subscribe messaging models. In both cases, messages have the same format as in the inter-controller case. The Arcadio library performs communication in external applications, hiding the implementation details. Non-dedicated networks are used and for security reasons, connections use authentication and encryption. During boot, the application loader establishes all communication links and they are managed later by the gateway element. Process interface communication is performed by a field bus (e.g., CanOPEN or EtherCAT).

Periodic review of variables is a common task and Arcadio offers a "bag" feature. The bag is a variable container with a set refresh period. When configured, it ensures an unlimited number of reads without further involvement of Arcadio. A *variable* in this context means a component input, output or internal variable.

After a bag is created and configured, COSME starts sending messages containing the desired values ot Arcadio at specified intervals of milliseconds. When a new bag is received, Arcadio triggers an event to the external ap-

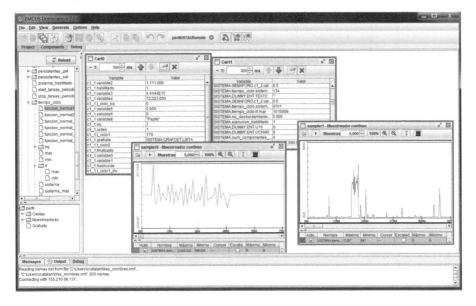

FIGURE 14.7
User interface debugging tool.

plication so that it can take appropriate action (usually making use of fresh data).

Figure 14.8 shows an external application (e.g., the machine tool HMI) which interacts with the Arcadio library in three ways: 1) The external application issues commands to Arcadio. This makes Arcadio create and send a message to COSME (e.g., create a new bag, add variables to the bag, set a new bag period, write a new value, ask for a single read, etc.). 2) Events are notified to the external application as the result of Arcadio receiving certain kinds of messages from COSME (e.g., bag arrivals, single read arrivals, and some echo telegrams). 3) The external application explicitly reads values from Arcadio which are taken from the buffer where received values are stored.

It may be necessary to block the external application execution until a value is received, or until it can be assured that a certain value has been written in the COSME application. For those cases, a synchronous model is available. Usually, external applications will be interested in reading component outputs and writing component inputs, but Arcadio also allows access to internal component variables (if they were marked as public or writeable at design time). Although it obviously breaks the encapsulation principle, this was decided with debugging and diagnosis tools in mind. The HMI developed for the case study machine is built on top of the Arcadio library. Given the networked nature of Arcadio, the HMI application can reside in the same computer where COSME and the control application are run (which is the solution adopted) or on different computers. Most of the HMI is built using the bag

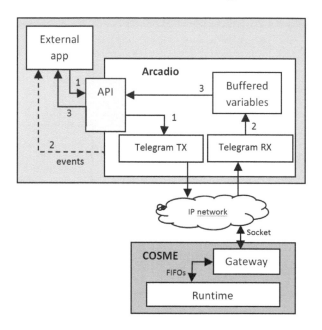

FIGURE 14.8
External applications can communicate with COSME using the Arcadio library. Arcadio prioritizes an asynchronous programming model, which is adequate for HMI, SCADA and MES/ERP integration applications; synchronous operations are also possible.

concept, receiving from COSME about 5 kB of data every 200 ms (considered adequate to HMI refresh rate). Tests have been successfully conducted with periods of up to 5 ms using conventional local networks.

14.5   Practical Issues and Conclusions

The COSME version used runs on a RTAI 3.7.1 Linux PC, and some middleware parts use Java SE6. RTAI modifies the Linux kernel to be able to execute real time tasks in user or kernel space. In the kernel space, RTAI applications are implemented with Linux kernel modules, which may be loaded or unloaded from memory during execution time. This solution has been adopted in the current COSME version, in which runtime elements and component types are kernel modules. COSME communicates the control software application and external applications by means of the gateway middleware element. This uses

RTAI FIFOs to communicate with the runtime and sockets for the external applications.

Hardware elements used in this glass cutting table are one B&R Automation PC 620, with two 100 Mbps network interfaces; two PCI-7841Adlink CANopen master cards; one s0700 SMC valve island; one X20 B&R remote I/O; nine Sdrive Power Stages standard motor drives; and one CISCO SLM200 Gigabit Smart Switch.

As previously mentioned, the component execution model is based on a daisy chain. The predefined chain types (in descending priority) are: initialization, finalization, external event, normal real time, normal non-real time and background. Each chain sequentially executes the corresponding component's stage functions. The CNC and PLC functionality are performed within the normal real time chain, which is sequentially executed every 1 ms, according to the component order predefined at design time. Functionalities related to preventive maintenance, external communications and data management are performed within the normal non-real time chain, which is sequentially executed every 20 ms. Also, CNC trajectories calculations are performed within the background chain, which is sequentially executed on demand. Finally, CANopen handling is performed within the external event chain, when an external event occurs.

The glass cutting machine prototype is shown on Figure 14.1. The control software has the following characteristics:

1. Thirty-five different FB types have been used, for example, digital I/O, CAN bus, gearbox (to rescale variables), PID, emergencies.

2. About 200 instances of these FB types are necessary to implement the FB network. So that, the model defined in Figure 14.3 can be implemented.

3. Seventy SFCs are defined.

4. Software control manages 9 axes on 2 ms cycle time using a Celeron PC.

COSME provides control designers a simplified model tailored to specific machine tools, so applications are easier to understand and maintain. It promotes more sophisticated products, adding features such as embedded preventive and predictive maintenance, integration with MES and ERP systems, remote diagnosis, increased software robustness and improved time to market.

This chapter shows an industrial application with real requirements. The application of well established techniques in software engineering (e.g., component-based programming) represents an advantage for a company by reducing design times, debugging and marketing of products.

Bibliography

[1] 4DIAC. www.fordiac.org/.

[2] R. W. Brennan, P. Vrba, P. Tichy, A. Zoitl, C. Sünder, T. Strasser, and V. Marik. Developments in dynamic and intelligent reconfiguration of industrial automation. *Computers in Industry*, Vol. 59, , 533–547, 2008.

[3] C. Catalán, F. Serna, A. Blesa, and J. Rams. Iec 61499 execution model based on life cycle of function blocks. In *Emerging Technologies and Factory Automation, IEEE Conference on*, pages 1–8, 2010.

[4] C. Catalán, F. Serna, A. Blesa, J. Rams, and J. Colom. Communication types for manufacturing systems: A proposal to distributed control system based on IEC 61499. In *Automation Science and Engineering, IEEE Conference on*, pages 767–772, 2011.

[5] C. Catalán, F. Serna, T. Civera, A. Blesa, and J. Rams. Control design for machine tools using domiciano, an IDE based on software components. In *Emerging Technologies Factory Automation, IEEE 16th Conference on*, pages 1–4, 2011.

[6] J. H. Christensen. Design patterns for systems engineering with IEC 61499. *Verteilte Automatisierung - Modelle und Methoden für Entwurf, Verifikation, Engineering und Instrumentierung*, 2000.

[7] M. Colla, T. Leidi, and M. Semo. Design and implementation of industrial automation control systems: A survey. In *Industrial Informatics, 7th IEEE International Conference on*, pages 570–575, 2009.

[8] N. Hagge and B. Wagner. Implementation alternatives for the OMAC state machines using IEC 61499. In *IEEE Conference on Emerging Technologies and Factory Automation*, 2008.

[9] K. Hall, R. Staron, and A. Zoitl. Challenges to industry adoption of the IEC 61499 standard on event-based function blocks. In *IEEE Conference on Industrial Informatics (INDIN)*, 2007.

[10] D. Hästbacka, T. Vepsäläinen, and K. S. Model-driven development of industrial process control applications. *Journal of Systems and Software*, Vol. 84, , 1100–1113, 2011.

[11] Holobloc. http://www.holobloc.com/doc/fbdk/index.htm, 2012.

[12] Iacocca Institute. 21st century manufacturing enterprise strategy: An industry-led view. 1991.

[13] International Electrotechnical Commission. *Programmable Controllers. Part 3: Programming Languages, IEC 61131-3*, 2.0 edition, 2003.

[14] International Electrotechnical Commission. *Function Blocks. Part 1: Architecture IEC 61499-1*, 2.0 edition, 2005.

[15] ISaGRAF. www.isagraf.com.

[16] Y. Koren, U. Heisel, F. Jovane, T. Moriwak, G. Pritsshow, G. Ulsoy, and H. van Brussel. Reconfigurable manufacturing systems. *Annals of the International Institution for Production Engineering Research (CIRP)*, Vol. 48, No. 2, 527–539, 1999.

[17] R. Lewis. *Modelling Control Systems Using IEC 61499 Applying Function Blocks to Distributed Systems*. Number 59. Institution of Engineering and Technology, 2001.

[18] S. Mekid, P. Pruschek, and J. Hernández. Beyond intelligent manufacturing: A new generation of flexible intelligent NC machines. *Mechanism and Machine Theory*, Vol. 44, No. 2, 466–476, 2009.

[19] M. Minhat, V. Vyatkin, X. Xu, S. Wong, and Z. Al-Bayaa. A novel open CNC architecture based on STEP-NC data model and IEC 61499 function blocks. *Robotics and Computer-Integrated Manufacturing*, Vol. 25, No. 3, 560–569, June 2009.

[20] nxtControl. http://www.nxtcontrol.com.

[21] G. Pritschow, Y. Altintas, F. Jovane, and Y. Koren. Open controller architecture: Past, present and future. In *Annals of the CIRP*, volume 50, pages 463–470, 2001.

[22] F. Serna, C. Catalán, A. Blesa, J. Colom, and J. Rams. Predictive maintenance surveyor design pattern for machine tools control software applications. In *Emerging Technologies Factory Automation, IEEE 16th Conference on*, pages 1–7, 2011.

[23] F. Serna, C. Catalán, A. Blesa, and J. Rams. Design patterns for failure management in iec 61499 function blocks. In *Emerging Technologies and Factory Automation, IEEE Conference on*, pages 1–7, 2010.

[24] F. Serna, C. Catalán, A. Blesa, J. Rams, and J. Colom. Preventive maintenance manager design pattern for component-based machine tools. In *Industrial Informatics, 9th IEEE International Conference on*, pages 615–620, 2011.

[25] T. Strasser, M. Rooker, I. Hegny, M. Wenger, A. Zoitl, L. Ferrarini, A. Dede, and M. Colla. A research roadmap for model-driven design of embedded systems for automation components. In *Industrial Informatics, 7th IEEE International Conference on*, pages 564–569, 2009.

[26] Turomas Group. www.turomas.com.

[27] V. Vyatkin. *IEC 61499 Function Blocks for Embedded and Distributed Control Systems Design.* Instrumentation Society of Automation (ISA), 2011.

[28] V. Vyatkin and J. Chouinard. On comparisons of the ISaGRAF implementation of IEC 61499 with FBDK and other implementations. In *IEEE Conf. on Industrial Informatics*, 2008.

[29] V. Vyatkin, M. Hirsch, and H.-M. Hanisch. Systematic design and implementation of distributed controllers in industrial automation. In *IEEE Conference on Emerging Technologies and Factory Automation*, pages 633–640, 2006.

[30] H. Xuemei. Distributed and reconfigurable control of lower level machines in automatic production line. In *8th World Congress on Intelligent Control and Automation*, pages 2339–2344, 2010.

[31] A. Zoitl, T. Strasser, and A. Valentini. Open source initiatives as basis for the establishment of new technologies in industrial automation: 4DIAC a case study. In *Industrial Electronics, IEEE International Symposium on*, pages 3817–3819, 2010.

15

Distributed Intelligent Sensing and Control for Manufacturing Automation

Robert W. Brennan

University of Calgary

CONTENTS

15.1   Introduction

To remain competitive in today's global market, manufacturers require systems that are capable of quickly responding to change while maintaining stable and efficient operation. Increasingly, the manufacturing control system is viewed as being central to achieving this goal. The main barriers to success in this area however, result from the combination of increasingly stringent customer requirements (*e.g.*, high quality, customizable, low-cost products that can be delivered quickly) and inherent manufacturing system complexity (*i.e.*, these systems are, by nature, distributed, concurrent and stochastic). Although manufacturing technology has become increasingly sophisticated to deal with these issues (*e.g.*, through advanced robotics and computer numerical control), without adequate control, the result is often a collection of "islands of automation" that lack the necessary integration for truly responsive behaviour.

This chapter describes a distributed intelligent sensing and control systems (DISCS) approach that meets the primary needs of modern manufacturing systems (*i.e.*, disturbance handling, availability, flexibility, and robustness). Our approach focuses on design strategies for reconfiguration that are supported by appropriate system analysis and safety management techniques. The motivation for this work is to enable manufacturing systems to quickly respond to change while maintaining stable system operation and efficient use of available resources.

Given the proliferation of ubiquitous computing, smart sensors, and certified embedded computers in aircraft, medical equipment, and other high-integrity devices, there has been a considerable interest in the modelling and analysis of industrial computing systems in recent years [14]. In the manufacturing domain, the main tools and techniques to support automation and control have come from the areas of agent systems [34] and programmable logic controller (PLC) languages [27]. The IEC 61499 standard [28] bridges these approaches by using a component-based design that supports configurability, interoperability, and portability; in recent years, this has led to significant progress in the modelling of distributed intelligent automation systems [31].

A system wide deployment of sensing and actuating devices, especially wireless devices, is becoming more and more common in industrial applications. Those devices are spatially distributed and primarily connected wirelessly for collecting, monitoring, and controlling conditions of large, expensive, or hazardous facilities and infrastructures. In this chapter, we propose applying wireless sensor networks (WSNs) to this problem in the form of a DISCS. In a DISCS, sensing and control devices are distributed spatially over the infrastructure sites, and/or decision-making within the applications is distributed rather than centralized. Either for improved productivity and efficiency or increased safety, distributed sensing and control shows unprecedented potential.

In order to provide the basic services required to support distributed intelligent sensing and control, we propose an application-oriented middleware architecture that considers both application requirements and network resource constraints. The intention of our work on this system architecture is to build the foundation to support "intelligent" behaviour of our DISCS (for our purposes, "intelligence" is the system's ability to automatically reconfigure in response to change). More specifically, we plan to use our application-oriented middleware architecture [5] as a basis for a multi-layered DISCS model that combines high-level software agents with low-level IEC 61499 function blocks to achieve dynamic reconfiguration. Higher-level agents will be used to manage a reconfiguration process that will be supported by the low-level, real-time services of our middleware architecture.

In the next section, we provide an overview of the recent work on distributed sensing and control systems with an emphasis on system reconfiguration and middleware support. We follow this with a description of our proposed DISCS, first describing the general DISCS architecture in Section 15.3, the function block implementation in Section 15.4, then an example of the implementation for mobile node tracking in Section 15.5. The chapter concludes with a summary of our current and future work in this area.

15.2 Related Work

In this section, we provide a brief review of recent work on distributed sensing and control systems. Given that one of the key motivations for our work in this area is to develop systems that are capable of quickly responding to change, we begin with an overview of the work on reconfiguration of distributed systems. This is followed by a review of related work on intelligent middleware to support wireless sensor networks.

15.2.1 Reconfiguration of Distributed Systems

The early work on dynamic reconfiguration falls under the umbrella of what Guler *et al.* [9] refer to as "transition management": *i.e.*, reducing the impact of software reconfiguration at run time and maintaining system consistency and stability. This began with Kramer and Magee's work on dynamic change management in distributed systems [12], which focused on allowing changes to be specified at a high level, then executed by "configuration management" at the task level.

The concept of "dynamic reconfiguration" was introduced by Setchi and Lagos [24], and referred to systems that have "the ability to repeatedly change and rearrange [their components] in a cost effective way". Walsh *et al.* [32] classified dynamic reconfiguration into different change types that were based on how system integrity was managed. This approach involved building a domain model that included various fault tolerance modes.

The early work on reconfiguration of IEC 61499 function block-based systems [28] was performed by researchers at the University of Calgary [2] and the Technical University of Vienna, PROFACTOR, and University of Applied Sciences Wels [25], and addressed IEC 61499 architecture and the concept of down-timeless systems [23]. This work focused on component-level reconfiguration and the identification of key services to support intelligent reconfiguration (*i.e.*, services to support execution control, state interaction, queries, component libraries). Rockwell Automation took a different approach to intelligent reconfiguration with its autonomous cooperative system (ACS) [20], that separates the low-level, real-time control portion of the system from the high-level, intelligent control portion of the system; reconfiguration is then managed and performed exclusively at the high (software agent) level.

Despite the amount of work in this area, the mechanisms to create the dynamic and emergent structures required for intelligent reconfiguration at the device level need to be better addressed [31]. The trade-off at this level is between deliberative and reactive behaviour: *i.e.*, the processing overhead required to support intelligent decision making can hinder an agent's ability to satisfy a physical device's timing constraints.

15.2.2   Intelligent Middleware

Middleware of distributed sensing and control systems refers to software and tools that can help hide the complexity and heterogeneity of the lower level hardware systems, and ease information processing and control at distributed application level. The challenges of distributed sensing and control middleware are numerous, *i.e.*, frequently changing requirements from applications, limited network resources including bandwidth and battery life, dynamic network topology and data availability, as well as application-specific data-processing knowledge. In addition, the lack of design tools in this area is another big challenge at this moment.

Compared with the middleware of a general sensor network, the middleware for industrial distributed sensing and control system must focus more on application requirements, execution control, and how to optimize information processing proactively according to both QoS (Quality of Service) and resource constraints. Work in this area has attracted wide interests over the past two decades driven by the potential of WSNs in industrial applications. Because of unique characteristics and technical challenges from both the application and the basic network sides, research and development into distributed sensing and control system middleware has focused on general methodology, as well as application-specific issues.

In an early review of distributed sensing and control middleware research, Romer *et al.* [22] identified the challenges associated with successful WSN middleware both in terms of design concepts and system implementations. At the time of their review, most research was in an early stage, and results were based on simulations or small-scale experiments in laboratory settings.

Liu and Martonosi [16] proposed a middleware system to manage an autonomic parallel sensor system called Impala. Impala enables application modularity, adaptively, and repair ability in WSNs. It also allows software updates to be received via the node's wireless transceiver and applied to the running system dynamically. In addition, Impala also provides an interface for on-the-fly application adaptation to improve the performance, energy efficiency, and reliability of the software system.

In [11], the authors described different types of sensor network applications and discussed existing techniques to manage those types of networks. They also overviewed a variety of related middleware and argue that no existing approach provides all the management tools required by sensor network applications. To meet that need, the authors developed a new middleware called middleware linking applications and networks (MiLAN) that targeted environment surveillance, home and office security, and medical monitoring applications.

An EPC sensor network (ESN) architecture was proposed by Wang *et al.* [33] as an integration system of RFID and WSN. The core of ESN was the middleware part. In the proposed middleware, complex event processing technology was used which could handle large volumes of events from distributed

RFID and sensor readers in real time. Through filtering, grouping, aggregating, and constructing complex events, ESN middleware provided meaningful reports for clients and increase system automation.

Gungor and Hancke [10] analyzed challenges, design principles, and technical approaches in the industrial wireless sensor networks. Their work identified a number of issues to be addressed via middleware such as analytical models for efficient deployment of IWSNs in the real world, diverse industrial application requirements, large scale networks, optimal sensor node deployment, localization, security, and interoperability between different IWSN manufacturers.

Distributed sensing for quality and productivity improvements was discussed in [6]. The paper also discussed the state-of-the-art practice, research challenges, and future directions related to distributed sensing. The discussion included the optimal design of distributed sensor systems, information criteria, and processing for distributed sensing and optimal decision making in distributed sensing.

To tackle the interoperability of various types of sensors and actuators using different wireless communication technologies (WiFi, Bluetooth, or RFID), Ramamurthy *et al.* [21] developed a wireless smart sensor platform, which has a "plug-and-play" capability of supporting hardware interface, payload and communications needs, and the means to update operating and monitoring parameters, as well as sensor and RF-link specific firmware modules "over-the-air". The research targeted applications of instrumentation and predictive maintenance systems and focused on hardware and firmware levels.

A service-oriented architecture for QoS configuration and management of WSNs was proposed by Anastasi *et al.* [1]. The proposed middleware tried to hide the complexity of low-level physical devices and support the management of heterogeneous real-time data at enterprise level.

In [35], a fast and simultaneous data aggregation solution is proposed for advanced data analysis over multiple regions. A novel distributed data structure called distributed data cube is used in query processing for expanding WSN applications.

Martinez *et al.* [18] presented a design and validation method for wireless sensor and actuator networks to address the complexity of nondeterministic and concurrent behaviour of distributed systems and to improve the analysis and design performance in simulation and testbed scenarios. Instead of focusing on general methodology research, application-specific WSN approaches are investigated in [17] and [13]. The former was motivated by real business cases from the oil and gas industry and focused on interoperability between different firmware systems. The authors presented a three-layer, service-oriented architecture that accommodated different sensor platforms and exposed their functionality in a uniform way to the business application.

In summary, the deployment of WSNs in industry will occur incrementally, and middleware will be crucial to form general wireless sensor devices into real industrial solutions for distributed application systems. Although considerable effort has been put into middleware research, most of the work focused on the network issues or sensing-only applications have been considered. There is still no existing middleware with efficient tools that have the desired features for industrial distributed sensing and control system middleware; *i.e.*, node intelligence and low-level control support are especially weak. In the remainder of this chapter, we describe an intelligent middleware design based on the international standard IEC 61499 [30] for modelling distributed control systems and intelligent agents.

15.3   DISCS Architecture

Our proposed application-oriented intelligent middleware architecture for DISCS is illustrated in Figure 15.1. The proposed middleware has a two-level structure: an agent facilitated upper level management and a function block based lower level implementation.

At the upper level, the QoS of an application is taken as inputs to the device manager and the task manager, which work collaboratively to serve application needs by dynamically passing the required network configuration and data information to a lower level network. The configuration and data information are sent and received through universal plug and play (UPnP)-based gateway and basic network and event handling protocols.

At lower level, application-based intelligence is embedded into corresponding sensor nodes (*i.e.*, sink nodes). Device-dependent raw data are collected from a specific node or a node cluster and processed locally. As a result, only application-oriented key parameters output from embedded algorithms are transmitted to the upper level task manager. The key links among different modules within the system are bidirectional events, which trigger, reconfigure, or put nodes to sleep according to upper level needs. Meanwhile, bidirectional events are also used to discover, remove, or reorganize node topology based on physical node availability and network connection. With the proposed middleware architecture, higher level information and lower level data have been processed locally, which makes the communication between levels simpler and more efficient. Furthermore, the QoS can be guaranteed by this flexible middleware structure through fast network reconfiguration. There are four unique components in the proposed middleware: device manager, task manager, sink nodes, and bidirectional events.

1. Device manager takes care of bottom-up and top-down configuration of physical network devices. Bottom-up configuration involves configuring the network to reflect physical connection changes at lower level. Top-down configuration involves configuring the network to the application's needs, which may involve node clustering and re-clustering. In general, the device

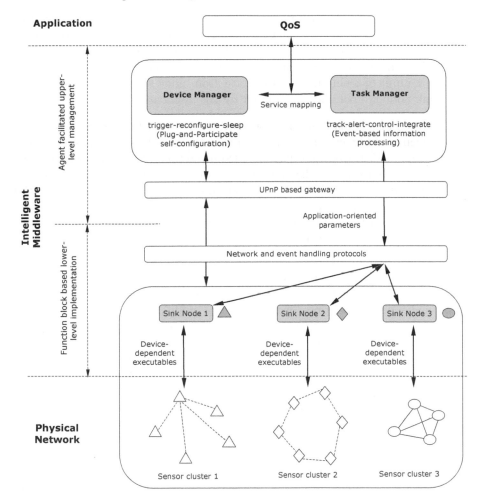

FIGURE 15.1
Distributed intelligent sensing and control system architecture.

manager stores information about the current status of the network connection, the node descriptions, and the running status of each node. More importantly, the device manager matches network connections to application needs.

2. Task manager is concerned with how to satisfy application requirements. For example, typical application tasks include moving object tracking, status alert, soft real-time control, and the integration of all these tasks. To obtain necessary information for each application task from the device level, collaboration with the device manager is the key.

3. Sink nodes function as data aggregators. When a sensing or control task is very complex, a large number of raw data are collected, and embedded data-processing algorithms need more efficient computation; a sink node will be deployed to separate data processing from basic data acquisition. The input of a sink node is raw data from node sensors; the outputs of a sink node are parameters required by the application.

4. Bidirectional events information flow inside our intelligent middleware is driven by bidirectional events, which are denoted as bidirectional arrows in Figure 15.1. Control events flow from the upper to the lower level, while sensing events flow in the opposite direction. This results in an event-based middleware where bidirectional events make taking upper level application needs as inputs for lower level efficient resource usage possible. They also form a path from upper level to lower level for execution control, not just a path from lower level to upper level for sensing.

As indicated in Figure 15.1, two enabling technologies for distributed application are adopted for our intelligent middleware. They are intelligent agents and IEC 61499 function blocks. At upper level, intelligent agents facilitate the management functions of the proposed middleware through their negotiation and decision-making capabilities. The desired management functions include dynamic device configuration, application task decomposition, and service mapping between device manager and application manager. At the lower level, function blocks will be used as modelling tools for event-based information processing, node intelligence embedding, and device-level execution control. Different from software agents, IEC 61499 function blocks are event-driven and can be internal algorithm-embedded. They are designed for complex low-level device execution and control. More details on the overall DISCS architecture can be found in [4] and [5]; the upper level multi-agent system implementation is described in detail in [7] and [8]. In the next section, we focus on the lower level, function block implementation of our proposed DISCS.

15.4 Function Block Implementation

IEC 61499 function blocks are modularized event-driven functional units that are executable, scalable, and distributable. As a result, IEC 61499 is well suited as a modelling tool for our proposed DISCS given its ability to achieve:

1. Event-driven execution

2. Diverse firmware encapsulation

3. Direct access to hardware

Low-level middleware requires explicit and efficient execution given its close proximity to physical devices. Since the *de facto* operating system for embedded WSNs uses event-driven execution [15], the event-based IEC 61499 function block model is a logical choice for this application. For example in our model, the upper level bidirectional events connected to task manager and device manager are inputs and outputs of a function block network (FBN) at application side, and triggers and I/Os of hardware are inputs and outputs of the FBN at physical device side. As defined in IEC 61499, execution of the FBN is managed by execution control chart inside each function block, which is associated with corresponding data inputs and data outputs.

Industrial distributed sensing and control applications typically involve a combination of diverse types of hardware and their associated firmware. To hide the complexity and heterogeneity of lower level hardware systems, different function block models and types can be employed to encapsulate different firmware systems into different level function block components, each with unified interfaces and clear inputs and outputs. To accommodate different levels of intelligence that various devices and firmware possess, necessary embedded intelligence can also be developed using function blocks to leverage these differences. Even a complete function block-based middleware with direct hardware access is feasible.

Finally, given the heterogeneity of the low-level hardware systems, direct access to hardware when required is one of the most important features of a middleware. IEC 61499 is based on the widely adopted standard IEC 61131-3 [29], which is a direct device-level control tool. As a result, it already defines the architectural elements necessary for hardware access. There is considerable work on function block technology in an execution environment for the low-level real-time control [3, 26, 36].

Figure 15.2 shows the subsystem architecture of our function block-based lower level implementation. The core of this implementation is the mapping of upper level application tasks and network requirements to lower level function-block applications through bidirectional events. The ultimate goal of this subsystem is to support upper level optimization and lower level dynamic network configuration. Considering power drain from increased internode communication, most function block applications (see FBA 1 and FBA 3 in Figure 15.2) will be mapped to individual sensor nodes, such as condition monitoring and individual machine execution control.

15.5 Example: Mobile Object Tracking

To illustrate the IEC 61499 model's ability to map application tasks across multiple nodes (*i.e.*, see FBA 2 in Figure 15.2), we provide an example of mobile object tracking in this section. Moving object tracking is a typical

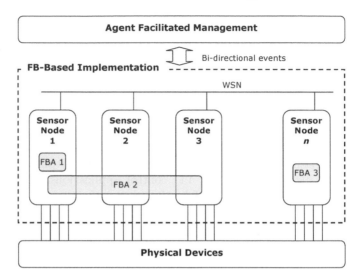

FIGURE 15.2
Function Block based subsystem architecture.

mobile sensor localization problem, in which several static nodes are used as anchors. The anchor nodes work together with a moving node equipped for this object tracking. The communication at different levels of function-block applications is achieved by events and their associated data through network protocols.

For this application, we identify three types of wireless sensor nodes and five software agents as illustrated in Figure 15.3. The basic agent model is built on the mediator architecture [19], and relies on sink nodes as the management and localization nodes given their higher processing capability. Mobile nodes and anchor nodes are represented by mobile node agents and anchor node agents, respectively, and for the purposes of simulating the factory environment, the model includes an environment agent.

For simplicity, only one instance of each of the node types is depicted in Figure 15.3. However, in the factory environment, various anchor nodes are dispersed across the geographic area of the factory floor, while one or more mobile nodes move through the environment along established pathways between equipment. Sink nodes manage the mobile node tracking process by enlisting nearby anchor nodes that receive distance estimates to mobile nodes and by employing locator agents to triangulate the distance data to determine mobile node locations.

Figure 15.4 displays how the resulting wireless *ad hoc* network tracks a mobile node by forming, re-forming and dissolving *ad hoc* clusters on the fly as the mobile node traverses its path. The IEC 61499 FBA is shown in Figure 15.5. Like the agent class diagram shown in Figure 15.3, we only show

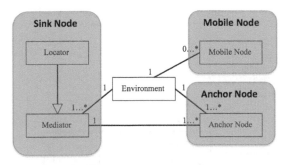

FIGURE 15.3
Mobile object tracking agent class diagram.

one instance of the nodes. To simplify the figure, we do not show the mobile node, which involves publishing beacon signals used by the anchor node to estimate distance using time difference of arrival (TDoA).

An example of moving object tracking with our DISCS simulation is shown in Figure 15.6. In this example, mobile object tracking is managed by FBAs distributed across multiple sensor nodes (*i.e.*, sink nodes and anchor node clusters in Figure 15.6). The group of sensor nodes that share a mobile object tracking task form a wireless sensor node cluster in our model. In order to balance resource load, we pass the tracking task from one cluster to another as the mobile object traverses the sensor field. More details on this approach can be found in [7] and [8].

Since function blocks are modular design tools, our lower level network implementation can utilize a library of predefined function block templates. For a specific application, these templates can be quickly customised according to task requirements and node configurations and built into application-specific FBNs. By encapsulating node intelligence and node configurations into FBNs, lower level implementation bridges the gap between high-level application requirements and low-level hardware devices. Through bidirectional events, agent-facilitated decision making and function block-based event-driven implementation work together, which makes the whole middleware for DISCS intelligent and adaptive from application requirements to network constraints and *vice versa*.

15.6 Future Work

In this chapter, we described a distributed intelligent sensing and control systems (DISCS) approach intended to enable manufacturing systems to quickly respond to change while maintaining stable operation and efficient use of re-

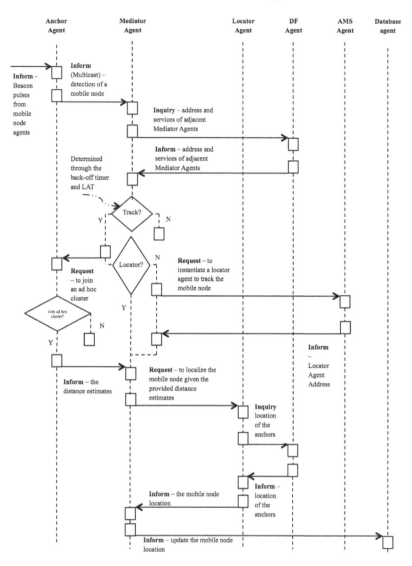

FIGURE 15.4
Mobile object tracking agent interaction diagram.

sources. The focus of this work has been on the need for a distributed sensing and control approach based on embedded WSN technologies. Although research and development on general wireless sensors network protocols has grown significantly in recent years and the technology has become more and more mature, adoption of the technology in industry is much slower relative to the need. Middleware is one of the key reasons. Effective middleware should consider application domain knowledge and network resource constraints, as

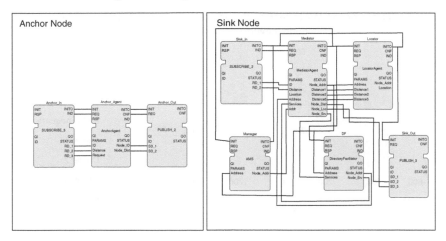

FIGURE 15.5
Function block application for mobile object tracking.

well as building complex application knowledge into networks with proper tools.

An application-oriented intelligent middleware approach is proposed in this chapter to address these issues that comprises a two-level structure: agent facilitated upper level management and function block-based lower level implementation. The former is responsible for optimal mapping between application task and network topology by considering both efficient usage of network resources and QoS of the application. The latter focuses on event-driven device-level execution, and function blocks are adopted as distributed modelling tools.

Our recent work in this area has focused primarily on the overall DISCS architecture and the upper level on the software agent-based management layer. Future work will focus on further refinement of the lower level, function block-based implementation layer. More specifically, we plan to extend our work on dynamic reconfiguration to tackle automatic reconfiguration. We anticipate that the layered DISCS architecture will provide a strong basis for this work given its focus on low-level timing and synchronization and the higher-reasoning capabilities of the upper, software agent layers. A systems approach will be used to solve the problem of automatic reconfiguration: *i.e.*, the nature and severity of the disturbance will determine how and where the DISCS responds. For example, unrecoverable faults will be handled by low-level middleware layer function blocks and services through the use of fail-safe modes; when faults can be managed by swapping hardware and/or software with exact copies (*i.e.*, homogeneous redundancy) or equivalent copies (*i.e.*, diverse redundancy), upper level agents will be involved in developing contingency plans that will be executed by lower level function block applications.

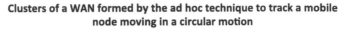

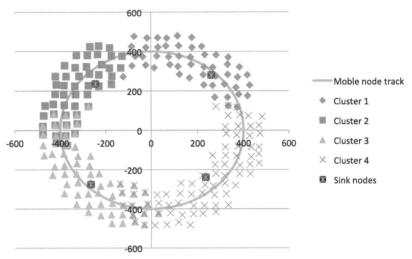

FIGURE 15.6

Mobile object tracking using the DISCS simulation.

Bibliography

[1] G. Anastasi, E. Bini, A. Romano, and G. Lipari. A service-oriented architecture for qos configuration and management of wireless sensor networks. In *Proceeding of IEEE Conference on Emerging Technologies for Factory Automation*, 2010.

[2] R. Brennan, M. Fletcher, and D. Norrie. An agent-based approach to reconfiguration of real-time distributed control systems. *IEEE Transactions on Robotics and Automation*, Vol. 18, No. 4, 444–451, 2002.

[3] R. Brennan, P. Vrba, P. Tichy, A. Zoitl, C. Sünder, T. Strasser, and V. Marik. Developments in dynamic and intelligent reconfiguration of industrial automation. *Computers in Industry*, Vol. 59, , 533–547, 2008.

[4] N. Cai and R. Brennan. Distributed sensing and control architecture for automotive factory automation. In *Holonic and Multi-Agent Systems for Manufacturing*, volume 5696 of Lecture Notes in Computer Science, pages 165–174. Springer, 2009.

[5] N. Cai, M. Gholami, L. Yang, and R. Brennan. Application-oriented intelligent middleware for distributed sensing and control. *IEEE Transactions*

on *Systems, Man, and Cybernetics–Part C: Applications and Reviews*, Vol. 42, No. 6, 947–956, 2012.

[6] Y. Ding, E. Elsayed, S. Kumara, J. Lu, F. Niu, and J. Shi. Distributed sensing for quality and productivity improvements. *IEEE Transactions on Automation Science and Engineering*, Vol. 3, No. 4, 344–359, 2006.

[7] M. Gholami and R. Brennan. Dynamic cluster formation and management in a factory wireless sensor network. In *Proceedings of IEEE International Conference on Systems, Man, and Cybernetics*, 2013.

[8] M. Gholami, M. Taboun, and R. Brennan. Comparing alternative cluster management approaches for mobile node tracking in a factory wireless sensor network. In *Proceedings of IEEE International Conference on Systems, Man, and Cybernetics*, 2014.

[9] M. Guler, S. Clements, L. Wills, B. Heck, and G. Vachtsevanos. Transition management for reconfigurable hybrid control systems. *IEEE Control Systems Magazine*, Vol. 23, , 36–49, 2003.

[10] V. Gungor and G. Hancke. Industrial wireless sensor networks: Challenges, design principles, and technical approaches. *IEEE Transactions on Industrial Electronics*, Vol. 56, No. 10, 4258–4265, 2009.

[11] B. Heinzelman, A. Murphy, H. Carvalho, and M. Perillo. Middleware to support sensor network applications. *IEEE Networks*, Vol. 18, No. 1, 6–14, 2004.

[12] J. Kramer and J. Magee. Dynamic configuration for distributed systems. *IEEE Transactions on Software Engineering*, Vol. 11, , 424–436, 1985.

[13] S. Lee, T. Jeon, H.-S. Hwang, and C.-S. Kim. Design and implementation of wireless sensor based-monitoring system for smart factory. In *Lecture Notes in Computer Science*, volume 4706, pages 584–592, 2007.

[14] N. Leveson. *Engineering a Safer World: Systems Thinking Applied to Safety*. MIT Press, 2011.

[15] P. Levis, S. Madden, J. Polastre, R. Szewczyk, K. Whitehouse, A. Woo, D. Gay, J. Hill, M. Welsh, E. Brewer, and D. Culler. Tinyos: an operating system for sensor networks. In E. A. Werner Weber, Jan M. Rabaey, editor, *Ambient Intelligence*, pages 115–148, 2006.

[16] T. Liu and M. Martonosi. Impala: A middleware system for managing autonomic parallel sensor systems. In *Proceedings of 9th ACM SIGPLAN Symposium on Principles Practice Parallel Program*, pages 107–118, 2003.

[17] M. Marin-Perianu, N. Meratnia, P. Havinga, L. DeSouza, J. Muller, P. Spiess, S. Haller, T. Riedel, C. Decker, and G. Stromberg. Decentralized enterprise systems: A multiplatform wireless sensor network approach. *IEEE Wireless Communications*, Vol. 14, No. 6, 57–66, 2007.

[18] D. Martinez, A. Gonzalez, F. Blanes, R. Aquino, J. Simo, and A. Crespo. Formal specification and design techniques for wireless sensor and actuator networks. *Sensors*, Vol. 11, No. 1, 1059–1077, 2011.

[19] F. Maturana and D. Norrie. Multi-agent mediator architecture for distributed manufacturing. *Journal of Intelligent Manufacturing*, Vol. 7, No. 4, 257–270, 1996.

[20] F. Maturana, R. Staron, and K. Hall. Methodologies and tools for intelligent agents in distributed control. *IEEE Intelligent Systems*, Vol. 20, , 42–49, 2005.

[21] H. Ramamurthy, B. Prabhu, R. Gadh, and A. Madni. Wireless industrial monitoring and control using a smart sensor platform. *IEEE Sensors Journal*, Vol. 7, No. 5, 611–618, 2007.

[22] K. Romer, O. Kasten, and F. Mattern. Middleware challenges for wireless sensor networks. *ACM SIGMOBILE Mobile Computing and Communications Review*, Vol. 6, No. 4, 59–61, 2002.

[23] M. Rooker, C. Sünder, T. Strasser, A. Zoitl, O. Hummer, and G. Ebenhofer. Zero downtime reconfiguration of distributed automation systems: The εcedac approach. In *Proceedings of 3rd International Conference on Industrial Applications of Holonic and Multi-Agent Systems*, pages 326–337, 2007.

[24] R. Setchi and N. Lagos. Reconfigurability and reconfigurable manufacturing systems: state-of-the-art review. In *Proceedings of 2nd International Conference on Industrial Informatics*, 2004.

[25] T. Strasser and R. Froschauer. Autonomous application recovery in distributed intelligent automation and control systems. *IEEE Transactions on Systems, Man, and Cybernetics, Part C: Applications and Reviews*, Vol. 42, No. 6, 1054–1070, 2012.

[26] T. Strasser, A. Zoitl, J. Christensen, and C. Sünder. Design and execution issues in iec 61499 distributed automation and control systems. *IEEE Transactions on Systems, Man, and Cybernetics, Part C: Applications and Reviews*, Vol. 39, No. 5, 534–546, 2009.

[27] TC 65/SC 65B. IEC 61131: Programmable controllers – Part 3: Programming languages. International Electrotechnical Commission, Geneva, 1993.

[28] TC 65/SC 65B. IEC 61499: Function blocks – Part 1: Architecture. International Electrotechnical Commission, Geneva, 2005.

[29] TC 65/SC 65B. IEC 61131: Programmable controllers – Part 3: Programming languages. International Electrotechnical Commission, Geneva, 2nd ed., 2003.

[30] V. Vyatkin. *IEC 61499 Function Blocks for Embedded and Distributed Control Systems Design.* Instrumentation, Systems, and Automation Society, 2007.

[31] V. Vyatkin. IEC 61499 as enabler of distributed and intelligent automation: state-of-the-art review. *IEEE Transactions on Industrial Informatics*, Vol. 7, No. 4, 768–771, 2011.

[32] J. Walsh, F. Bordeleau, and B. Selic. Domain analysis of dynamic system reconfiguration. *Software and Systems Modeling*, Vol. 6, No. 4, 355–380, 2007.

[33] W. Wang, J. Sung, and D. Kim. Complex event processing in EPC sensor network middleware for both RFID and WSN. In *Proceedings of 11th IEEE International Symposium on Object Oriented Real-Time Distributed Computing*, pages 165–169, 2008.

[34] G. Weiss. *Multi-Agent Systems.* MIT Press, 1999.

[35] D. Wu and M. Wong. Fast and simultaneous data aggregation over multiple regions in wireless sensor networks. *IEEE Transactions on Systems, Man, and Cybernetics, Part C: Applications and Reviews*, Vol. 41, No. 3, 333–343, 2011.

[36] A. Zoitl. *Basic real-time reconfiguration services for zero down-time automation systems.* PhD thesis, Vienna University of Technology, 2007.

16

Model-Driven Design of Cardiac Pacemaker Using IEC 61499 Function Blocks

Yu Zhao

University of Auckland

Partha S. Roop

University of Auckland

CONTENTS

16.1 Introduction

The use of cardiac rhythm-management devices such as implantable defibrillators and pacemakers has rapidly increased over recent decades. To improve the quality of performance, the complexity of such devices has grown. However, safety issues caused by the software still challenge development. Based on the database of the U.S. Food and Drug Administration (FDA) in 2010, at least 6 of the 23 recalls of defective devices were related to software errors [13].

Response time analysis [15] is crucial for cardiac pacemaker software. Existing techniques use timed automata (TA)-based closed-loop models [3] for conducting such analysis. However, these models make idealised assumptions regarding the underlying execution platform, which could affect verification fidelity. In this chapter we propose a model-driven design approach starting with modeling the software using IEC 61499 function blocks. We perform low-level timing analysis to compute the worst case execution time (WCET) [20] of the generated C code from these high-level models. The obtained WCET values are used for the creation of high-fidelity TA models needed for response time analysis. Earlier response time analysis frameworks [11, 13] perform analysis without considering the low-level timing of the underlying architecture. Unlike these, the proposed approach enables high-fidelity response time analysis of implantable devices such as pacemakers.

16.2   Pacing System in a Nutshell

For patients who suffer from *bradycardia* (lower than normal heart rate), cardiac pacemakers are implanted to regulate the heart beat. A pacemaker senses the intrinsic pulses from the heart and decides whether to stimulate the heart while obeying the underlying timing sequences. A pacemaker should not accelerate pacing for patients with *tachycardia* (higher than normal heart rate). Figure 16.1 depicts a closed-loop system consisting of the heart and a pace-

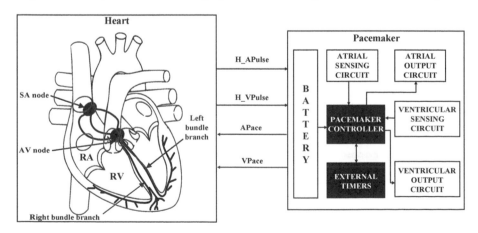

FIGURE 16.1
Closed-loop system of heart and pacemaker.

maker. We term this a pacemaker-aided system. Four interactions are observable:

- **H_APulse** denotes an atrial pulse generated by the heart to be sensed by the pacemaker.

- **APace** denotes the atrial stimulus generated by the pacemaker to regulate the heart rate.

- **H_VPulse** denotes a ventricular pulse generated by the heart to be sensed by the pacemaker.

- **VPace** denotes the ventricular stimulus generated by the pacemaker to regulate the heart rate.

H_APulse and **APace** are atrial events, while **H_VPulse** and **VPace** are ventricular events.

In this chapter, we focus on designing the pacemaker controller by the model-driven approach, and also provide models of the heart and timers. This section starts with an introduction to the models of the electrical system in the heart, followed by a detailed presentation of the DDD mode pacemaker. Finally, the section summarizes related work.

16.2.1 Electrophysiology of Heart and Its Modeling

Figure 16.2 shows the structure of the electrical system in the heart, and the electrocardiogram (ECG) signal created by the electrical conduction in this system. In the electrical system, all the pulses start at the SinoAtrial (SA)

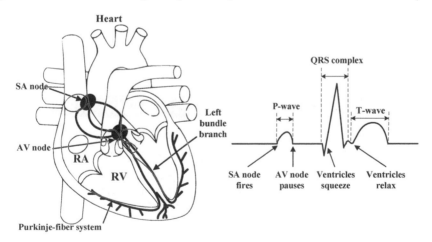

FIGURE 16.2
Heart electrical system and observed ECG.

node in the right atrium and are spontaneously and periodically generated electrical signals. Such signals propagate along the conduction pathway until they reach the `AtrioVentricular` (AV) node. This process leads to the contraction of the right atrium, which squeezes the blood to the right ventricle. The left atrium performs an identical process nearly simultaneously. In the ECG, a P-wave signifies such a conduction process. The role of the AV node is for synchrony between atrium and ventricle, where the propagation of electrical signals will be deferred. Physiologically, during the deferred time, blood squeezed from the atrium flows entirely into the right ventricle. Following the delay, signals traverse through the Purkinje fiber system, which renders the contraction of the ventricles. In the ECG, this is represented by the QRS complex. The following T wave implies relaxation of the ventricular tissue, which is known as the refractory period. A regularly beating heart usually maintains a rate from 60 beats per minute (bpm) to 100 bpm. On the ECG, a fixed sequence of the P-wave, QRS complex and T-wave will be periodically observed.

Jiang et al. and Chen et al. [7, 12] present models of the electrical system, where [12] is grounded on timed automata, and hybrid automata is used [7]. The timed automata models of the heart capture the timing behaviour of the electrical system and close the loop with the pacemaker models. Then, temporal properties are verified based on the models of such a closed-loop system. As suggested by [12], different levels of abstraction could be achieved for the heart model based on the requirements of reality. This, however, could require a long time for verification if more realistic heart models are created. The work presented in [7] also includes the studies of electrical behaviour of the ECG signal. By using hybrid I/O automata, the heart model could detect whether the accumulated electrical voltage is able to trigger an action.

Figure 16.3 presents a simple heart model consisting of the right atrium (RA) model and the right ventricle (RV) model using timed automata [3]. Timed automata facilitates a set of real-valued clocks (e.g., `Clock1` and `Clock2`). All the clock values increase at the same speed, and may be com-

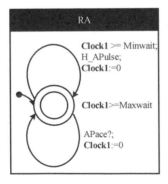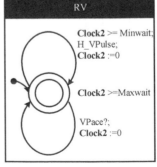

FIGURE 16.3
An example of a heart model using timed automata [13].

pared to integers (e.g., `Maxwait` and `Minwait`). Since no synchronisations are considered, these models randomly produce `H_APulse` and `H_VPulse`, respectively, to the pacemaker, within lower time bound `Minwait` and upper time bound `Maxwait`. When the clock value is greater than or equal to the lower time bound, a transition is enabled, during which a pulse (i.e., `H_APulse` or `H_VPulse`) is generated for the pacemaker and the clock value is reset. Moreover, the model is also able to synchronise with the pacemaker. `APace` or `VPace`, produced by the pacemaker, could trigger a transition and meanwhile reset the clock value. Since only one state exists in the RA and RV models, all the transitions lead to this state. Next, the clock value starts to increase instantaneously.

16.2.2   Timing Requirements and Diagram of Pacemaker

Modern pacemaker manufacturers have implemented various operating modes in these devices. In this chapter, we chose one of the commonly used modes, called the DDD [5]. Such a mode is capable of sensing and pacing both the atrium and ventricle. Also, to prevent pacing the ventricle too rapidly, the DDD mode facilitates an upper rate bound and a lower rate bound. Once the ventricular rate reaches the upper bound, stimuli stop, while stimuli are produced if the rate does not reach the lower bound. Moreover, a pacemaker represents a hard real-time and safety-critical device, in which timing requirements must be satisfied. Such requirements in the DDD mode pacemaker are elaborated below. We use an example trace to explain the different timing constraints related to the DDD mode pacemaker.

- AEI: `Atrial Escape Interval`. This describes the maximum time interval between a ventricular event and the corresponding atrial event.

- AVI: `Atrial Ventricular Interval`. This describes the maximum time interval between an atrial event and the corresponding ventricular event.

- VRP: `Ventricular Refractory Period`. This defines the resting period for the ventricle after each ventricular event. Physiologically, ventricular events are not detectable by a pacemaker during a resting period.

- PVARP: `Post Ventricular Atrial Refractory Period`. This defines the resting period for the atrium after each ventricular event. Physiologically, atrial events are not detectable by a pacemaker during a resting period.

- URI: `Upper Rate interval`. This indicates the minimum time interval between two consecutive ventricular events. Such an interval is used in dual chamber modes (such as the DDD mode) pacemaker, and it serves as the upper bound for the heart rate.

- LRI: `Lower Rate Interval`. This indicates the maximum time interval between two consecutive ventricular events. In dual chamber mode, the LRI time interval equals the summation of AVI and AEI. It serves as the lower bound for the heart rate, and hence the value of LRI is greater than URI.

Additionally, we deploy three blocking periods in the DDD mode based on the specification in [2]. Such periods prevent cross-talk during sensing when each ventricular event or APace occurs. They are usually shorter than the refractory periods.

- PVPB: Post Ventricular Pace Blocking. This signifies the blocking period after each generated VPace, during which H_APulse will be considered an invalid event.

- PVSB: Post Ventricular Sense Blocking. This signifies the blocking period after each ventricular event, during which H_APulse will be treated as an invalid event.

- PAPB: Post Atrial Pace Blocking. This indicates the blocking period after each APace, during which H_VPulse will be considered an invalid event.

Five ECG waveforms representing the electrical conduction of the heart are shown as examples in Figure 16.4. Waveform ① indicates a normal heart behaviour. We could observe that the related AVI timers stop before expiry, which shows a normal synchrony between an atrial event and the following ventricular event. Likewise, the behaviour of AEI shows a normal synchrony between ventricular event and the coming atrial event.

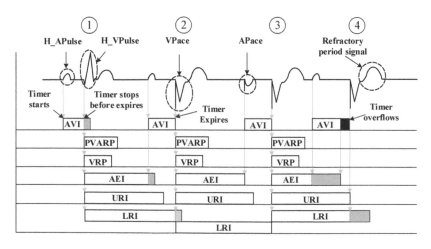

FIGURE 16.4
Timing diagram of DDD mode pacemaker. Adapted from [5].

Waveform ② presents the situation in which an artificial pulse (VPace) will be generated by the pacemaker if an intrinsic ventricular pulse is not sensed within the AVI interval. Also, the LRI requirement is not violated.

Waveform ③ features a situation where both AEI and AVI are expired. Consequently, to achieve the LRI requirement, both APace and VPace are produced.

Waveform ④ presents an atrial event that is earlier than usual. It may cause a temporary fast ventricular rate if there is no upper limitation. The existence of URI defers the generation of VPace by extending the *AVI* interval to the length of the *URI* interval.

16.2.3   Related Work

Electrical heart models are of two types. There are many computational models of the heart [8, 10, 18, 22], functioning as "virtual organs" [19] that replicate heart behaviours at scales from the cell to the organ. These high-fidelity models are used for computer simulation, but are unsuitable for mathematical verification (model checking) [4] or real-time simulation of pacemakers. In contrast, abstract models of cardiac electrical behaviour have been developed using timed automata [3] and hybrid automata [9]. These models have lower computational complexity and hence are ideally suited to closed-loop, real-time simulations, and analysis for correctness of pacemaker design. Hybrid automata-based models have been primarily used for testing-based validation [7]. We are interested in timed automata models as they can be used for formal response time analysis.

In [11, 16] two approaches for response time verification of the pacemakers have been developed using an abstracted model of the heart with timed automata. This approach makes an idealised assumption regarding the time automata transitions that are considered instantaneous (i.e., take zero time). In order to overcome this limitation, we use the model-driven approach. We combine information from low-level BCET and WCET [20] analysis to precisely compute the delay bounds on transitions. The results are then used to create timed automata models for response time analysis.

16.3   Overview of Proposed Approach

As shown in Figure 16.5, the proposed approach starts with model creation using IEC 61499 function blocks (Step ①). The execution of the function block models is based on synchronous semantics [21]. For simplicity, we term them as synchronous function blocks. Using FBC [21], we compile each function block into its own C function. Step ② statically analyzes the execution of the generated code and provides BCET and WCET for the following step. In Step

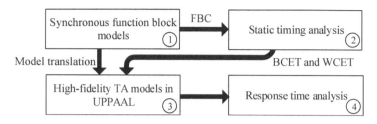

FIGURE 16.5
Overview of proposed approach.

③, to overcome the unrealistic assumption that transitions consume no time, delays are inserted for certain transitions. We designate such models with delayed transitions as high-fidelity models. The approach ends with Step ④ in which the response time analysis is conducted.

Synchronous function blocks execute in a lock-step fashion, using a global logical clock that ticks. During a tick, all the input events are sampled once, followed by a reaction of each ECC in the function block network based on the input events. The tick ends with an emission of all the output events after the execution of the ECCs. A reaction of an ECC could be: (1) the ECC stays at the same state if no transition occurs; (2) the ECC proceeds to the next state if an input event triggers the transition. During a reaction, the algorithm associated with the state is executed, and the output event is emitted at the end of the tick. All ECCs in a network execute in a logically paralleled fashion as all producer-consumer dependencies rely on a unit delay.

PTARM [14] is chosen as the underlying platform of the pacemaker controller. The precision timed ARM (PTARM) architecture facilitates four threaded-interleaved pipelines and a local scratch-pad memory. Figure 16.6 features the time progression of a threaded-interleaved pipeline from t1 to t11. The following situation demonstrates that no control and data dependen-

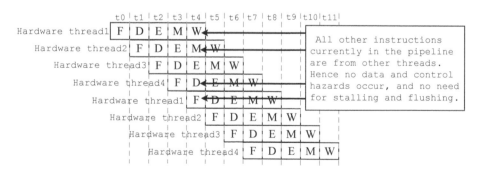

FIGURE 16.6
Task allocation on the PTARM.

cies occur in the PTARM. Time step t4 shows that all other stages are filled with instructions from other threads when the instruction from the hardware thread 3 is in the execution stage (E). Therefore, there are no control and data dependencies between instructions, hence eliminating control and data hazards and facilitating simplicity for timing analysis.

Timed automata create high-fidelity models of the pacemaker controller. UPPAAL [6] is the tool that supports the TA framework, which also facilitates useful functions. We use the committed locations to model the states in which time is not allowed to elapse and the output event is produced instantaneously. Also, the broadcast channels are used for synchronisations among timed automata in a TA network.

16.4 Modeling Using IEC 61499 Function Blocks

Based on the timing requirements (see Section 16.2.2) and the diagram of the DDD mode pacemaker, we created IEC 61499 function block models for the pacemaker controller. It starts with a detailed presentation of basic function blocks (BFBs) for each timing requirement, followed by a description of the network of basic function blocks. Such a network is encapsulated in a composite function block (called *DDD_Mode_PacingLogic*). The hierarchical layout of the models is presented at the end. The layout includes a network that integrates a ventricular pulse detector, an atrial pulse detector, an event log function and the DDD mode pacing logic. With the encapsulation of the network, a composite function block is created as the pacemaker controller model, and is ready to be exposed to the environment consisting of the heart and the external timers (see Figure 16.1).

16.4.1 DDD Mode Pacing Logic Model

Seven basic function blocks are created to model the aforementioned timing requirements, each of which has a Logic suffix. We illustrate how different basic function blocks are used to capture the timing requirements in Figure 16.7. For example, AVI timing is modeled by the AVI_Logic basic function block. Such basic function blocks could interact with the external timers by sending stop or start signals or reading expiry signals. To maintain the timing requirements, seven timers are created and are allocated the values of time intervals of each timing requirement.

The URI_Logic is shown in Figure 16.8. This function block communicates with the rest of the function blocks in the network. The state machine starts with the initial state called S0, and proceeds to the S1 state when the VSense or VPace event occurs. During this state, an algorithm (namely aSetValue) sets the countdown value for an external timer corresponding to the URI in-

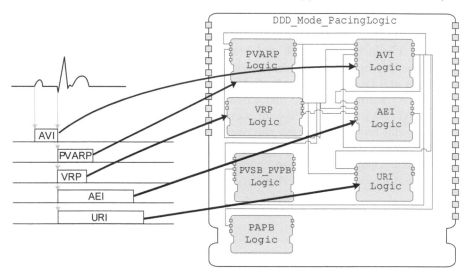

FIGURE 16.7
An overview of the timing diagram for function block models.

terval, followed by an output event (URI_TimerStart) that triggers the timer. Next, the state machine progresses to the S2 state due to the **true** statement. While in this state, the state machine can move to the S3 state or move back to the initial state based on VSense or URI_TimerExpired signals. When VSense is received, the URI timer is forced to stop using the URI_TimerStop output.

The AEI_Logic and AVI_Logic are shown in Figure 16.9. As an important timing requirement for the synchrony between the atrial pulses and the ventricular pulses, the AEI_Logic ECC receives ventricular events to progress from the initial state, and generate APace_Out when the time for the synchrony has expired. This is achieved by reading the AEI_TimerExpired signal. Also,

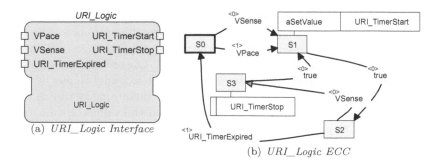

(a) *URI_Logic Interface*

(b) *URI_Logic ECC*

FIGURE 16.8
Basic function blocks for *URI_Logic*.

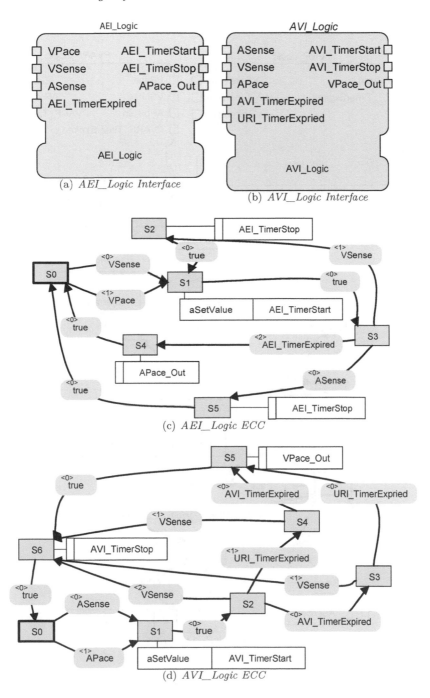

FIGURE 16.9
Basic function blocks for *AEI_Logic* and *AVI_Logic*.

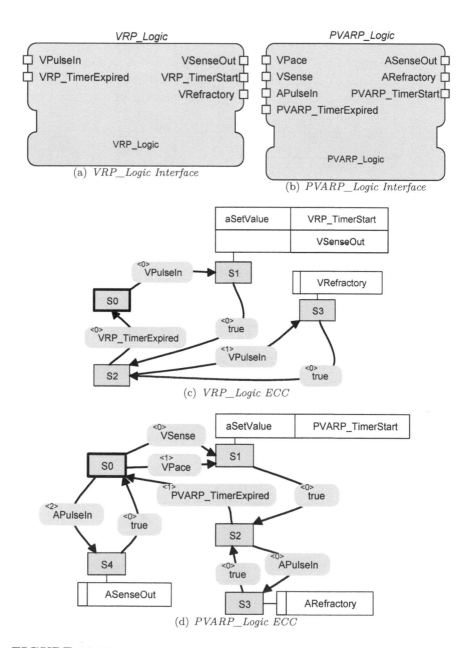

FIGURE 16.10
Basic function blocks for *VRP_Logic* and *PVARP_Logic*.

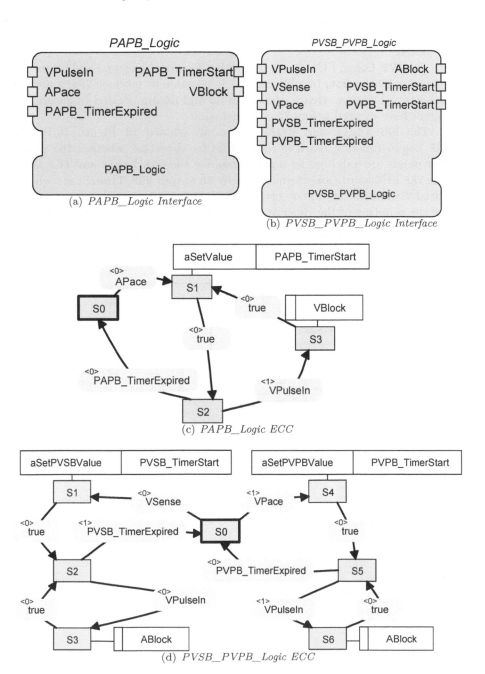

FIGURE 16.11
Basic function blocks for *PAPB_Logic* and *PVSB_PVPB_Logic*.

as shown in the timing diagram, the AEI timer may be stopped by ASense signals. This is interpreted in S1 state, where the AEI_TimerStop output is produced.

The AVI_Logic ECC starts by waiting for atrial triggers, and in response generates VPace_Out. In addition, AVI is not able to produce VPace until the URI interval expires. Hence, the states S3 and S4 are created to ensure both signals have expired before VPace generation.

The VRP_Logic and PVARP_Logic are shown in Figure 16.10. The VRP_Logic function block is responsible for detecting whether the ventricular pulses are valid as the input events for the URI_Logic and AEI_Logic. The VRP ECC starts an external timer by an output VRP_TimerStart when an initial VPulseIn signal is observed. All the VPulseIn signals that occur during the counting of the timer are considered invalid signals. They are termed

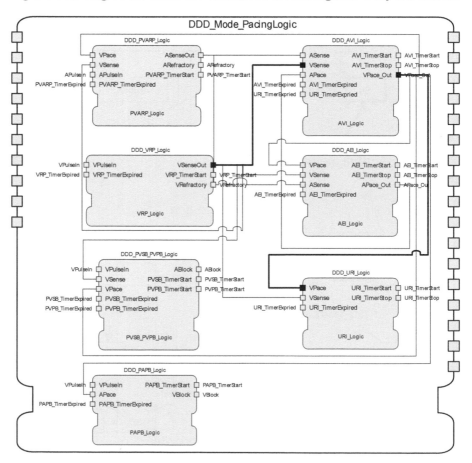

FIGURE 16.12
The network of basic function blocks of the timing requirements.

VRefractory signals. The VSenseOut output can only be generated after the S1 state is reached. The ECC of PVARP_Logic follows a similar process except that the initial triggers are VSense and VPace.

The PAPB_Logic and PVSB_Logic are shown in Figure 16.11. We integrate PVSB and PVPB timing requirements into one function block. The ECC of this function block starts with the S0 state, and proceeds to either the S1 state or the S4 state based on VSense or VPace signals. Once such a state is reached, the related external timer is started after the countdown value is assigned. Also, the ECC could record the blocked signal (i.e., ABlock) when VPulseIn occurs during the countdown periods. Likewise, The PAPB_Logic ECC treats VPulseIn signals that occur during the countdown periods as VBlock signals. The difference between these two function blocks is the initial trigger.

The network of basic function blocks is shown in Figure 16.12. After the creation of the basic function blocks, they are then composed to build a network. The network shows internal connections of these function blocks and external events consumed or produced from or to the outside. For instance, the detected ASense (i.e., ASenseOut) signal is generated by the PVARP_Logic function block and could only be observed by the connected function blocks such as AEI_Logic. In addition, the output signal could be observed within the network or produced to the outside of the network. For example, VPace (i.e., VPace_Out) is an input event for the URI_Logic ECC. It is also an external output to the ventricular pacing circuit (see Figure 16.1). The network is then encapsulated in a composite function block, namely DDD_Mode_PacingLogic for the creation of the pacemaker controller model.

16.4.2 Hierarchical Layout of Pacemaker Controller Model

We complete the pacemaker controller development by building a network of function blocks (presented in Figure 16.13). First, the VPulseDetector BFB is responsible for processing the ventricular pulses sensed by the ventricular sensing circuit (shown in Figure 16.1). If a valid ventricular event is detected, this function block produces an output to the input of the DDD_Mode_PacingLogic function block called VPulseIn. Likewise, the APulseDetector deals with the atrial pulses from the atrial sensing circuit, and is able to produce output events to the APulseIn interface of the DDD_Mode_PacingLogic. The DDD_Mode_PacingLogic CFB contains the logics of all the timing requirements. Finally, the Event_Log basic function blocks transfer all the recorded signals (including APace, VPace, ABlock, and ARefractory) to a physical device. This would help the caregivers to observe pacing conditions of the pacemaker-aided closed-loop system. Like the network in Figure 16.12, the network of the pacemaker controller contains external inputs and outputs. These inputs and outputs normally come from devices like the pacing circuits and the external timers.

The detailed ECCs of the VentricularPulseDetector, Event_Log and `AtrialPulseDetector` BFBs are shown in Figure 16.14. In the `AtrialPulseDetector` ECC, when an `H_APulseIn` is observed, the ECC progresses to the `S1` state, in which the signal processing algorithm (namely `aPulseDetection`) is initially executed. An output (i.e., `PulseDetectionOut`) indicates a valid atrial pulse has been detected. Moreover, such a detection process could be stopped once the blocking periods (`PVSB` and `PVPB`) are started by the `DDD_Mode_PacingLogic`. It waits for the expiry of such periods to process new incoming pulses. Analogously, the `VentricularPulseDetector` ECC is responsible for ventricular pulse detection, in which `PAPB` could disable the process. Finally, the `Event_Log` ECC simply reads the input events, and proceeds back to the initial state after generating the `TransOut`.

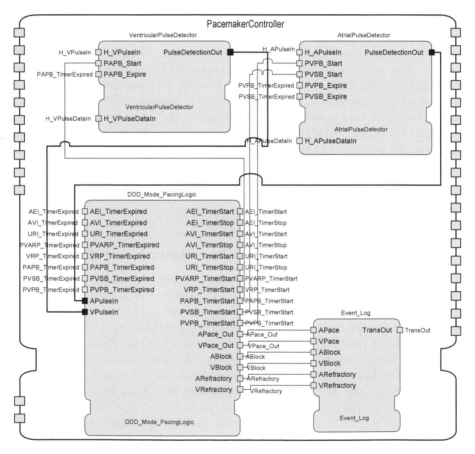

FIGURE 16.13
Function block network of pacemaker controller.

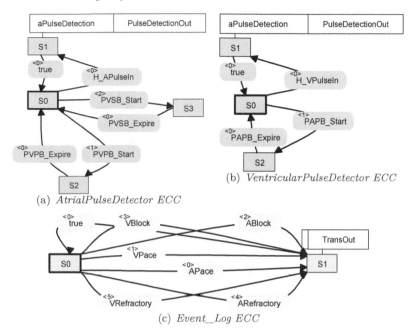

(a) *AtrialPulseDetector ECC*

(b) *VentricularPulseDetector ECC*

(c) *Event_Log ECC*

FIGURE 16.14
ECCs of *AtrialPulseDetector*, *VentricularPulseDetector* and *Event_Log* basic function blocks.

16.5 High-Fidelity Model Creation

UPPAAL provides a platform to conduct response time analysis by creating timed automata models. To take advantage of the TA framework, we design translation rules for converting the IEC 61499 function blocks to UPPAAL models. To remove the unrealistic assumption regarding instantaneous transitions (see Section 1.1.3) in UPPAAL models, we use the results from timing analysis of low-level code to ensure realistic values of transition timing. Consequently, high-fidelity TA models are created by inserting such bounds to transitions. On the basis of such high-fidelity models, the response time analysis is conducted to ensure high-assurance results. This section starts with an introduction to the static timing analysis of the generated code from function blocks. Then, we present the heart model and the external timer model using timed automata, which are plant models. Finally, the pacemaker controller model using TA is presented with model translation rules.

16.5.1   Timing Analysis of Synchronous Function Blocks

We manually allocate the function blocks to four hardware threads of the PTARM. The code generated from the `AtrialPulseDetector` function block is allocated to the hardware thread 1. The hardware thread 2 contains the code generated from the `VentricularPulseDetector`. The hardware thread 3 contains the `DDD_Mode_PacingLogic`, and the `Event_Log` code is implemented to the hardware thread 4. Ticks (see Section 16.3) may be illustrated using Figure 16.15 on the basis of the PTARM execution.

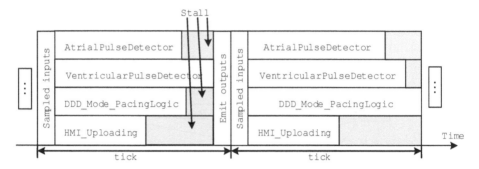

FIGURE 16.15
Execution of pacemaker controller based on PTARM.

Figure 16.16 presents an overview of the timing analysis approach. First, the pacemaker controller (implemented in the IEC 61499 function block) is compiled to executable C code by the function block compiler (see Section 16.3). Figure 16.17 presents the structure of the generated code. For each PTARM hardware thread a cyclic schedule is implemented to execute the function blocks allocated to that thread. For example, `AtrialPulseDetector_run` represents the C function for the `AtrialPulseDetector` function block, and this function will be implemented on hardware thread 1 of the PTARM.

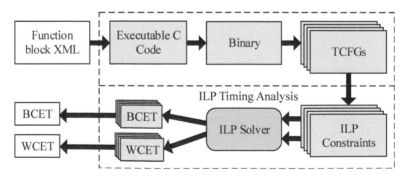

FIGURE 16.16
Static timing analysis overview.

```
for(;;) {
 [sampled inputs]

 AtrialPulseDetector_run();
 VentricularPulseDetector_run();
 DDD_Mode_PacingLogic_run();
 Event_Log_run();

 [emit outputs]
}
```

```
for(;;) {
 [sampled inputs]

 AEI_Logic_run();
 AVI_Logic_run();
 PAPB_Logic_run();
 PVARP_Logic_run();
 PVSB_PVPB_Logic_run();
 URI_Logic_run();
 VRP_Logic_run();

 [emit outputs]
}
```

FIGURE 16.17
Pacemaker controller code (left) and DDD mode pacing logic code (right).

Also, the DDD_Mode_PacingLogic consisting of seven basic function blocks is implemented on thread 3.

Next, the C code is compiled to binary using the ARM-GCC compiler. The binary is then converted to an intermediate format for timing analysis called the timed control flow graph (TCFG). The TCFG is a control flow graph with execution times annotated on each node, and is generated for each hardware thread. The process to generate the TCFG from binary is an adaptation of the approach presented in [17].

To conduct ILP-based timing analysis, we first define the BCET and WCET of the system, which are the maximum values of the BCET and WCET among all hardware threads. The BCET and WCET of a single hardware thread are, respectively, the shortest and longest times it takes to execute an iteration of the cyclic schedule (i.e., a tick). Therefore, the ILP objective function to determine the BCET and the WCET of one hardware thread is to find the minimum and maximum execution times, respectively. We then create objective formulas and constraints, and use an ILP solver such as Groubi [1] to solve the corresponding objective functions. At last, the outcomes are provided for the creation of high-fidelity models.

16.5.2 High-Fidelity Models of Pacemaker Controller

Due to synchronous executions, we derive two types of delays. First, the algorithm associated to states defers either the generation of output events, or the receiving of input events from the incoming tick. Second, due to the unit delay [21] in a function block network, a delay needs to be added to the internal transitions in a network. Figure 16.18 presents unit delays in the DDD_Mode_PacingLogic network. Both of the aforementioned types of transitions consume tick time. As previously shown using timing analysis, the BCET

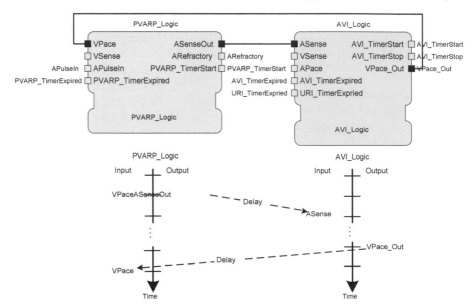

FIGURE 16.18
Unit delay in the DDD mode pacing logic network.

and WCET values may be calculated to bound the tick length. Therefore, the delays of the transitions could also be bounded to happen in the closed interval of [BCET, WCET]. After the consideration of the delays, we define some rules to translate function block models to TA models in UPPAAL. The rules are listed as below:

Rule 1. *Each ECC is translated into a timed automaton to form a network.* Since no encapsulation is allowed in the TA network, the encapsulations in the function block network no longer exist in UPPAAL models. For example, the DDD_Mode_PacingLogic encapsulates seven basic function blocks, and is not modeled in the TA network. Specifically, the TA network of the pacemaker controller is composed of ten timed automata due to ten ECCs in the function block models.

Rule 2. *Each state in an ECC is translated to a location of TA in UP-PAAL.* Specifically, each initial state is mapped to an initial location, and other states are mapped to normal locations. This is illustrated using #2 in Figure 16.19.

Rule 3. *Each input event and output event is translated to a broadcast channel.* Specifically, each input is mapped to *channel_name?*, indicating a consumer. Each output is mapped to *channel_name!*, indicating a producer. It is possible to model a synchronisation between one producer and multiple consumers using broadcast channels. This is illustrated using #3 in Figure 16.19.

Rule 4. *Each output event is generated from a committed location.* For ease

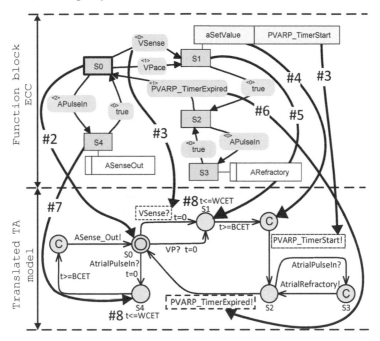

FIGURE 16.19
Model translation example of *PVARP_Logic* ECC.

of adding delays, a committed location is implemented before the generation of a related output event. Committed locations do not allow time to elapse. This is illustrated using #4 in Figure 16.19.

Rule 5. *Each state that is associated with an algorithm needs to insert a bounded delay time at the corresponding location in TA.* Specifically, to represent the bound [BCET, WCET] in UPPAAL models, we create a clock for such a state, using an invariant which is assigned to the value of WCET, and a transition which is triggered when the clock value is greater than BCET. We illustrated this rule using #5 in Figure 16.19.

Rule 6. *Delays are not inserted for each input event that is an interface with the environment.* Due to the synchrony hypothesis, the pacemaker controller is able to capture the events from the environment (i.e., the heart and external timers) whenever the environment produces such events. We illustrated this rule using #6 in Figure 16.19.

Rule 7. *Each state that generates an output event for internal communication needs to insert a bounded delay time.* This defines the place to add a unit delay. As with the fifth rule, a clock is created for the delay time. This is illustrated using #7 in Figure 16.19.

Rule 8. *Each timed automaton implements a maximum of one clock for bounding the delay time.* This simplifies the process that resets the clock, and may also shorten the verification time. Specifically, all clocks need to be reset before entering or leaving the state with a bounded delay. This rule is illustrated using #8 in the TA model of Figure 16.19.

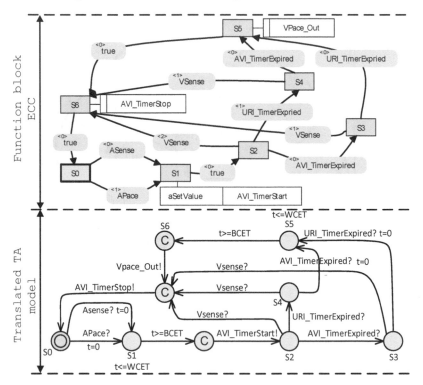

FIGURE 16.20
Translated model of *AVI_Logic* ECC.

We apply the aforementioned rules to the pacemaker controller models and illustrate the translated PVARP_Logic and AVI_Logic TA models using Figure 16.19. As defined by **Rule 8**, only one local clock t is defined. We also present the translated AVI_Logic ECC in Figure 16.20. The delay added in the location S1 is based on the algorithm associated with this state in the ECC. The delay value is not doubled because the unit delay is inserted by the producer PVARP_Logic timed automaton. This is observable in Figure 16.19, where **Rule 6** is applied.

16.5.3   Environment Model Creation for Closed-Loop System

For a closed-loop system, models of the environment need to be created. By setting an invariant and a clock guard for a clock in UPPAAL, an external timer model can generate an expiry signal (using broadcast channels) if the clock guard detects that the transition condition is achieved. Also, it could be

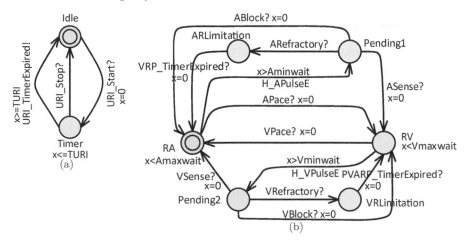

FIGURE 16.21

TA models for the heart (left) and the external timer (right).

started or stopped through signals from the pacemaker controller. An example of the external timer is presented in Figure 16.21(a). As for the heart model, the clocks may also be used. The heart can randomly generate signals using broadcast channels within a time bound, which could be realised by wisely setting invariants and clock guards.

To remove unrealistic traces of the model in Figure 16.3 while keeping the closed-loop system deadlock-free, we build connections between RA and RV and present the heart model in Figure 16.21(b). Specifically, APace? induces the transition from RA to RV, while VP? renders the transition in an opposite way. Since a local clock t is declared in this automaton, each transition resets the clock for the next random event generation. The benefit of such synchronisations is that the state space with respect to the heart model is reduced during the verification, which could render a more scalable modeling of the pacemaker. The same transitions are applied to the spontaneous pulses (i.e., H_VPulse and H_APulse). However, such spontaneous pulses should be blocked during the blocking periods. Hence, the transition to the RV location could not happen directly if the guard (i.e., x is greater than the Aminwait) suffices. Instead, a location (i.e., Pending2) is inserted to enforce a synchronisation with PVARP, since PVARP decides whether the spontaneous pulse is valid. If ABlock is present, the automaton will go back to the initial state (i.e., RA). In addition, the pulses that occur during the refractory period should also be captured in the heart model. To mimic such a behaviour, we first assume that only one pulse is allowed during the refractory period. This is shown as only one VR triggers the transition to the VRLimitation location. Multiple pulses could be set according to the needs.

16.6 Response Time Analysis Using High-Fidelity Models

We use the UPPAAL verifier to conduct the response time analysis. Before the analysis, we need to specify the values for all the timing requirements. We also need to build monitor models in TA for the related TCTL properties. Such values could be specified in the declarations of UPPAAL. Figure 16.22 features such declarations. Specifically, in the details of the timing requirements, the values serve as the hard deadlines for each timing requirement used in the properties.

```
// Place global declarations here.
/* BCET & WCET calculation */
const int clock_period = 200; // time unit: ns.
const int scalar = 1000; // 1 us = 1000 ns.
const int BCET_Cycle = 244; // unit: clock cycles.
const int WCET_Cycle = 6373;  // unit: clock cycles.
const int BCET = BCET_Cycle * clock_period / scalar; //unit: us.
const int WCET = WCET_Cycle * clock_period / scalar; //unit: us.
/* details of the timing requirements */
const int TAEI = 800000; // 800 ms = 800000 us.
const int TVRP = 150000; // 150 ms = 150000 us.
const int TAVI = 150000; // 150 ms = 150000 us.
const int TPVARP = 200000; // 200 ms = 200000 us.
const int TPVPB = 125000; //For atrial blocking,125ms=125000us.
const int TPVSB = 45000; //For atrial blocking,45ms=45000us.
const int TPAPB = 65000; //For ventrical blocking,65ms=65000us.
const int TURI = 900000; // URI should be greater than AEI and
 // less than LRI. 900 ms = 900000 us.
const int TLRI = TAEI + TAVI;
/* parameters for heart pulses */
const int Vmaxwait = 1200000; // 1.2s = 1200 ms = 1200000 us.
const int Vminwait = 1000; // 1 ms = 1000 us.
const int Amaxwait = 1200000; // 1.2s = 1200 ms = 1200000 us.
const int Aminwait = 1000; // 1 ms = 1000 us.
```

FIGURE 16.22
Declarations in UPPAAL.

To verify response time properties, monitor models must be built. Based on the method presented in [13], we create such models and present two examples using Figure 16.23. In 16.23(a), the monitor checks whether the PVARP interval has expired. By creating a clock in the automaton, it measures the time starting from a ventricular event to the instance when the expiry signal is received. Such a measured time is recorded in the `Ex_Get` location. The value is used to compare with the value declared. The comparison is conducted instantaneously, and hence a committed location is implemented. Likewise, the LRI monitor (shown in Figure 16.23(b)), uses a committed location `Detected` to record the interval between two consecutive ventricular events. It keeps monitoring such an interval since the `Wait2` location is always reached after the `Detected` location.

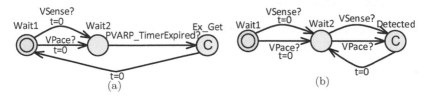

FIGURE 16.23
The TA monitors for PVARP (left) and LRI (right).

As a consequence, the related properties could be created as:

- `A[] (Monitor_PVARP.Ex_Get imply Monitor_PVARP.x>=TPVARP).`

- `A[] (Monitor_LRI.Detected imply Monitor_LRI.x<=TLRI).`

To check all times such properties must be satisfied, `A[]` is used. The word `imply` means that if the previous condition is satisfied, the following statement should also be satisfied. We extend such a method to all the related timing requirements, and create additional properties as below:

- `A[] not deadlock.`

- `A[] (Monitor_URI.interval imply Monitor_URI.x>=TURI).`

- `A[] (Monitor_VRP.Ex_Get imply Monitor_VRP.x>=TVRP).`

- `A[] (Monitor_PVPB.Ex_Get imply Monitor_PVPB.x>=TPVPB).`

- `A[] (Monitor_PVSB.Ex_Get imply Monitor_PVSB.x>=TPVSB).`

- `A[] (Monitor_PAPB.Ex_Get imply Monitor_PAPB.x>=TPAPB).`

There is no need to create properties for AVI and AEI, since LRI monitors a more significant interval which includes both the AVI and AEI intervals. The result of response time analysis shows the properties related to LRI, VRP, PVARP, and PVPB are **not** satisfied. Because delayed transitions are introduced in high-fidelity models, the specified declarations for the timing requirements are no longer able to serve the analysis. We set new deadlines for the declarations by adding tolerances [2] shown in Figure 16.24. We accordingly modify the properties by adding those tolerances. The new properties are as below:

- A[] not deadlock.

- A[](Monitor_LRI.Detected imply Monitor_LRI.x<=
  (TLRI+TLRI_tolerance)).

- A[](Monitor_VRP.Ex_Get imply Monitor_VRP.x>=
  (TVRP+TVRP_tolerance)).

- A[](Monitor_PVARP.Ex_Get imply Monitor_PVARP.x>=
  (TPVARP+TPVARP_tolerance)).

- A[](Monitor_PVPB.Ex_Get imply Monitor_PVPB.x>=
  (TPVPB+TPVPB_tolerance)).

- A[](Monitor_PAPB.Ex_Get imply Monitor_PAPB.x>=TPAPB).

- A[](Monitor_PVSB.Ex_Get imply Monitor_PVSB.x>=
  (TPVSB+TPVSB_tolerance)).

- A[](Monitor_PAPB.Ex_Get imply Monitor_PAPB.x>=
  (TPAPB+TPAPB_tolerance)).

The response time analysis is then conducted using the modified properties. The result shows that all of the properties are satisfied. By conducting the response time analysis on high-fidelity models, the approach benefits the developers in two ways. First, it helps a developer locate model problems by diagnostic traces. Second, it informs the developer that the generated code is not able to meet the hard deadlines before its implementation. The developer needs to either redesign models or lower the clock frequency of the platform to pass the analysis.

```
/* tolerance values for the timing requirements */
const int TLRI_tolerance =5000; // 5ms = 5000 us.
const int TVRP_tolerance =5000; // 5ms = 5000 us.
const int TPVARP_tolerance = 5000; // 5ms = 5000 us.
const int TAVI_tolerance =5000; // 5ms = 5000 us.
const int TPVSB_tolerance =5000; // 5ms = 5000 us.
const int TPAPB_tolerance =5000; // 5ms = 5000 us.
const int TPVPB_tolerance =5000; // 5ms = 5000 us.
```

FIGURE 16.24
Tolerances in declarations.

16.7 Conclusions

In this chapter, we provide a framework for response time analysis of cardiac pacemakers that combines static timing analysis at code level with high-level modeling. We chose IEC 61499 function blocks to achieve high-level modeling and synchronous semantics for the execution of such models. At the code level, static timing analysis is conducted using integer linear programming. The calculated best case and worst case execution times are used to bound the execution of certain transitions in the TA models, to which the function block models are translated. The related translation rules are designed to create high-fidelity models using timed automata. Then, response timing analysis is conducted on those models. To the best of our knowledge, the proposed approach, for the first time, develops a unified framework for response time analysis that considers the code level execution time while producing the results from high-level response time analysis.

There are several limitations of the proposed approach. First, our heart model is based on an abstract design called the random heart system [13]. While the model is good for scalability of our analysis, it lacks fidelity compared to recent models such as [7]. Our work improves the model fidelity of the pacemaker, and we also need to consider the high-fidelity model of the heart as both models consist of closed-loop systems. Second, issues like power consumption [7] should be considered in addition to the response time during the analysis, due to the fact that a pacemaker must work continuously for a few years once it is implanted.

Bibliography

[1] Gurobi. http://www.gurobi.com/.

[2] Boston scientific. ADVANTIO pacing system specifications, 2012. http://www.bostonscientific.com/.

[3] R. Alur and D. L. Dill. A theory of timed automata. *Theoretical Computer Science*, Vol. 126, No. 2, 183–235, 1994.

[4] C. Baier and J.-P. Katoen. *Principles of model checking*. MIT Press, 2008.

[5] S. S. Barold, R. X. Stroobandt, and A. F. Sinnaeve. *Cardiac Pacemakers Step by Step*. Futura Publishing, 2004.

[6] G. Behrmann, A. David, and K. G. Larsen. A Tutorial on Uppaal. Department of Computer Science, Aalborg University, Denmark, 2004.

[7] T. Chen, M. Diciolla, M. Kwiatkowska, and A. Mereacre. Quantitative verification of implantable cardiac pacemakers over hybrid heart models. *Information and Computation*, Vol. 236, No. 0, 87–101, 2014.

[8] G. D. Clifford, S. Nemati, and R. Sameni. An artificial vector model for generating abnormal electrocardiographic rhythms. *Physiological Measurement*, Vol. 31, No. 5, 595, 2010.

[9] T. A. Henzinger. The theory of hybrid automata. In M. K. Inan and R. P. Kurshan, editors, *Verification of Digital and Hybrid Systems*, volume 170, pages 265–292. Springer, Heidelberg, 2000.

[10] J. R. Hussan, P. J. Hunter, and M. L. Trew. A clustering method for calculating membrane currents in cardiac electrical models. *Cardiovascular Engineering and Technology*, Vol. 3, No. 1, 3–16, 2012.

[11] E. Jee, S. Wang, J. K. Kim, J. Lee, O. Sokolsky, and I. Lee. A safety-assured development approach for real-time software. In *Embedded and Real-Time Computing Systems and Applications, IEEE 16th International Conference on*, pages 133–142, 2010.

[12] Z. Jiang, M. Pajic, R. Alur, and R. Mangharam. Closed-loop verification of medical devices with model abstraction and refinement. *International Journal on Software Tools for Technology Transfer*, Vol. 16, No. 2, 191–213, 2014.

[13] Z. Jiang, M. Pajic, S. Moarref, R. Alur, and R. Mangharam. Modeling and verification of a dual chamber implantable pacemaker. In C. Flanagan and B. König, editors, *Tools and Algorithms for the Construction and Analysis of Systems*, volume 7214, pages 188–203. Springer, Heidelberg, 2012.

[14] I. Liu. *Precision Timed Machines*. PhD thesis, University of California, Berkeley, May 2012.

[15] S. Malik, M. Martonosi, and Y.-T. S. Li. Static timing analysis of embedded software. In *Proceedings of 34th Annual Design Automation Conference*, pages 147–152, New York, 1997. ACM.

[16] M. Pajic, Z. Jiang, I. Lee, O. Sokolsky, and R. Mangharam. From verification to implementation: A model translation tool and a pacemaker case study. In *18th Real-Time and Embedded Technology and Applications Symposium, IEEE*, pages 173–184, 2012.

[17] P. S. Roop, S. Andalam, R. von Hanxleden, S. Yuan, and C. Traulsen. Tight WCRT analysis of synchronous C programs. In *Proceedings of International Conference on Compilers, Architecture, and Synthesis for Embedded Systems*, pages 205–214, New York,, 2009. ACM.

[18] R. F. Schulte, G. B. Sands, F. B. Sachse, O. Dossel, and A. J. Pullan. Creation of a human heart, model and its customisation using ultrasound images. *Biomedizinische Technik/Biomedical Engineering*, Vol. 46, No. s2, 26–28, 2001.

[19] N. Trayanova. Your personal virtual heart. *Spectrum, IEEE*, Vol. 51, No. 11, 34–59, November 2014.

[20] R. Wilhelm, J. Engblom, A. Ermedahl, N. Holsti, S. Thesing, D. Whalley, G. Bernat, C. Ferdinand, R. Heckmann, T. Mitra, et al. The worst-case execution-time problem-overview of methods and survey of tools. *ACM Transactions on Embedded Computing Systems (TECS)*, Vol. 7, No. 3, 36, 2008.

[21] L. Yoong, P. Roop, V. Vyatkin, and Z. Salcic. A synchronous approach for IEC 61499 function block implementation. *Computers, IEEE Transactions on*, Vol. 58, No. 12, 1599–1614, 2009.

[22] J. Zhao, T. D. Butters, H. Zhang, A. J. Pullan, I. J. LeGrice, G. B. Sands, and B. H. Smaill. An image-based model of atrial muscular architecture effects of structural anisotropy on electrical activation. *Circulation: Arrhythmia and Electrophysiology*, Vol. 5, No. 2, 361–370, 2012.

17

Smart Grid Application through Economic Dispatch Using IEC 61499

Srikrishnan Jagannathan

Pennsylvania State University

Peter Idowu

Pennsylvania State University

CONTENTS

17.1   Introduction

Smart grid automation, in its preliminary and evolving state, is achieved through centralized control, where a central control system monitors the entire grid automation and communications. These systems are more commonly known as supervisory control and data acquisition (SCADA) systems which are essentially centralized in nature. SCADA systems are implemented using large software programs that are custom developed based on the application and are difficult to reuse for other applications. Modern control systems require flexibility for advanced automation. To achieve this high level of flexibility in these systems, a new software technology based on the interaction of distributed objects and that aims at decentralizing control is required.

The IEC 61499 is a novel method of software development that enables modeling control applications in a distributed manner. The standard presents guidelines for automation protocols in various applications ranging from industrial to smart grid.

The objective of this chapter is to highlight and demonstrate the benefits of microgrid generation through an innovative economic dispatch application implemented in IEC 61499. The contribution of this research is the use of the function block (FB) concept to implement an economic dispatch application considering levelized cost of energy (LCOE), the availability of distributed generators (DG) and the load forecast to balance loads. The LCOE has been calculated based on studies conducted by the Energy Information Administration (EIA). Previous work in this field has been to perform load balancing using FBs with a focus only on the availability, not the economic aspect. The economic modeling of renewable energy systems is implemented in the economic dispatch application in combination with a load forecasting model for each load. A saving in cost of energy production is expected to be achieved through the implementation of different test behaviors. Hence the effectiveness of each behavior is depicted by comparing the power supplied by the utility under the different applications and by comparing the cost of energy production of each of the generating sources for a single day. This cost value can help policy makers in deciding laws for inter-operation of different utilities.

17.1.1   Features of Smart Grid

A smart grid can be viewed as an electricity grid with decision making abilities. It is an electricity network that can intelligently combine power generation and the consumer needs to ensure delivery of sustainable power in an efficient, economic and reliable manner. The core idea of a smart grid vision is to transform the electric grid from a centralized network controlled by the producer to one that is decentralized and consumer-interactive [10]. The smart grid technology can use sensors and meters to send and receive data about the grid. This can help in solving the problem of power disturbances in a grid and also increase the efficiency and security of the supply. State-of-the-art micro-controllers implemented with suitable software can enable distributed generation and help in avoiding grid congestion, thereby improving the implementation of renewable energy resources. A smart grid is developed with the intention of transforming the traditional grid into a modern system that keeps pace with the developments in the generation and distribution technologies.

Real-time two-way communications are available in a smart grid that allow customers to communicate with the grid. Consumers are able to save energy and sell back energy to the grid. Communication is carried out through advanced metering technologies. Installing distributed generating units like residential solar panels and small wind turbines will improve the efficiency of the smart grid by generating power close to the loads, avoiding distribution losses, and peak shaving. Peak shaving is the process of reducing the amount of energy purchased from the utility company during peak hours when the rates are highest. It will allow small domestic customers and businesses to sell power to their neighbors or even back into the distribution grid. The same concept can be applied to larger commercial organizations that have renewable power systems that can transfer excess power back into the grid during peak demand hours.

With a smart grid it is possible to improve the reliability of the power system and the power quality and reduce carbon emissions by integrating renewable-energy sources and engaging distributed storage to allow the penetration of these renewable generation resources in a cost-effective manner. With the provision of advanced infrastructure, it is possible to develop new automated transmission and distribution systems capable of implementing various power system protection and control functionalities such as managing reactive power to maintain system voltages, voltage regulation, enabling reducing cost of operations based on marginal production costs, ramping and load following, and reducing transmission loads when congestion costs are high. As a consequence of using a smart grid concept, a power system network may be equipped with various features worth discussing.

The features of the smart grids include: (i) integration of renewable energy resources; (ii) implementation of various distributed energy resources (DERs) and the concept of a micro-grid; (iii) optimized implementation of distributed generation by considering factors such as fuel cost, operation and maintenance cost; (iv) use of robust communication protocols as an efficient way of data exchange between different intelligent electronic devices developed by different vendors in a plug and play fashion; and (v) distributed control and automation, necessary for the operation and maintenance of a grid.

17.1.2 Research Motivation

With appropriate strategies to implement a smart grid at a practical level, a traditional power system can be transformed into an intelligent grid that makes it efficient and economically beneficial to operate; this creates a win-win situation for the power producers and consumers. The motivation of this research is to implement intelligent smart grid control at the application level in a smart grid, that has the characteristics discussed previously. Traditional control systems such as the PLC and DCS, lack the flexibility and design and fail to offer the desired levels of integration [5]. The IEC 61499 automation standard is being considered as a novel automation solution to overcome these limitations [2]. The advantage of this standard is that it provides a solution that is flexible and can be used to model control systems by breaking down the control process into basic FB instances. It is currently implemented in a variety of industrial applications. Its range of applications is now extending to include smart grids. The functionalities implemented using this standard have been in the fields of protection coordination and fault isolation and restoration [13, 14].

17.1.3 Contribution

The research described in this work proposes an innovative economic dispatch application implemented through the FB concept of IEC 61499. Previous work in this field has been load balancing using FBs with a focus only on the availability of power generation, without including the economic aspect. The previous work also does not consider the ability of loads to predict their need or load forecasting [7, 12]. The work carried out in this chapter implements an economic dispatch application with factors that have not been included in the previous work. The economic modeling includes studies conducted by the EIA and National Renewable Energy Laboratory (NREL) to estimate the parameters of the cost function. The inputs to the control system performing the economic dispatch application will include the cost of generation for renewable and conventional energy systems based on real world economic constraints such as capital cost, fuel rate, heat rate, along with the availability of generation and the forecasted load profile. The final goal is to achieve a saving in cost of energy production through the implementation of each behavior. Hence the effectiveness of each behavior is depicted by comparing the power supplied by the utility under the different applications and comparing the cost of energy production of the various distributed generators for a single day.

17.1.4 Benefits Achieved by Using IEC 61499

Modeling the real world: While designing an application, it is advantageous to break down complex real-world problems into basic software components. This reduces the complexity of the problem and simplifies the process of application development. Economic dispatch has been considered as a practical area of focus. Through the implementation of IEC 61499, the economic dispatch has been broken down into small steps by creating FBs that perform certain parts of the overall application.

Reusability: In object-oriented programming applications, developers are able to re-use the same object classes in a variety of applications through the creation of instances. This feature can be extended to FBs. For example, as a part of this application, a FB was created to perform power comparison. Once this FB was developed and tested, it became a part of the database or library of the engineering tool. Instances of this FB were then used multiple times in the application.

Reduce complexity: While it is good practice to understand the behavior of a FB, the FB can nevertheless be used without understanding the internal behavior as long as its associated task is known. The application was built by creating and linking FBs provided in the library of the software tool used. This eased the development and maintenance of various control schemes. The time required to build the application was greatly reduced. The system has modularity of code and easy reconfiguration; This could be a benefit for engineers who do not have a background in automation but are interested in developing control applications.

Easy reconfiguration: IEC 61499 provides a method of graphical design of the system and also an easy way of distributing the functions in the automation process. Also there is the benefit of easy changing of existing applications.

17.2 Essential Concepts

Before describing the application, the essential concepts relevant to understanding the control behavior are discussed.

17.2.1 Economic Dispatch

Economic dispatch is defined as the reliable operation of generating units to produce energy at the lowest cost to consumers, while considering the constraints that exist on the system. The classical economic dispatch approach essentially involves allocating generation to loads in an economical manner. The four major focus areas in the field of economic dispatch are [1]:

Optimal power flow focuses on an minimizing costs of generation in a system by considering real power (P), reactive power (Q) and voltage constraints in a system.

Economic dispatch in relation to automatic generation control (AGC) provides optimum dispatch with load frequency constraints.

Dynamic dispatch considers minimizing operating costs and providing generation based on forecasts of load.

Dispatch with renewable energy is an economic dispatch solution with financial constraints on costs of modeling facilities with renewable energy generation.

At the core, an economic dispatch aims at optimizing an objective function while considering the different constraints for a system. These constraints can be the voltage and frequency set points, cost of generation, or load forecast parameters. Each of these constraints can be affected by various parameters. For example, the load forecast variable could be modeled as a stochastic process, or the cost of generation can be affected by parameters such as location, fuel cost and capacity factor.

17.2.2 Load Forecasting

Load forecasting is an important aspect of power system planning and operation. Forecasting implies predicting the load ahead of the delivery of the load. The forecasting intervals are basically the time intervals for which the load is forecasted. Based on the forecast intervals, load forecasting can be classified as very short term (few seconds to several minutes), short term (half an hour to a few hours), medium term (days to weeks) and long term (months to years) [4]. In order to maximize financial benefit, a good load forecasting method must be capable of reflecting the current and future trends. Based on historical data and statistical techniques short term load forecasting (STLF) can be performed.

17.2.3 Levelized Cost of Energy

This section describes the economic model used to evaluate the system. For every developing technology a key aspect that researchers and industries look into is the feasibility of the technology. Hence engineering economics is a key area of focus. The feasibility of the potential solutions that engineers develop is evaluated along with the technical aspects. In the field of power generation, the LCOE is the mathematical quantity that can be used to compare the relative cost of energy production from a generating source.

LCOE can be considered a measure summarizing the overall competiveness of the various generating technologies. It is the cost, per kilowatt-hour in dollars, of constructing, operating and maintaining a generating plant over its assumed lifetime. Essential inputs to calculating LCOE include capital costs, fixed and variable operations and maintenance (OandM) costs, financing costs, plant factors, and fuel costs. These factors vary with each plant type. The influence of each of the factors on LCOE varies for different generating technologies. Technologies such as hydro and solar have no fuel costs and hence have relatively small variable OandM costs. In such cases the LCOE is affected by the estimated capital cost and generation capacity. For technologies with significant fuel cost, such as conventional generating units, the capital cost and fuel cost affect LCOE. The availability of various incentives, including state or federal tax credits, can also impact the calculation of LCOE. Each of these factors and their values varies with location and time as technologies evolve and fuel prices change.

The LCOE equation allows a comparison of alternative technologies for different scales of operation, operating time periods or investment [9]. Using this motivation, the LCOE has been used as a reference model for the economical evaluation of the potential solutions proposed in this work.

17.2.3.1   Factors Affecting LCOE

As discussed in the previous section, there are several factors affecting the LCOE. This section describes them in detail.

1. Capital Investment Costs: The major cost components affecting the generating technologies are:

   (a) Construction costs: These are normally the major costs of any project. The civil construction costs are determined by the trends in prices of the country where the project is to be developed. In countries undergoing economic transition, the civil construction costs are generally lower than in their developed counterparts. This is due to deployment of local construction materials and local labor. Construction costs are affected by the inherent characteristics of the topography, geology weather conditions and the construction design of the project. Hence the construction costs are site specific. This could lead to variation in investment cost and consequently the LCOE, even for projects having the same capacities.

   (b) Cost of power system equipment: The costs of electromechanical equipment follow world market prices for these components rather construction costs. The type of technology used for generation is a significant component. For example, in a solar power plant the type of solar module affects the investment cost since it requires the installation of specific equipment. Additionally the investment costs also cover planning, licensing, environmental impact analysis, wildlife mitigation, historical and archaeological mitigation, and recreation mitigation.

Studies [8] show a general tendency of increasing investment cost as the capacity increases and that there is also a wide range of cost for projects of the same capacity. For plants with lower installed capacity, for example 5 MW, the costs of electromechanical equipment tend be more significant. However, with an increase in plant capacity, the costs of construction are more influential. The same overall generating capacity can be achieved through a combination of several smaller or large generating units. Plants implementing multiple smaller generating units have higher costs per kilo-watt (kW) as compared to plants using fewer larger units. The reason of having higher costs per kW associated with a higher number of generating units is because of a greater flexibility and efficiency achieved by the integration of plants into the electric grid. For plants that have fewer units with larger generating capacity, the investment costs tend to be reduced. For example, for a larger generating unit, a hydropower project can be set up to use less volume flow and consequently smaller hydraulic conduits. The sizes of the equipment and their related costs are also lower. Results [8] have shown the characteristic distribution of investment costs based on geographic areas.

2. Operation and Maintenance Cost: Renewable energy technologies, after commissioning, generally require minimal operation and maintenance. A major reason is that plants operating with such technologies do not incur the cost of fuel continuously. OandM costs can be calculated as a percentage of the investment cost per kW. In contrast conventional generating units such as fuel base units have a higher OandM cost due to recurring cost of fuel. This again varies with location and market conditions.

3. Capacity Factor: The capacity factor of a power plant is the ratio of its actual power to the output it could produce if operating at full nameplate or rated capacity continuously over a given period. The parameter affecting capacity factors is mainly the availability of the renewable resources. This in turn depends on the location and the type of technology used. For hydropower, the capacity factor is generally designed during the planning and optimization stages of the project by considering both the market demand for power and the statistical distribution of flow. Reservoirs can be designed to increase the stability of flow for base-load production and supply highly variable and reliable flow to a power plant. For solar power, the main parameter affecting capacity factor is the annual solar radiation and the type of panel or module used. The type of module affects the efficiency of conversion of light to electricity which affects the actual output of the plant and hence the capacity factor. A low capacity factor implies low production and higher LCOE. Overall the capacity factors of renewable energy technologies are lower compared to the those of conventional generating units. This is due to the intermittency or unavailability of the renewable energy sources at all times.

4. Lifetime: For renewable energy plants, the structures like dams, tunnels, canals, powerhouses for hydropower or the components in concentrated solar power (CSP) technology for solar are the major cost components that must have long lifetimes. Electrical and mechanical equipment, with comparatively shorter lifetimes, contribute less to the cost. Hence it is common to use longer lifetimes for hydropower plants as compared to other electricity generation sources. Typical lifetimes are in the range of 40 to 80 years.

5. Discount Rate: The discount rate is basically a percentage that can influence the LCOE based on the trends of expenditures and revenues that occur over the lifetime of the investment. Investors generally choose discount rates based on the risk-return characteristics of available investment options. A high discount rate is beneficial for technologies that have low initial capital investment and high running costs. On the other hand, a low discount rate is conducive for renewable energy sources such as hydropower and solar power which have comparatively higher capital investment and lower maintenance costs. Typical values for renewable energy technologies are between 7 to 10.

17.2.3.2 LCOE Calculations

The LCOE equation depends on the capital investment cost, discount rate, OandM cost, capacity factor, fuel cost, heat rate and plant lifetime. The discount rate may be real or nominal. Using the values of discount rate and lifetime periods, the capital recovery factor (CRF) is calculated. CRF is the ratio of a constant annuity to the present value of receiving that annuity for a given period of time. Using the discount rate i, and plant lifetime n, in years, CRF is calculated as shown in Equation (17.1).

$$CRF = \frac{i * [1 + i]^n}{[1 + i]^n - 1} \tag{17.1}$$

Based on definitions in [6] LCOE is calculated as shown in Equation (17.2):

$$LCOE = \frac{CI * CRF + FOM}{8760 * CF} + (FC * HR) + VOM \tag{17.2}$$

where:
CI: capital investment costs
FOM: fixed OandM
CRF: capital recovery factor
CF: capacity factor
FC: fuel cost
HR: heat rate
VOM: variable OandM costs

17.3 Software Tools

The software tool used to develop the economic dispatch application is Function Block Development Kit (FBDK). The software used to develop the power system network is MATLAB®/Simulink. FBDK has the ability to be programmed in JAVA and MATLAB®/Simulink supports JAVA programming as well. This feature is used to achieve a synchronized simulation between the two environments.

17.4 Application Development

This chapter depicts the implementation of two economic dispatch applications. Each application has a different control behavior in the FB environment. This section describes the system configuration and each of the economic dispatch behaviors in detail.

17.4.1 Description of Power System Network

The power system configuration and parameters have been extracted from the benchmark system of the IEEE Standard 399-1997 with some modifications to allow for autonomous micro-grid operation [3]. The system, shown in Figure 17.1, consists of a three-feeder distribution subsystem operating at 13.8 kV. This subsystem is connected to the main network or grid through a 69 kV radial line. The large network at the end of the radial line is represented by a 69 kV, 1000 MVA short circuit capacity bus. Each of the three feeders is connected to the large network at one end and a DG source and a combination of variable loads at the other. Since the focus is on observing the behavior of the load balancing applications during normal conditions, each of the DGs is modeled to operate at steady state and are designated DG 1, DG 2, DG 3, each operating at 13.8 kV. DG 1 is a conventional unit running on natural gas capable of supplying 4 MW. DG 2 is modeled as a PV plant of capacity 4 MW. DG 3 is a hydro power plant capable of supplying 3 MW. Loads L1 to L6 are supplied through the radial feeders of the subsystem. The loads vary over time and generate a variable load profile. The simulation time is 12 seconds representing 24 hours in a day. Thus a micro-grid collectively has a DG, two loads and corresponding distribution network components.

Each DG has a LCOE that is calculated using an economic model. The factors used in calculating the value of the LCOE are described in Section 17.2.3.1. The values of LCOE are shown in Table 17.1.

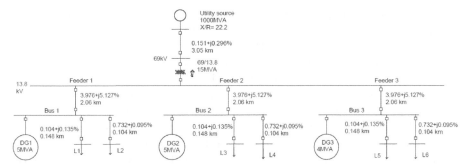

FIGURE 17.1
Single line diagram of system network.

TABLE 17.1
LCOE for Generating Technologies under Consideration

Factor	DG 1	DG 2	DG 3
Period (years)	30	30	30
Rate of Interest(percent)	10	7	7
Capital recovery factor	0.10	0.08	0.08
Capital cost (USD/kW)	1090	2600	2359.3
Capacity factor (percent)	75	25	57.5
Fixed O and M (USD/kW-yr)	11.96	7.56	11.3975
Variable OandM (USD/kWh)	0.003	0.001	0.0036
Heat rate (Btu/kWh)	8185	0	0
Fuel cost (USD/MMBtu)	5.03	0	0
LCOE (cents/kWh)	6.1	10.1	4.4

17.4.2 Stage 1 – Normal Grid Operation

In this case the loads are supplied power by their respective micro-grid DGs and the utility without load balancing consideration. The power supplied to loads is shared by generators based on their impedances.

17.4.3 Stage 2 – Single Microgrid Economic Dispatch

The application developed in this stage involves interaction between the utility and the DG associated with loads to be balanced. Any increase in load above the threshold limits, in a specific micro-grid is shared only between the utility and the DG in that micro-grid. For example, if there is an increase in the load L1, the necessary control action chooses between the utility or a combination of the utility and DG 1 to balance the load. The application is developed on one device with one resource. The control architecture is shown in Figure 17.2.

The criteria for load balancing are based on only the availability of a DG to match power and the LCOE of the generating sources. Each grid has two loads associated with it. The power consumed by each of the loads is designated P(1), P(2), P(3), P(4), P(5), P(6). The utility is connected to the grids through CBs labeled CB(1), CB(2), CB(3) respectively for grids 1, 2 and 3. Each load has a pre-determined threshold value (Pth1, Pth2, ... Pth6). The individual threshold values are calculated as the mean of the varying load profile exhibited by each of the loads. The attributes of a load are received at a central control center. The value of the forecasted load in a grid is compared with the threshold for that particular grid. If the power consumed exceeds the threshold for even one of the loads in the micro-grid, the CB between the utility and the micro-grid containing that load is closed. This ensures that the necessary power is supplied to the load in an efficient way and the load is hence balanced. The described algorithm is executed on one device.

17.4.3.1 List of Function Blocks Used in Stage 2

CLIENT FBs for data exchange: The FB on the left in Figure 17.3 acts as the interface in accepting data from the smart metering device(SMD) in Simulink. The input to this block is the IP address of the metering device and the guard condition which has a boolean value of 1. The FB execution gets triggered only when it receives an event input trigger at INIT. The values LREAL, TIME are the data types of the values received from the SMD. They are essentially the power value, start time and duration of load, respectively.

The FB on the right in Figure 17.3 sends the CB signals back to the utility in Simulink through the inputs SD_1, SD_2 and SD_3. The FB on the left

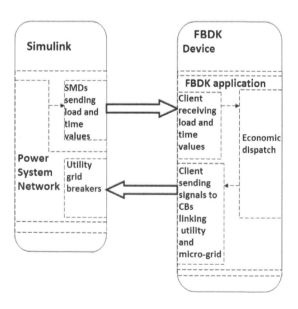

FIGURE 17.2
Reference architecture for Stage 2.

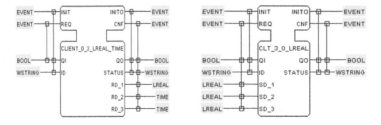

FIGURE 17.3
Function block for receiving and sending data between Simulink and FBDK.

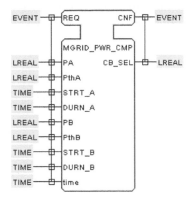

FIGURE 17.4
Function block for power comparison and time synchronization.

receives data from Simulink. Again both the SIFBs act as links for exchanging data from the application in FBDK to the network in Simulink.

Power Compare FB: This FB, shown in Figure 17.4, essentially executes the comparison of the loads with their threshold for a single grid in a time synchronized manner. The output of this FB is a signal communicated to the CB connecting the grid to the concerned micro-grid. If the loads are above the threshold, the output has a value of 1 and if they are less than the threshold, the output of this FB has a value of 0.

17.4.4   Stage 3 – Intermicrogrid Economic Dispatch

This subsection focuses on an economic dispatch application and involves lateral interaction between the micro-grids. Any increase in load in a specific micro-grid can be shared by the utility and the other micro-grids in the network. The application has been divided into multiple sub-applications and implemented on multiple devices. The criteria for economic dispatch are based on availability of power, the cost of generation and the forecasted load. This stage is an example of a distributed application, i.e., a single application distributed into multiple sub-applications on multiple devices and hence can be considered as an application implemented on remote devices. A block diagram is shown in Figure 17.5.

17.4.4.1 Mathematical Formulation

The mathematical formulation of the economic dispatch application is discussed. At the core of every economic dispatch application is an objective function that aims at supplying loads in a cost-effective manner. This objective function is then implemented in the FB environment. The objective function and the constraints used in this work can be formulated as in Equations (17.3) to (17.7).

$$[CB_{signal}, C_{min}] = f(P_{load}, Time_{Start}, Time_{Duration}, LCOE, P_{Gi})$$

$$\text{subject to} \qquad L_{(min,i)} <= P_{Gi} <= P_{(Gi,max)} \qquad (17.3)$$

$$\sum P_{Gi} = \sum P_{load}$$

i represents the ith generating source

$$P_{Gi,max} = min(P_{Gi,rated}, P_{th}) \qquad (17.4)$$

$$P_{th} = \sum_{j=1}^{K} mean(P_{load,j}) \qquad (17.5)$$

K: the number of loads in a micro-grid.
$P_{Gi,rated}$: the rated capacity of generator i

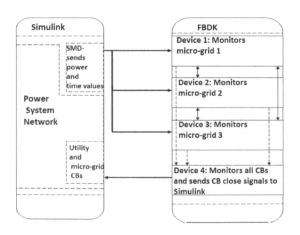

FIGURE 17.5
Reference architecture for Stage 3.

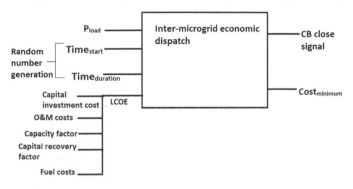

FIGURE 17.6
Economic dispatch application.

$L_{min} = min(P_{load})$
where

$$C_{min} = min \sum_{j=1}^{N} LCOE * E_{Gi} \qquad (17.6)$$

N: the total number of generating sources
E_{Gi}: energy output of generator based on forecasted P_{load}

$$E_{Gi} = P_{Gi} * 1hr \qquad (17.7)$$

P_{load}: load demand (MW)
$Time_{start}$: start time of load
$Time_{duration}$: duration of load
$LCOE$: levelized cost of energy (USD/MWh), see Equation (17.2)

Figure 17.6 shows a block diagram of the economic dispatch application implemented in this work. The proposition is to minimize the cost of inter-operation of micro-grids. Based on the economic, load and availability constraints of the system, which go as inputs to the control system, the output of the system is the minimum cost of operation and the appropriate control signals for the circuit breakers to the micro-grid DGs for supplying loads. Here the inputs to the system are the forecasted load demand (P_{Load}), load start time and duration ($Time_{Start}$, $Time_{Duration}$), and the LCOE. The load start time and duration are generated by a random number generator following uniform distribution to create a load profile that acts as the forecast to the control system. The cost function for the renewable energy systems is based on the LCOE which considers the investment cost of equipment, capital cost, OandM costs, and capacity factor. Thus the economic dispatch solution combines dynamic dispatch and load forecasting to provide a cost-effective solution. The LCOE values are based on statistical analysis studies conducted

by Energy Information Administration, National Renewable Energy Laboratories and U.S.DOE.

Similar to Stage 2, each load has a pre-determined threshold value (Pth1,
Pth2, ... Pth6). Each Micro-grid is controlled by a local control center (LCC)
and is responsible for performing the economic dispatch application in respective grids. The LCC control applications are developed in FBDK. Hence there
are three local control centers in total. These local control centers then communicate their decisions, to open or close the breakers, to a central control
center (CCC) which sends the trip signal appropriate CB. The three grids are
connected to each other through a circuit breaker. The CB between grid 1
and grid 2 is CBgrid1-grid2, CB between grid 2 and grid 3 is CBgrid2-grid3,
and CB between grid 1 and grid 3 is CBgrid1-grid3. In addition to these CBs
there are also three CBs connecting each of the grids to the main utility. These
are labeled CBUtility1, CBUtility2 and CBUtility3, respectively, for grids 1,
2 and 3. These CBs are controlled by a CCC after receiving signals from the
LCC. Each load has a particular threshold value. When a load in a micro-grid
exceeds the threshold value, that grid looks for generating sources that offer
power at the lowest price. If any of the micro-grids has excess capacity, it supplies it to the grid in deficit. If no DG is able to supply the required amount of
power entirely, the power is supplied through a combination of DGs and the
main utility. Only when load(s) cannot be supplied by the micro-grid DGs is
the deficit supplied by the main grid entirely. Based on which DG is scheduled
to supply the power, the corresponding breaker is closed, thus ensuring the
necessary amount of power supply to the load.

17.4.4.2 List of Function Blocks Used in Stage 3

In addition to the FBs used in Stage 2, two more FBs are created in Stage 3.

CB close decision signal: This FB is present in device 4 of the reference
architecture for Stage 3 and is shown in Figure 17.7(a). It takes as input the
CB signals from the first three devices 2 at a time. The CB will close if even one
of the inputs goes to 1. The algorithm for the FB is shown in Figure 17.7(b).

FB for cost comparison: The FB in Figure 17.8 performs the comparison algorithm. The inputs CostX, CostY and CostUTIL are the LCOEs of
the DGs from which the concerned grid might take power.

After performing the power comparison algorithm, the FB sends out three
outputs for three CBs. For example, if we consider grid 1 as the grid demanding
power, CB_A_B is the CB connecting grid 1 and grid 2. CB_B_C is the CB
connecting grid 1 and grid 3. CB_UTIL is the CB connecting grid 1 and the
utility. The 1 implies CB close and the 0 implies CB open. The other inputs
are the instantaneous load values, their duration and threshold limits.

The implementation is shown in Figure 17.9. The output of the FB shown
in Figure 17.4 triggers the cost comparison FB. This FB is responsible for
performing the economic dispatch application based on the cost and load
inputs. The outputs of the cost comparison FB are basically signals to the most

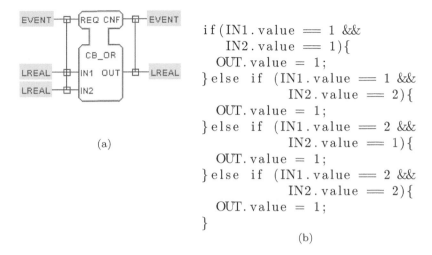

```
if (IN1.value == 1 &&
    IN2.value == 1){
OUT.value = 1;
}else if (IN1.value == 1 &&
             IN2.value == 2){
OUT.value = 1;
}else if (IN1.value == 2 &&
             IN2.value == 1){
OUT.value = 1;
}else if (IN1.value == 2 &&
             IN2.value == 2){
OUT.value = 1;
}
```

(b)

FIGURE 17.7
Decision making function block (left). Decision making function block algorithm (right).

cost effective distributed generator to increase its generation to compensate the load.

The behavior implemented in this stage is an example of a distributed application. To successfully achieve distributed behavior, there must be proper communication of relevant data between the different devices. In FBDK, this has been implemented through the publish and subscribe blocks: PUBL1 and SUBL1, respectively. The data exchanged between different devices using the publish and subscribe blocks are the load values. The publish and subscribe FB along with the SERVER and CLIENT FBs discussed in the subsequent sections are collectively called the service interface FBs implemented in this application. They form the basis of developing a distributed application.

17.5   Co-Simulation between MATLAB® and FBDK

The execution of the application has been executed through a co-simulation model between MATLAB® and FBDK. To achieve successful co-simulation both the software tools must be able to exchange data in a language easily understood by both sides. Before sending data from MATLAB® into FBDK, the data and variables have to be encoded into data types supported and understood by FBDK. This has been the underlying principle of co-simulation followed in this work. The co-simulation has been carried out through a simple

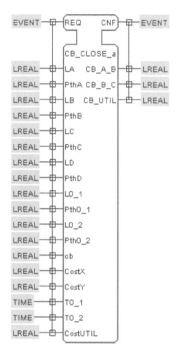

FIGURE 17.8
Cost comparison function block.

approach using the class files used to create the FBs in FBDK. Co-simulation
has been performed using the feature that FBDK is a Java-based program.
Hence each FB consists of a class file. These class files can be compressed
into jar files. A unique feature of MATLAB is that it allows the addition of
Java library files when they are compressed into a jar file. The jar files can be
added into MATLAB® through the command window. The addition of the jar
files implies the addition of the class files into MATLAB®. Hence instances
or objects of these classes can be included in programs in MATLAB® and
Simulink. Co-simulation is done using TCP/IP. Since MATLAB® is the side
initiating the communication, it is the SERVER and FBDK is the CLIENT.
Hence class files of SERVER FBs are included in MATLAB®.

The procedure to include the class file of a SERVER FB exchanging time
values is as follows. In the folder or package where the file is stored, open
command prompt. The compiling and running of the Java program under
consideration are done by typing the following commands in the command
window:

```
javac $-$cp fbrt.jar $-$d . TIMESENDER.java
java $-$cp fbrt.jar;. time.server.TIMESENDER
```

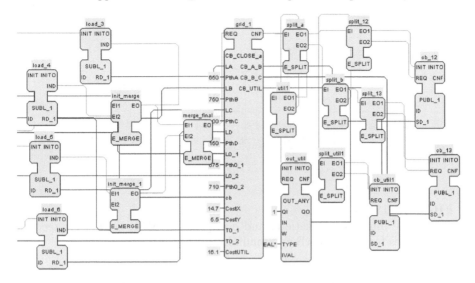

FIGURE 17.9
Implementation of cost comparison function block.

Here the name of the Java file is TIMESENDER. On executing the above two commands, the file TIMESENDER gets compiled, we then create a JAR file with the class files and include them in the MATLAB® directory.

The TIME SENDER is essentially a FB of type SERVER_1_0 that sends time values to FBDK. The basic class files are inherited from fbrt.jar. Hence TIME SENDER can be viewed as a virtual FB created at the MATLAB® end to enable co-simulation. At the receiving end, the FB will be of type CLIENT FB. In this application it has been named CLT_0_1_TIME. The composite FB encapsulates a CLIENT FB that exchanges time values. The composite FB looks as shown in Figure 17.10(a) and the SIFB looks like Figure 17.10(b). The data value RD_1 is of data type TIME in this case.

In order to successfully achieve co-simulation between Simulink and FBDK time synchronization needs to be achieved between the two platforms. Alternatively, time step in Simulink must be matched with the time step in FBDK. The method implemented in this application is adapted from the method described in [11] The MGRID_PWR_CMP block is the FB that performs the time comparison and consequently the synchronization. For convenience it is shown again in Figure 17.11(a) along with FB. The inputs labeled STRT_A, DURN_A, STRT_B, DURN_B, and time perform the time synchronization. This enables the execution of a particular event at a specified instant of time.

The input labeled TIME receives its input from the SERVER FB created in MATLAB® using the procedure described in Section 17.5. The ECC for the FB above looks as shown in Figure 17.11(b).

Having described the ECC and state transitions, the part of the appli-

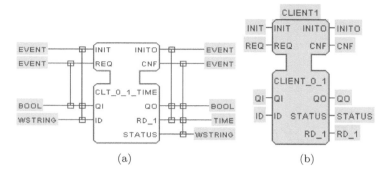

(a)   (b)

FIGURE 17.10
Composite function block (left) and encapsulated SIFB (right) for receiving time values from MATLAB®/Simulink into FBDK.

cation performing the time synchronization is shown in Figure 17.12 to provide a better understanding. The data from MATLAB® is received using CLIENT_0_3_LREAL_TIME. The values received denote load demand, load start time and load duration. These are the essential parameters required for the implementation of the economic dispatch application. They are then sent to the FB labeled MGRID_PWR_CMP. The time synchronization and load comparison are performed using this FB. This part of the application common to Stage 2 and Stage 3.

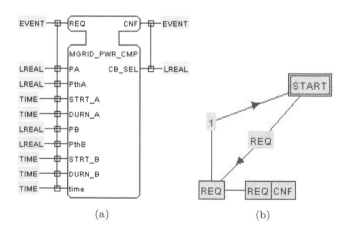

(a)   (b)

FIGURE 17.11
Function block for power comparison and time synchronization (left) and corresponding ECC (right).

17.6   Simulation Results, Conclusions, and Future Work

As discussed in the previous sections, the application development is divided into three stages. Each stage has a different load balancing behavior run for the same variation in load. The time variation for each load is developed using a random number generator. This logic ensures a randomly varying load profile. The total simulation is run for 12 seconds to simulate 24 hours in a day. Thus an environment has been created for comparing the effectiveness of each behavior. The effectiveness of each application is decided by comparing the power supplied by the DGs in each of the micro-grids and utility. Based on the power generated, the total cost of operation for one day is calculated and compared. The results show the power generated by the DGs and the utility along with the calculation for cost of production of power generated by the DGs and utility.

17.6.1   Simulation Results for Power Outputs

Stage 1

The results for this stage are shown in Figure 17.13. The utility and the DGs are supplying the loads based on the generator impedances. This stage represents the typical operation of a power network without any form control. It forms the basis for the development of the subsequent stages.

Stage 2

Figure 17.14 shows the power supplied by the utility and the DGs under the load balancing application in Stage 2.

In comparison with the power supplied in Stage 1 there is a reduction in power supplied by the utility and a corresponding increase in the power supplied by micro-grid DG 3. This result is based on the comparison of the

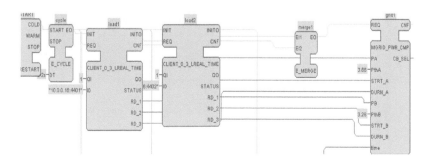

FIGURE 17.12
Application for time synchronization and power comparison.

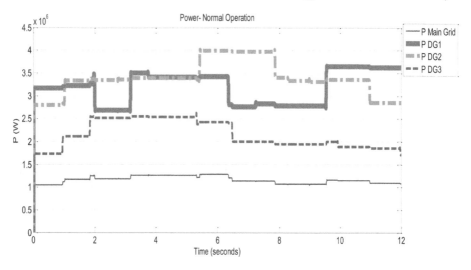

FIGURE 17.13
Power supplied for 24 hours: Stage 1.

availability and LCOE of DG 3 and the utility. For the model under consideration, since cost of DG 3 is less than the cost of electricity of the utility network,
DG 3 is the primary choice for supplying the excess demand. The reduction in
utility power can be noticed in the time window starting 0.5 seconds to about
6 seconds. In stage 1, the power supplied by the utility is approximately 1.25

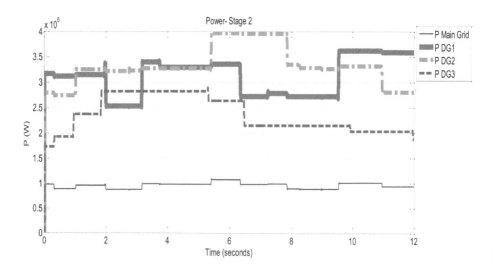

FIGURE 17.14
Power supplied for 24 hours: Stage 2.

MW. In stage 2, for the same time window, the power supplied by the utility reduces to approximately 1 MW. This reduction in power supplied by the utility needs to be compensated for the load to continually receive power. This reduction is compensated by DG 3. An increase in generation can be observed for DG 3 by the same amount. The power generated by DG 1 and DG 2 is almost unchanged, indicating that the loads associated with these DGs are under the threshold. The benefit of the application at this stage is that it reduces the cost of generation as compared to the cost of generation in Stage 1 and is straightforward to implement.

Stage 3

The load balancing application in Stage 3 is more efficient than the application in Stage 2 in terms of cost of operation. This is because the economic dispatch application also considers economics of generating sources in adjacent micro-grids. The application considers availability and LCOE as the factors when calculating the power generation of each sources in Stage 3. This cost factor assists policy makers in deciding the cost of selling power. Figure 17.15 shows the power generated by the DGs and the utility network.

The difference between the power outputs in this stage and the power outputs in Stage 2 can be noticed for the utility and the other DGs as well.

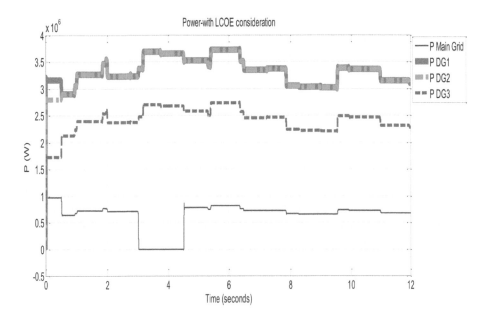

FIGURE 17.15
Power supplied for 24 hours: Stage 3.

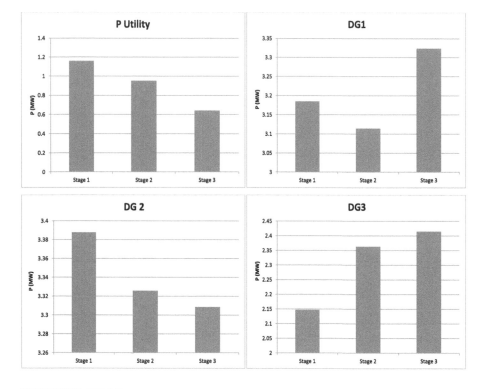

FIGURE 17.16
Average power supplied by each of the generating units for each Stage for 24 hours.

During zero power for the utility, the DGs compensate by increasing their generation.

Figure 17.16 is a group of four bar charts showing the power supplied by the four generating units for each stage. On observing the average power supplied by generating units in each of the stages, it can be concluded that economic dispatch application aims at choosing the generating unit capable of supplying the loads at the lowest price. A decrease in power generated by one or more generating units is compensated by the other more economically feasible generating units.

17.6.2 Daily Cost of Generation for Different Stages

This section tabulates the total cost of generation of each of the sources for a single day. The cost of energy generation forms a good basis for the effectiveness of the application developed in the FB environment.

Table 17.2 shows that the values in the total cost column reduce with the development of successive stages. The total is the sum of the cost of energy

TABLE 17.2
Cost of Generation for 24 Hours in U.S. Dollars per day

	Utility	DG 1	DG 2	DG 3	Total Cost
Stage 1	03741	04441	08630	01938	18750
Stage 2	03071	04340	08471	02133	18015
Stage 3	02060	04674	08388	02170	17292

of the individual generating units and the utility network. This indicates the effectiveness of each load balancing application. The reduction in this application between Stages 3 and 1 is approximately 7.7%. This is the saving achieved when these sizes of generating units are considered. For larger generating units, the savings would be proportional. These applications are developed as global solutions. Hence depending on the locations the individual cost of generation (DG 1, DG 2, DG 3 or utility) will vary. This is evident from the graphical representation of the costs of the individual generating units in Figures 17.17 through 17.20 shown below.

However, the overall goal of achieving a reduced total cost is achieved. A reduction in power supplied by the one generating unit is compensated by an increase in the generation of the other DGs or the large utility network. This aspect is also reflected in the cost of energy production for these DGs. For the scenario considered, the levelized cost of generation for the DGs is lower than the cost of electricity from the utility network. Hence there is a reduction in power supplied for the utility network. This in turn reflects the cost of electricity distributed by the utility network for a single day. Also, based on the LCOEs of the individual generating units, there will be a increase or decrease in the cost of operation based on the economic dispatch algorithm.

The total cost of operation for 24 hours for the entire network is shown in Figure 17.21. A gradual decrease is observed in the total cost of energy production for a day.

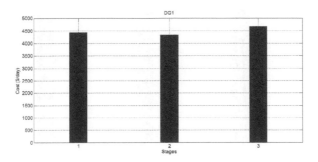

FIGURE 17.17
Cost of energy production of DG 1 for 24 hours.

The effectiveness of each scheme can be better demonstrated by comparing the percentage saving achieved by load balancing applications at all stages with the normal operation in Stage 1. This is shown in Figure 17.22.

17.6.3   Conclusion

On observing the results for each stage, it can be pointed out that with the development of each stage using the FB concept, there is a reduction in the cost of generation. The economic dispatch application uniquely combines the features of load forecasting, economic modeling of renewable energy systems and the concept of LCOE in performing the economic dispatch application. The benefit of using LCOE for financial calculations is that the input variables of the LCOE equation are modeled using data available, thus more accurately reflecting the cost uncertainty associated with a specific generating technology. The economic dispatch application is a global solution and depending upon the location, capacity and generating technology, the results will vary. Thus in locations where renewable energy technologies are more expensive than conventional generating technologies, the federal and state governments can provide subsidies and incentives to help bring down the cost of these technologies. Economic dispatch can be performed to guide policy makers, regulators and investors in making sound financial decisions for optimal power system operation. It can help utility operators decide on policies relating to inter-operation of power system networks. The LCOE index provides a platform for assessing different renewable energy resources by providing an economic ranking system that considers the initial capital cost, OandM cost and the environmental impacts of the renewable sources of generation. Policy makers consistently make an effort to create a mutually beneficial environment for the inter-operating utility networks. This can be achieved by developing and implementing smart and intelligent control applications that give the regulators an economic benefit. This project combines the aspects of intelligent

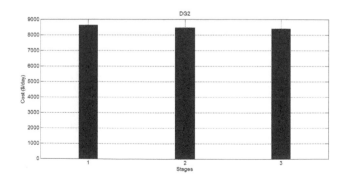

FIGURE 17.18
Cost of energy production of DG 2 for 24 hours.

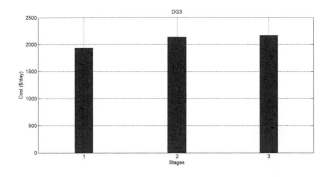

FIGURE 17.19
Cost of energy production of DG 3 for 24 hours.

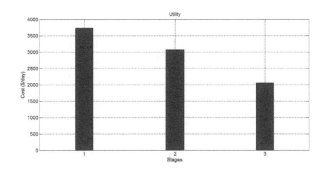

FIGURE 17.20
Cost of energy production of utility for 24 hours.

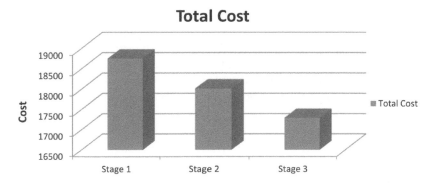

FIGURE 17.21
Total cost of energy production for 24 hours.

control and economic benefit by implementing the FB control to develop load balancing applications that prove to be economically beneficial.

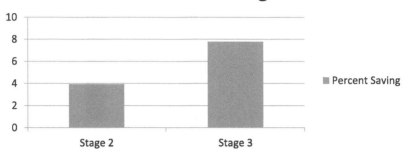

FIGURE 17.22
Percentage saving achieved through function block applications.

In conclusion it can be said that IEC 61499 is certainly an effective and reliable method for the development of control applications in smart grid. It can also be said that any complex process can be considered as a collection of simple functional modules. The method of co-simulation used in this work can be used to show that the FB can be viewed as a concept that can implemented using a programming language supporting object- oriented programming as well. The possibility of using IEC 61499 for other control applications such as demand response and generator excitation control in a smart grid should be looked into in depth.

17.6.4 Future Work

The focus of this chapter has been on the control of distributed generation of the grid. Future work can include the control of load demand on the grid, also called demand side management, by including load queuing. This extends the area of applications of IEC 61499 into demand response. Also the model considered a single DG within each micro-grid. A future extension would be to modify the economic dispatch behavior to accommodate multiple DGs within the same micro-grid.

Acknowledgments

Sincere gratitude to Jim Christensen, Alois Zoitl and the other members of the IEC 61499 research team for their continued support and guidance.

Bibliography

[1] E. Chowdhury and S. Rahrnan. A review of recent advances in economic dispatch. *IEEE Transactions on Power Systems*, 1990.

[2] T. Hussain and G. Frey. Migration of a PLC controller to an IEC 61499 compliant distributed control system: Hands-on. In *International Conference on Robotics and Automation*, pages 3984–3989. IEEE, April 2005.

[3] F. Katiraei. Micro-grid autonomous operation during and subsequent to islanding process. In *Power Engineering Society General Meeting*, volume 2. IEEE, June 2004.

[4] D. Kothari and I. Nagrath. *Modern Power System Analysis*. Tata McGraw Hill., 2011.

[5] R. Lewis. *Modeling Control Systems Using IEC 61499: Applying Function Blocks to Distributed Systems*. The Institution of Electrical Engineers, 2001.

[6] National Renewable Energy Laboratory, Colorado. *A Manual for the Economic Evaluation of Energy Efficiency and Renewable Energy Technologies*. U.S. Department of Energy, Golden, Colorado, March 1995.

[7] S. Patil, V. Vyatkin, and B. McMillin. Implementation of FREEDM smart grid distributed load balancing using IEC 61499 function blocks. In *Industrial Electronics Society, 39th Annual Conference of IEEE*, pages 8154–8159, Nov 2013.

[8] R. Pichs-Madruga, O. Edenhofer, and Y. Sokona. Renewable energy sources and climate change mitigation: Special report of the Intergovernmental Panel on Climate Change. Technical report, Cambridge University Press, 2011.

[9] U.S. Energy Information Administration. Levelized cost and levelized avoided cost of new generation sources. Technical report, Office of Integrated and International Energy Analysis, U.S. Department of Energy, Washington, 2014.

[10] J. Vasudevan, K. Sekkappan, K. Yashwant, and K. S. Swarup. Research issues in a smart grid-application to automation, distributed energy sources and demand response. *IIT, Madras*, 2012.

[11] C.-H. Yang, G. Zhabelova, C.-W. Yang, and V. Vyatkin. Cosimulation environment for event-driven distributed controls of smart grid. *IEEE Transactions on Industrial Informatics,*, 2013.

[12] G. Zhabelova, S. Patil, C.-W. Yang, and V. Vyatkin. Smart grid applications with IEC 61499 reference architecture. In *Industrial Informatics 11th IEEE International Conference on*, pages 458–463, July 2013.

[13] G. Zhabelova and V. Vyatkin. Intelligent logical nodes of IEC 61850 and IEC 61499 for multi-agent smart grid automation. *IEEE Transactions on Industrial Electronics*, 2011.

[14] G. Zhabelova, V. Vyatkin, and N. Nair. Standards-based intelligent fault management system for FREEDM green hub model. In *IEEE International Conference on Industrial Electronics,*, 2011.

Part IV

Laboratory Automation Examples

18

Workspace Sharing Assembly Robots: Applying IEC 61499 to System Integration and Application Development

Matthias Plasch

PROFACTOR GmbH

Gerhard Ebenhofer

PROFACTOR GmbH

Michael Hofmann

PROFACTOR GmbH

Martijn Rooker

PROFACTOR GmbH

Sharath Chandra Akkaladevi

PROFACTOR GmbH

Andreas Pichler

PROFACTOR GmbH

CONTENTS

18.1   Introduction

Future production systems are characterized by highly distributed and modular architectures, which have to be highly adaptable to satisfy customer and
market needs. These systems usually consist of heterogeneous software and
hardware components, based on different platforms and frameworks, thus leading to high levels of complexity and development effort during system development and integration. Complex robotic assembly systems share many of the
previously mentioned characteristics. In order to tackle these difficulties, we
present a collection of concepts which aid system integration and application
development based on functional independent and highly modular software
and hardware system components. The aim is to reduce development time
and increase stability of applications by relying on proven, reusable software
components and development approaches. This work aggregates experiences
gained in various industrial and research projects in the field of robotics, and
is founded on previous work [7, 25, 26].

The applicability of the concepts presented have been applied to a complex robotic system consisting of two workspace sharing robots, which was
designed to solve an assembly task by cooperation. For system integration
and application modelling, the 4DIAC framework for distributed automation
and control [34], according to the industrial standard IEC 61499 [13], was
used. Section 18.2 provides a state of the art overview on system integration
frameworks, scripting languages and technologies used in the presented development approaches. Section 18.3 describes the application use case and the
workspace-sharing robotic system as well as its system components; the development approaches presented in Section 18.4, were applied. In Section 18.5
the development patterns are applied to the application example. Section 18.6
summarizes and concludes the work.

18.2   Related Work

This section gives an overview of related work and state-of-the-art topics. The first part focuses on frameworks that support the integration of heterogeneous services in the domains of robotics and industrial automation. In the second part, supporting technologies and software components for the concepts presented in this chapter are presented.

18.2.1   Integration Frameworks

Various approaches targeted the integration of services based on different platforms into one system built on service-oriented architecture (SOA) concepts. SOA-developed components are characterized by functional availability and are designed to push reuse and composition of existing functionality to enhance services [8]. The major advantages of SOA are the treatment of services as independent resources and their accessibility without the need of platform-specific knowledge [37]. Within the field of robotics, Kononchuk et. al. [17] developed an interconnection protocol, RoboCop, to combine robotic engineering tool chains and programs to a scalable control system structure. Frameworks building on SOA-driven development for integrating devices including their sensors and actuators into a robotic system are proposed by [9] and [37]. In both cases, devices are treated as ubiquitous services. Yang and Lee [37] additionally use a hierarchical approach to express complex services (components) as compositions of simple services. A comparison of implemented SOA approaches adapted to robotic work-cell programming was performed by Veiga et. al. [35]. Emphasis here is on methods for service description and service orchestration. The analyzes proved that SOA-driven development of service components, combined with graphical development environments, reduces the integration effort significantly [35].

In the domain of robotics, Orocos is one of the most mature and established open source software frameworks [21]. More recent work extends this data flow-oriented framework with Java technology resulting in JOrocos, influenced by service component architecture (SCA) concepts [31] and additionally bridging the gap between component-based approaches (e.g., using real-time control functionality inside functional components) and modern information system technologies (e.g., using state-of-the-art Java toolkits for graphical user interfaces) [4].

YARP (yet another robotic platform) [19] is an open source software library that enables module-based software development for robotic platforms. YARP modules run as separate executables and communicate over an extensible set of communication protocols such as TCP, UDP, MPI, etc. The communication patterns provided by YARP include publish and subscribe as well as client and server architectures. YARP was started in 2002 to support software development for humanoid robots such as the iCub [20].

To simplify large-scale software integration, the ROS (robot operating system) [30] was developed in 2007. It was not designed for a specific target application (as a wide variety of frameworks before) but designed to support the philosophy of modular tools-based software development [28]. A heavily growing ROS community proves the success of ROS' design principles, although it focuses mostly on Linux platforms and hardly provides information concerning the development approach.

In this context, the project BRICS (Best Practices in Robotics) proposed initial guidelines for component design resulting in better integratable components [5] according to the model-driven engineering (MDE) approach. Recently, a reuse-oriented development process for component-based robotic systems [3] was published following the developed design philosophy using the BRICS Integrated Development Environment (BRIDE) [5], which is a prototypical software tool chain. It includes a graphical user interface to support their approach for robotic system engineering.

Investigating the domain of industrial automation, wherein particular embedded devices have been pushing the technology more and more toward distributed control systems and also toward more service-oriented architectures [16], one comes across the established 4DIAC initiative based on IEC 61499 [13], which proposed an open source framework for distributed industrial automation and control since 2007 [34]. The 4DIAC framework consists of an IEC 61499-compliant automation and control environment for distributed systems. Currently, a modular runtime environment, an integrated engineering environment, a library of function blocks (FBs) and a set of example projects are available. The runtime environment FORTE supports the Windows operating systems, all Posix-compliant systems, NET+OS 7, eCos and VxWorks. Moreover, several evaluation boards (e.g., Digi Connect ME, Pandaboard) and PLC systems (e.g., Bachmann M1, Siemens EC31 with S7 IO modules) are supported.

The 4DIAC system has been applied in different domains. Examples are the control of an inverted pendulum where a state-space controller was dynamically replaced by a more complex state-space controller [38]. The Austrian Institute of Technology presented an example application of the usage of 4DIAC for smart grids applications where IEC 61850 was integrated into IEC 61499 FBs [33]. Our development approach, described in the following section builds on the standard IEC 61499, rather than its predecessor IEC 61131. This is because the programming paradigm of IEC 61499 is more suitable for distributing control intelligence to autonomous devices, leading away from centralized control structures. Moreover, the event-driven execution structure of FBs eases the development of applications to coordinate functional independent components, as required by the proposed integration methodology.

18.2.2   Integrated Technologies

Scripting languages are used in various domains, especially for rapid proto-typing and connecting software and web applications. The resulting interop-erability of interconnected heterogeneous systems or programming languages can be seen as a major advantage of the usage of scripting languages [22]. LUA is a C based, lightweight, and embeddable scripting language which is mainly used as extension language for software components [14]. Application fields of LUA are video games, mobile and embedded devices and automation applications. As an example for an industrial application, Girder [27], a soft-ware toolbox for industrial and home automation, uses LUA as base language. Software integration is eased by the provided LUA C-API. The execution of LUA is sped by using LuaJIT [23], a just-in-time interpreter library which extends the LUA base library package. A brief overview on many well-known scripting languages, including interesting facts about their history, is provided by [2].

ZeroMQ (ZMQ) is a high performance, asynchronous messaging library used in the fields of distributed computing and concurrency frameworks. Com-munication endpoints are based on common sockets, supporting protocols like inter-process, in-process, TCP and multicasts. The message distribution is based on a queue, which does not require a broker system to execute prop-erly. Based on basic communication patterns like publish-subscribe, request-response, and push-pull, a huge variety of approaches to implement a concur-rency framework is described in the user guide [11]. ZMQ is an open source project and its integration is based on an application programming interface (API), which is available in a wide range of programming languages [11].

CANDELOR SDK [10] is a software library for 3D scene interpretation. It contains an object localization algorithm to detect 3D CAD objects in a point cloud. CANDELOR works on full 3D point clouds. Configuration is performed automatically based on the CAD model. CANDELOR is a scalable solution for object localization. It can be used for realtime recognition as well as for high precision offline recognition.

18.3   Description of Robotic System

This section gives an overview on the hardware layout of the robotic sys-tem and its functional software components. The robotic system is used to demonstrate the applicability of the concepts presented in Section 18.4. These concepts were developed to support the automation and coordination of sev-eral robots performing an assembly task.

18.3.1   Hardware Layout

The tested robotic system consists of two independent, industrial manipulators of type Stäubli RX130B and Stäubli TX90L [29], each providing six degrees of freedom. Each robot is provided with a robot controller. The robot controller CS8 series in the VAL3 environment controls the robots for running various applications. To enable the manipulation of objects, each robot is equipped with a servo-electric two-finger parallel gripper PG70 from Schunk [32]. These grippers provide highly precise gripping force control and long stroke (about 34 mm per finger). Active feedback about the distance between the gripping fingers is also provided, to help recognize a successful grip by the robot. To enable the acquisition of current scene information, the system is equipped with a low-cost depth sensor of type Asus Xtion PRO [1]. The acquired scene data is represented as depth image based on a 3D point cloud. An overview on the hardware setup is given in Figure 18.1.

FIGURE 18.1
Hardware setup of the robotic system.

18.3.2   Experiments and Demonstration Use Case

The system setup as described in Section 18.3.1 was chosen to execute an assembly task in a number of variations. The main goal is to assemble fuse boxes by inserting a number of fuses into the appropriate boxes. By using two different types of object configurations (labelled type A and type B), given in Figure 18.2, the flexibility and adaptability of the system will be shown. Additionally to performing the assembly task, the robotic system had to meet the conditions given below.

- The robots should act in parallel to fulfill the task, thus sharing the workspace and avoiding robot-to-robot collisions and robot-to-object collisions.

- Fuses can be located in the workspace in arbitrary positions (lying, standing or upside-down), but need to be located on the workspace tray (see Figure 18.1).

- In case a robot cannot insert a fuse due to its initial orientation, the robots should cooperate to finish the insertion. As an example, robot A is able to pick a lying fuse. For insertion, robot A hands the fuse to robot B and it is able to insert the fuse.

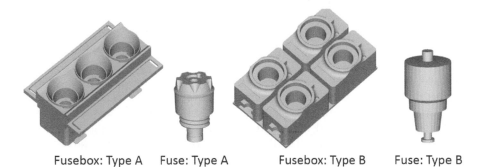

Fusebox: Type A   Fuse: Type A   Fusebox: Type B   Fuse: Type B

FIGURE 18.2
Assembly object types.

18.3.3   Required Functional System Components

This section gives an overview on the functional system components that are necessary to realize the execution of an assembly task as described in Section 18.3.2.

18.3.3.1   3D Vision System

The 3D vision module is founded on 3D object recognition and localization, based on CAD models of the objects of interest. This process builds on feature-based matching, which tries to detect correlations of the acquired depth image and pre-processed CAD models of the objects to be detected. The functionality is realized by the randomized global object localization algorithm

(RANGO) [10], which is an extension of the random sampling consensus algorithm (RANSAC) [24]. The RANGO procedure returns a list of transformation matrices, while each matrix describes the location and orientation of a detected object within the workspace. Additionally to object localization, the vision module provides complementary functionalities as listed below.

- Dealing with false positive detections by rejecting e.g., floating objects.

- Calculation of potential deposit and fixation positions for assembly operations.

As a prerequisite, the depth sensor needs to be calibrated relative to the robot workspace, thus sharing the same coordinate system as the two industrial robot systems.

18.3.3.2 Manipulation Planner: Collision-Free Path Planning

The results of the vision module are used to plan and calculate collision-free robot manipulation paths to enable handling of the detected objects. Based on predefined grasp as well as deposit points on the CAD model of the objects, the manipulation planner determines how the object can be grasped. The path planning algorithm is based on the rapidly-exploring random trees (RRT) approach [18][6]. In order to enable collision-free cooperation of both robots, a virtual wall to separate the workspace into two areas is dynamically inserted into the working area. Using the limited workspace, the first robot tries to find a collision-free path for grasping and manipulating objects. Among both robots, one is considered the master and is allowed to shift the wall until it finds a valid collision-free path. The other robot (slave) may use the remaining workspace for path planning.

Dependent on the selected grip point, the manipulation planner component decides whether a deposit path can be directly calculated or a hand-over is necessary. For performing an object hand-over, collision-free paths to predefined hand-over positions are planned. Dependent on the current position of the robots, the paths are executed in parallel or sequentially to avoid robot collisions.

18.3.3.3 Robot Control Module

To control the industrial robots, a ModbusTCP connection is established to send commands and receive status information from the Stäubli robot controllers. Thus, sending a set of joint configurations in sequence, resulting in a movement path, is possible. The two-jaw grippers of type Schunk Power-Cube, used for grasping operations, are controlled via their CAN interface, using the provided PowerCube API. By using force (current)-based gripper movements, an active feedback from the grippers can be interpreted, which is used to determine whether a grasp operation was successful.

18.4 Development Approach

One of the main tasks of system development is the establishment of an extendable system architecture, which includes heterogeneous and functional independent hardware and software system components. These components are based on different technologies and frameworks, thus leading to increased complexity during system integration and application development. The application development approach for the described robot team system that executes assembly tasks is founded on the standard IEC 61499. Previous work [7] showed the applicability of the main concepts of IEC 61499 to coordinate heterogeneous system components to achieve a certain system behavior. This section gives an overview on the concepts used for the integration of heterogeneous system components. Moreover, design patterns and methods which greatly aided the development of complex control applications will be presented. The concepts discussed in this section will be applied to the exemplary robotic assembly system.

18.4.1 Integration of Heterogeneous Software Components

The functional modules described in Section 18.3.3.1 and Section 18.3.3.2 are external high performance software components, which need to be integrated for application development, using the 4DIAC toolchain. The C++-based FB runtime environment FORTE can be extended by external software libraries. This is accomplished by using the build system of FORTE, which is based on CMake, to include the required library files and source code files. The included library functions and algorithms can be referred within a function block type. Figure 18.3 depicts a scheme explaining the extension process that is summarized below.

- By using the 4DIAC function block type editor, the interface of the service FB is designed. Moreover, entry points for the algorithms which should be invoked by the FB are specified. An execution control chart (ECC) can be defined to ensure that service functions can only be called in a defined sequence, e.g., initialization before algorithm execution.

- In the next step, source code stubs are generated for the FB type and included into the FORTE by configuring the build system accordingly. The algorithms, which also invoke functions from the included libraries, are now implemented within the generated source files.

- In the last step, the FORTE needs to be recompiled to include the external library functions.

Because these algorithms require large computational resources, a solution that avoids disturbing the default runtime behaviour was required. A threaded

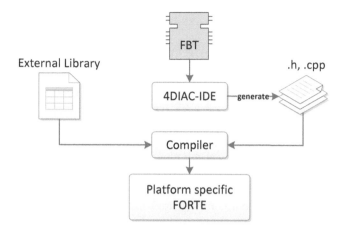

FIGURE 18.3
Extension of FB source code with external library functions.

function block type as described in Section 18.4.2 will be introduced to solve this issue.

Alternatively, the flexible communication infrastructure of the FORTE can be extended to establish a connection to an external component or facilitate distributed control architectures. In addition to the ASN.1 communication protocol, as described in the Compliance Profile for Feasibility Demonstration [13], the FORTE currently supports the fieldbus and communication protocols ModbusTCP, Ethernet Powerlink and OPC-DA. However, especially when dealing with large amounts of data (e.g., data produced by 3D sensors), which needs to be transferred in distributed systems, efficient messaging concepts are required. Therefore, support for ZMQ which is an asynchronous messaging library suitable to serve as scalable messaging protocol in distributed system architectures is realized. The protocol is especially designed to support parallel computing, without restricting the network topology, while having very low communication overhead.

18.4.2 Threaded Function Block Types

The execution of events by the FB runtime environment FORTE is performed by a separate event execution thread. If an event is raised, it is put into an event queue that is processed continuously. As the FORTE needs to respond to raised events as fast as possible, the main thread should not be blocked during the execution of an FB's algorithm. Especially in case of algorithms that require large computational resources, the event execution thread is very likely blocked.

To prevent blocking or suspension of the event execution thread due to time consuming algorithms, threaded function block types should be used to invoke such functionalities. Threaded FBs run another working thread, which is in charge of executing the FB's algorithms. After an algorithm is executed the working thread is suspended, and resumed when the next algorithm execution event is triggered. Figure 18.4 shows an example of an interface of a threaded Function Block type. The input event *REQ_LONG_RUNNING_WORK* is used to start a resource-intensive algorithm. The event outputs *STARTED_LONG_RUNNING_WORK* and *FINISHED_LONG_RUNNING_WORK* signal that the algorithm has been started, and has finished execution, respectively.

FIGURE 18.4
Threaded function block interface example.

18.4.2.1 Distribution of Services

Complex algorithms like image processing operations usually require extensive computational resources. To provide high overall performance, these algorithms can be distributed among the network. Figure 18.5 shows an example of a sequence of resource-intensive algorithms. The FBs model the invocation of a vision algorithm and a path planning algorithm. Sensor data serves as input for the vision algorithm.

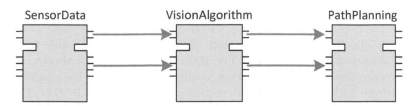

FIGURE 18.5
Application with algorithms encapsulated as FBs.

The extraction of algorithm FBs to a service can be done by inserting client and server FBs which act like a remote procedure call. Figure 18.6 depicts the extracted algorithm as a service. This enables the mapping of the service to a different resource. At this stage, only IEC 61499 features are required to model a distributed application.

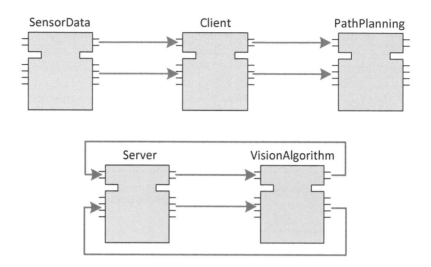

FIGURE 18.6
Encapsulating the algorithm FBs into a service.

To leverage the advantage of this concept, the algorithm (service) has to run on several servers. Every client is able to access the vision algorithm. The main feature is that the client does not know which server is actually invoked. For this, a client server and a router dealer with scheduler pattern provided by ZMQ [12] were added to the IEC 61499 runtime environment. For the client server architecture, ZMQ will automatically distribute all requests to the servers in a round robin scheduling. This pattern supports the addition of more clients, but the round robin scheduling will not guarantee equal distribution of the workload. Figure 18.7 shows the client server approach.

A more sophisticated pattern is the router dealer pattern where a central broker, as depicted in Figure 18.8, distributes all service requests to available servers. The main advantage is that clients and server are able to connect and disconnect from the broker. The broker has to deal with aborted requests and reschedule requests if a server is disconnected. The load balancer can be custom-made according to the requirements and available data. The main disadvantage of the pattern is that the broker is a single point of failure.

18.4.2.2   Service Controller with Lazy Connection Pattern

As explained in Section 18.4.2.1, the encapsulation of algorithms as a service is an effective development pattern. A major drawback of the approach is the resulting initialization chain of service and client function blocks. This chapter will introduce an extension to this pattern to prevent long initialization chains by adding a function block with the functionality to monitor and establish a connection.

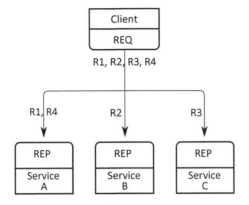

FIGURE 18.7
Client-server architecture with round robin distribution.

The main idea is to establish a connection to the service only if required, which is also known as the lazy initialization pattern. This means the connection established only if a request is triggered. This will be handled by the function block *servicecontroller*. The function block monitors the current connection state and decides whether the connection has to be established or the request has to be dropped because of a pending request. Additionally, the function block is monitoring the current request. If the request times out, the connection is disconnected and re-established if a new request occurs. The logic is shown as ECC in Figure 18.9.

The typical usage is shown in Figure 18.10, which depicts a composite function block that encapsulates a service as described in Section 18.4.2.1, as well as the introduced *servicecontroller* FB. As the *servicecontroller* handles the initialization routine, dedicated INIT and INITO events are not necessary. The example includes an additional *E_DELAY* function block to monitor the connection and requests. If the connection or the request exceeds the specified time-out, the connection is closed and the status QO will be set to false.

The pattern can be expanded with additional functionalities, like automatically disconnecting the connection if it is not needed for a given time or automatically retrying a request upon failure.

18.4.3 Application Design Patterns

During the implementation of the generic control application logic, a number of design patterns for IEC 61499 function block networks were applied. Moreover, certain FB types for the evaluation of logical conditions were implemented. This sections deals with an overview and a description of the techniques and patterns used.

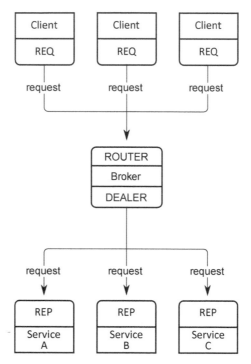

FIGURE 18.8
Request replay broker.

18.4.3.1   Flexible Service Parameterization Pattern

Data multiplexers can be effectively used in combination with simple event multiplexers, to perform a flexible parameterization pattern for services, as depicted in Figure 18.11. An event multiplexer (E_MUX) is used to determine the parameterization case for the subsequent service function call of the service FB. Moreover, a data multiplexer (F_MUX) is required to select the parameterization values forwarded to the service. The execution confirmation events are split according to the given parameterization case using event demultiplexer (E_DEMUX) for every service confirmation event.

Apart from enabling flexible parameterization of a service, this pattern realizes a one-to-one mapping of a service to a certain hardware or software component. Such a mapping is required if a functional component can only be accessed through one connection at a time.

18.4.3.2   LUA Scripted FB Types for Rapid Prototyping of Algorithm and Coordination FBs

Based on the work in [25], generic FB types as well as FBs that execute LUA scripts can be used to enable rapid prototyping of algorithm and service FBs.

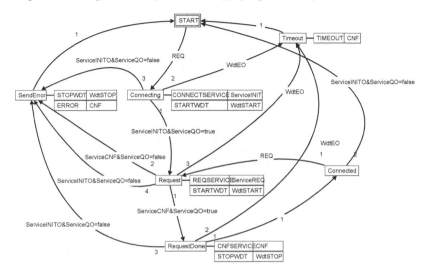

FIGURE 18.9
Logic of the *ServiceController* as ECC.

LUA Scriptable Basic FB types

In [25] it was shown that LUA scripts can be effectively used to realize online interpretable ECCs as used in basic FB types. Online interpretability of basic FB types implies the requirements as follows.

- Guard conditions of ECC transitions and algorithm expressions within an ECC action need to be interpreted while loading the FB.

- The execution of the ECC needs to be configurable based on the extracted ECC information and the interpreted algorithms and conditions.

To realize the online interpretation of expressions and conditions, the scripting language LUA [14] was integrated into the FORTE in our previous work [25]. As a basic FB's algorithms and conditions are usually coded in IEC 61131-compliant structured text language (ST) [15], they need to be exported into LUA expressions using a ST-TO-LUA-exporter. Figure 18.12 depicts the approach to make the ECC directly configurable based on LUA condition expressions and algorithms. ST expressions are transformed into LUA code and returned as C-String. An executable LUA script is created by putting the resulting LUA expressions into a LUA function, which can be pushed onto the LUA stack and run during the FB's execution. After construction, the script is compiled and in case of success, a function pointer referring to the script's execution method is created.

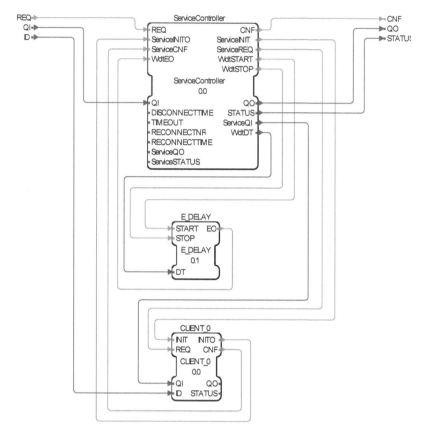

FIGURE 18.10

Typical usage of the *ServiceController* with lacy initialization.

During script execution, internal variables, data input and data output values are registered as global variables containing their current values and then the script is run by calling *lua_pcall*. Once the script execution is complete, the updated values are fetched from the globally registered LUA variables and assigned to the FB's variables. LUA-based FB types also need to be stored in the type library of the FORTE. A dynamic FB type library as described in [25] is used to include interpreted FB types during runtime startup. The dynamic type library approach leads to reduced development time, since recompiling of the FORTE is not necessary after changes in an FB's logic.

Extension of Generic FBs using LUA Scripts

FBs which perform simple algorithms or logical operations often vary only in their interfaces, e.g., having different numbers of equally treated data in-

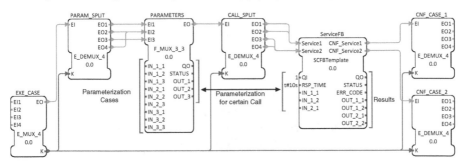

FIGURE 18.11
Flexible parameterization of service FBs.

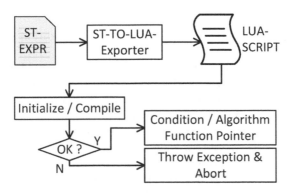

FIGURE 18.12
Structured Text to LUA script transformation and compilation.

puts/outputs or event inputs/outputs. This would require that every FB type having the same functionality and little difference in the interface needs to be compiled into the FORTE runtime environment. To overcome this problem, generic FB types whose interfaces are directly derived from the corresponding FB type names as described in [25] are used.

The flexibility of generic FBs can be combined with LUA scripts to facilitate changes of the execution logic, e.g., for the evaluation of logical conditions or simple expressions. Figure 18.13 depicts a generic IF FB type, which is used to evaluate logical conditions, similar to an if-statement. The ANY data inputs are used to pass parameter values, which are included in the condition string (LUA syntax required) specified using the COND input. The condition $\# IN\_1 / 2 < IN\_2$, where the unary $\#$-operator returns the length of its operand IN_1, is not satisfied and the ELSE event output is triggered. The LUA script is constructed by including the specified condition into a prede-

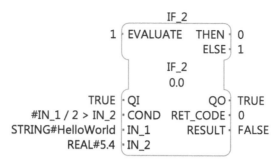

FIGURE 18.13
Generic IF FB type.

fined LUA code stub, and recompiled only if the condition is changed during runtime.

As a main advantage, the condition expression can be changed during runtime, since the resulting script just needs to be recompiled through the LUA interpreter. Comparable to the generic IF FB type, Figure 18.14 shows the implementation of the same behavior using a discrete FB application logic. In terms of execution speed, the discrete FB application executes about three times faster than the scripted, generic IF FB. This is due to the LUA script execution call as well as the data transfer to and from global LUA variables. However, the discrete FB network lacks flexibility as the condition logic is fixed. Changes in the condition logic directly imply changes of the FB network.

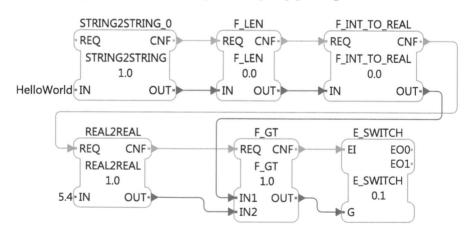

FIGURE 18.14
IF realized as FB network.

Benefits of Using Scriptable Function Blocks

The usage of scriptable function blocks in combination with a dynamic type library, as discussed in [25], enables rapid prototyping of logic and algorithm FB types without requiring continuous re-compilation steps of the runtime environment FORTE. During the development phase, changes in an FB's execution logic are adopted by simply restarting the FORTE. As the LUA FBs are dynamically loaded based on standard FB type files, they can be easily exported as C++ files and compiled into the FORTE, once development and tests are done.

As discussed in Section 18.4.3.2, generic, scriptable FB types reduce the complexity of FB applications, since the evaluation of simple expressions or conditions can be realized using a single FB instance, instead of discrete application logics. Moreover, scriptable logic FB types provide flexibility regarding their parameterization, as scripts can be updated during runtime.

18.5 Resulting System Architecture

By integrating the functional components introduced in Section 18.3.3 according to the methodologies described in Section 18.4, a system architecture, as depicted in Figure 18.15, was derived.

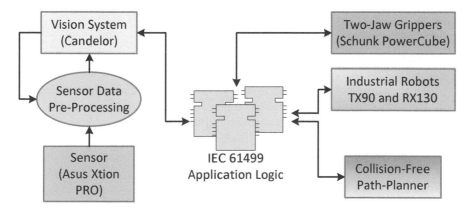

FIGURE 18.15
Resulting system architecture of workspace sharing robotic assembly system.

To realize an assembly task as described in Section 18.3.2, the system components depicted in Figure 18.15 need to be coordinated through an application logic. The goal is to develop a generic application logic based on IEC 61499 concepts, which enables the robotic system to perform manipulation and assembly operations. Figure 18.15 shows the communication relations between the logic and the functional components. The execution sequence of functions within the generic application logic is summarized below.

- Scene data acquisition and preprocessing of the acquired point cloud (including noise reduction)

- Object recognition and localization to find positions and orientations of objects of interest

- Planning of collision-free manipulation paths for object manipulation and assembly; robot interaction (e.g., handover of a grasped object) possible

- Execution of manipulation paths and gripper coordination

Dependent on the specific assembly task, various instances of the generic application logic that are parameterized differently are executed and coordinated by a task-specific management application (controller). Figure 18.16 depicts an example in which two instances of the generic logic are used to manipulate two different types of objects. Additionally, a logic used to control a simple graphical user interface (GUI), based on WxWidgets [36], enables the human operator to interact with the system. The management application is responsible for the coordination of the logic instances and for data exchange. The major advantage of this architecture is its continuously hierarchical structure, thus enabling application extension with reasonable effort in time.

Assembly Task Use Case

With reference to Section 18.5, a specific assembly task use case is described in this section, based on the application architecture in Figure 18.16.

Two instances of the generic application logic capable of manipulating objects, are created: one to handle fuses and the second to manipulate fuse boxes (see Figure 18.2). The controller application is use case-specific and coordinates the object-manipulating application instances to fulfill the assembly task. An assembly scenario as follows is assumed.

- One fuse box and three fuses need to be assembled

- The fuse box and one fuse are located close to robot A and the remaining fuses are close to robot B

- As the second and the third fuse are not located within the reachable area of robot A and the fuse box cannot be reached by robot B, cooperation between both robots is required to perform the task

The required functionality of the controller application is summarized in Figure 18.17.

During system startup, all FB applications according to Figure 18.16 are initialized. Once the generic application instances to manipulate fuses and fuse boxes have been initialized, the assembly procedure is started by triggering the

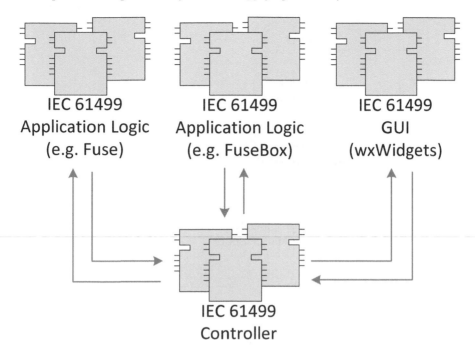

FIGURE 18.16
Multiple application logic instances handling different object types are coordinated by the main controller.

Start signal of the GUI. The controller application receives the initialization states as well as the signal from the GUI, and triggers the fuse-application to perform a fuse assembly task. Assembling fuses is repeated until the insertion fails for the first time, for the possible reasons as follows.

- There are no fuses left to be inserted.

- There is no empty slot.

- The manipulation planner cannot calculate collision-free grasping and insertion movement paths for the robots.

The last potential reason includes objects that are not reachable by a robot. An example of such a situation is explained at the beginning of Section 18.5. In this case, the assembly problem could also not be solved e.g., through handing over a fuse robot A to robot B. Consequently, according to Figure 18.17, the fusebox-application is triggered to move the fuse box into a common work area that is reachable by both robots. To complete this operation successfully, the fuse insertion procedure is repeated until all fuses are inserted and the assembly task is finished. Note that for simplification reasons, Figure 18.17 does not cover additional handling routines, e.g., for object localization errors.

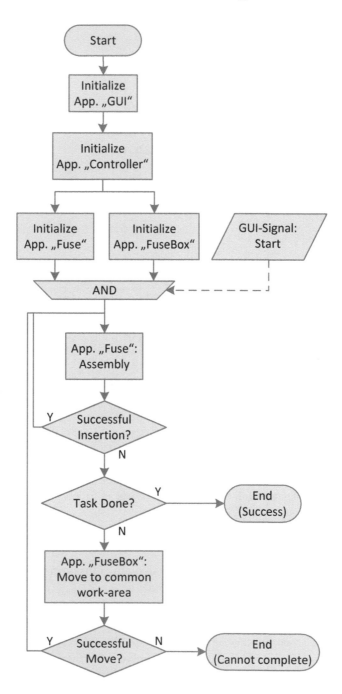

FIGURE 18.17
Assembly task procedure.

18.6   Summary and Conclusion

In this chapter, a collection of system integration and application development concepts are presented. These concepts were applied to a specific robotic application scenario where two robots sharing a common workspace cooperate to perform an assembly task. To reduce the complexity during application development, every control application builds on services that encapsulate component functionality. Hence, to ensure correct behavior and stability of a control application, thorough tests and revision cycles are necessary to be performed for every service. Reusing well-tested and proven software components, greatly reduces developing and debugging effort during integration and application development. The high level of flexibility of the 4DIAC framework enables the fast integration of components and application development based on the introduced development patterns. During the integration and development work, the authors identified a number of helpful features for increasing the development flexibility and reducing work effort. Examples are a monitoring functionality for composite FB types and an FB type export filter for threaded FB types. Additionally, the 4DIAC tool functionality to extend the interface of generic FB types dynamically, within an application is required. This could reduce the number of type files for each interface configuration of a generic FB type within the tool's type library and improve clarity.

Bibliography

[1] ASUSTeK Computer Inc. ASUS Xtion PRO. http://www.asus.com/Commercial_3D_Sensor/Xtion/.

[2] N. Bezroukov. Criptorama: A Slightly Skeptical View on Scripting Languages. http://www.softpanorama.org/Scripting/index.shtml.

[3] D. Brugali, L. Gherardi, A. Biziak, A. Luzzana, and A. Zakharov. A reuse-oriented development process for component-based robotic systems. In *Simulation, Modeling, and Programming for Autonomous Robots*, pages 361–374. Springer, 2012.

[4] D. Brugali, L. Gherardi, M. Klotzbücher, and H. Bruyninckx. *Service Component Architectures in Robotics: The SCA-OROCOS Integration.* Springer, 2012.

[5] H. Bruyninckx, M. Klotzbücher, N. Hochgeschwender, G. Kraetzschmar, L. Gherardi, and D. Brugali. The BRICS component model: A model-based development paradigm for complex robotics software systems. In

Proceedings of 28th Annual ACM Symposium on Applied Computing, pages 1758–1764. ACM, 2013.

[6] J. Capco, M. Rooker, and A. Pichler. RRT planner for the binpicking problem. In *Robot Motion and Control, 9th Workshop on*, pages 154–160. IEEE, 2013.

[7] G. Ebenhofer, H. Bauer, M. Plasch, S. Zambal, S. C. Akkaladevi, and A. Pichler. A system integration approach for service-oriented robotics. In *Emerging Technologies & Factory Automation, 18th Conference on*, pages 1–8. IEEE, 2013.

[8] T. Erl. *Service-Oriented Architecture: Concepts, Technology, and Design.* Prentice Hall, Upper Saddle River, 2005.

[9] Y.-G. Ha, J.-C. Sohn, Y.-J. Cho, and H. Yoon. A robotic service framework supporting automated integration of ubiquitous sensors and devices. *Information Sciences*, Vol. 177, No. 3, 657–679, 2007.

[10] P. Hintjens. CANDELOR: Computer vision library for 3D scene interpretation. http://candelor.com/.

[11] P. Hintjens. Zguide: ZeroMQ Guide. http://zguide.zeromq.org/ page:all.

[12] P. Hintjens. *Code Connected Volume 1: Learning ZeroMQ.* iMatrix Corporation, 2013.

[13] IEC 61499-1. *Function Blocks. Part 1: Architecture.* International Electrotechnical Commission, Geneva, 2005.

[14] R. Ierusalimschy, W. Celes, and L. Figueiredo. What is LUA?. http://www.lua.org/about.html.

[15] International Electrotechnical Commission. IEC 61131-3. *Programmable Controllers. Part*, Vol. 3, , 1993.

[16] F. Jammes and H. Smit. Service-oriented paradigms in industrial automation. *IEEE Transactions on Industrial Informatics*, Vol. 1, No. 1, 62–70, 2005.

[17] D. Kononchuk, V. Kandoba, S. Zhigalov, P. Abduramanov, and Y. Okulovsky. Robocop: A protocol for service-oriented robot control systems. In *Research and Education in Robotics*, pages 158–171. Springer, 2011.

[18] J. J. Kuffner and S. M. LaValle. RRT-connect: An efficient approach to single-query path planning. In *Proceedings of IEEE International Conference on Robotics and Automation*, volume 2, pages 995–1001. IEEE, 2000.

[19] G. Metta, P. Fitzpatrick, and L. Natale. Yarp: yet another robot platform. *International Journal on Advanced Robotics Systems*, Vol. 3, No. 1, 43–48, 2006.

[20] L. Natale, F. Nori, G. Metta, M. Fumagalli, S. Ivaldi, U. Pattacini, M. Randazzo, A. Schmitz, and G. Sandini. The ICUB platform: a tool for studying intrinsically motivated learning. In *Intrinsically Motivated Learning in Natural and Artificial Systems*, pages 433–458. Springer, 2013.

[21] OROCOS. Open Robot Control Software. http://www.orocos.org.

[22] J. K. Ousterhout. Scripting: Higher level programming for the 21st century. *Computer*, Vol. 31, No. 3, 23–30, 1998.

[23] M. Pall. The LuaJIT Project. http://luajit.org/index.html.

[24] C. Papazov and D. Burschka. An efficient RANSAC for 3D object recognition in noisy and occluded scenes. In *Computer Vision–ACCV 2010*, pages 135–148. Springer, 2011.

[25] M. Plasch, M. Hofmann, G. Ebenhofer, and M. Rooker. Reduction of development time by using scriptable IEC 61499 function blocks in a dynamically loadable type library. In *Emerging Technology and Factory Automation*, pages 1–8. IEEE, 2014.

[26] M. Plasch, A. Pichler, H. Bauer, M. Rooker, and G. Ebenhofer. A plug & produce approach to design robot assistants in a sustainable manufacturing environment. In *22nd International Conference on Flexible Automation and Intelligent Manufacturing, Helsinki*, 2012.

[27] Promixis. Girder: Home and Industrial Automation Software. http://www.promixis.com/index.php.

[28] M. Quigley, K. Conley, B. Gerkey, J. Faust, T. Foote, J. Leibs, R. Wheeler, and A. Y. Ng. ROS: an open-source robot operating system. In *ICRA Workshop on Open Source Software*, volume 3.2, page 5, 2009.

[29] S. Robotics. Staeubli Robotics. http://www.staubli.com/en/robotics/.

[30] ROS-Project. ROS âĂŞ Robot operating System. http://www.ros.org/.

[31] SCA. Service Component Architecture. http://www.osoa.org/.

[32] Schunk. Schunk: Superior Clamping and Gripping. http://www.schunk.com/.

[33] T. Strasser, F. Andren, and M. Stifter. A test and validation approach for the standard-based implementation of intelligent electronic devices in smart grids. In *Holonic and Multi-Agent Systems for Manufacturing*, pages 50–61. Springer, 2011.

[34] T. Strasser, M. Rooker, G. Ebenhofer, A. Zoitl, C. Sunder, A. Valentini, and A. Martel. Framework for distributed industrial automation and control (4DIAC). In *6th IEEE International Conference on Industrial Informatics*, pages 283–288. IEEE, 2008.

[35] G. Veiga, J. Pires, and K. Nilsson. Experiments with service-oriented architectures for industrial robotic cells programming. *Robotics and Computer-Integrated Manufacturing*, Vol. 25, No. 4, 746–755, 2009.

[36] WxWidgets. WxWidgets: A Cross-Platform GUI Library. https://www.wxwidgets.org/.

[37] T.-H. Yang and W.-P. Lee. A service-oriented framework for the development of home robots. *International Journal of Advanced Robotic Systems*, Vol. 10, , 2013.

[38] A. Zoitl. *Real-Time Execution for IEC 61499*. ISA, 2008.

19

Hierarchically Structured Control Application for Pick and Place Station

Monika Wenger

fortiss GmbH

Milan Vathoopan

fortiss GmbH

Alois Zoitl

fortiss GmbH

Herbert Prähofer

Johannes Kepler University Linz

CONTENTS

19.1   Introduction

A key point for automation system development is to reduce the control software engineering effort. With its strong focus on encapsulation and reuse, IEC 61499 claims to improve the situation for control engineers. The IEC 61499 defines only means and models to develop control systems in a component-oriented way. The key point, however, how to use the models of IEC 61499 to achieve the goal of modular reusable control applications, is beyond the current scope of the standard.

In previous works [3, 4] we developed a structural design pattern guiding control engineers for structuring control applications for mechatronical machines. The core idea behind the design pattern is a strong hierarchical approach where higher levels are coordinating underlying levels and receive commands from a higher level coordinator. The design patterns require that the control software hierarchy strictly follow the mechatronical hierarchy. This has the great advantage that each mechatronical component of a machine has a counterpart in the software structure, which makes the control software easier to navigate. Additionally changes in the mechatronic setup of a machine can be easily re-done in the control software and the changes are isolated to the area of the mechatronic change.

In addition to the hierarchy, the design pattern also defines that IEC 61499 adapter interfaces are to be used for defining the component's interaction interfaces. This further decouples the elements of the control application, which increases reusabilty.

In this chapter we show how to utilize this hierarchical design pattern by applying it to an educational pick and place station. Section 19.2 will recap the design pattern. In Section 19.3 we describe the resulting hierarchical structure of our control application. This is followed by a detailed description of the components and their interactions in Section 19.4. A conclusion wraps up this chapter.

19.2   Principles for Hierarchical Structured Control Applications

The principles for the design of hierarchically structured control applications [4] are summed up within this section. These principles for layered design of control applications are applied on control applications set by International Electrotechnical Commission (IEC) 61499.

19.2.1 Layered Structure

The layered structure of a control application arranges components at different hierarchy levels that provide services at various design depths. One layer offers services to the layer above and receives services from the layer below. Each component provides a clearly defined interface and resembles the hierarchical structure of the machine. Every software component is designed to be dependent on the mechanical structure of the plant and represent a specific hardware component. The first step for the design of a layered structure is to examine the mechanical structure of the desired plant. To implement a hierarchical control architecture, a bottom-up strategy is advisable since it prevents the creation of unnecessary layers. It also helps to leverage the modular architecture, and facilitates reuse of components and the flexibility of the system in general.

The components of a lower layer are coordinated and controlled by components of a higher layer and perform lower control tasks, whereas components of a higher layer perform higher control tasks. Components for the higher layer are determined based on the application and its complexity. The interaction between the components is achieved by defined interfaces between components of higher and lower layers. The commands are send from components of the higher layer to components of a lower layer and feedback is sent in the reverse direction. The coupling of the different components is hierarchical, but the behavior is implemented by the components of the lowest layer. Basically there are four types of layers:

- Layer 0: components of this layer implement the interface to the hardware like sensors or actuators, and can also implement elementary continuous control loops.

- Layer 1: components of this layer directly control one or more layer 0 component to combine elementary actions within elementary control operations and provide them as self-contained control operations to the next higher layer. These components set actuators and turn respective control loops on or off, supervise sensor values from the lower level components to determine whether the process correctly works, and report errors to the next higher layer.

- Layer 2 to layer n: components of these layers coordinate the activities of their subcomponents, synchronize potentially parallel activities, shut down any subcomponent process safely, check for the occurrence of errors, and report to connected components of a higher layer.

- Layer n: components of this layer provide overall process control for a plant.

19.2.2 Layering with IEC 61499

IEC 61499 [2] provides an application-centric engineering approach featuring a device-independent application model and a late hardware binding through its distribution model. Zoitl and Prähofer [4] present some design guidelines and rules for structuring component-oriented hierarchical automation solutions implemented with IEC 61499. The following guidelines are based on this work.

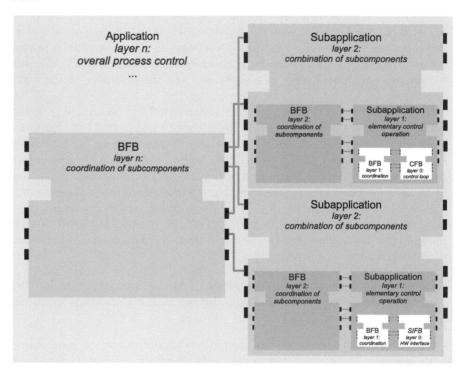

FIGURE 19.1
Hierarchical layers applied on IEC 61499 and based on [4].

The highest layer n is represented by an IEC 61499 application, as illustrated in Figure 19.1. Every layer has to provide a coordination function block (FB) for its subcomponents. Such a coordination FB should be realized by the execution control chart (ECC) of a basic function block (BFB), since an ECC coordinates the behavior for different events by transitions to a specific state that performs a proper action. When considering error handling, ECCs can become rather complex. To reduce the complexity, one error transition from each state to a common error state can be used.

Lower level components can be realized by any IEC 61499 FB type, like composite function blocks (CFBs), BFBs, or service interface function blocks (SIFBs), whereas the atomic components of layer 0, which usually provide hardware access, have to be realized by SIFBs. Any component is therefore realized by one or more IEC 61499 FBs.

In order to organize the components hierarchically CFBs or subapplications are provided by IEC 61499. FB networks inside CFBs are not distributable and therefore not sufficient to build hierarchical structures. The groupings of components or FBs are performed by subapplications within IEC 61499. Subapplications help reduce the number of FBs and thus the complexity of the application in general. They model a hardware abstraction and hide the inner complexity of a system in general. The hierarchical organization requires platform-independent SIFBs for hardware access on layer 0, which can be configured within the resource after mapping.

Since component interfaces can become quite large, adapters provided by IEC 61499 are utilized to model the interactions between layers. Adapters group received and sent events and data to improve clarity and understandability of components. Due to their grouping functionality, they reduce the number of required connections. Any adapter represents services offered or required by a component. The two types of adapters within IEC 61499 are plugs and sockets. Sockets are used in any component to control the service in terms of starting or stopping or updating. Since ECC do not support timeout conditions, components of the first layer use a timeout adapter `ATimeout` connected to a `E_TimeOut` CFB which provides the timeout functionality.

19.3 Structure of Pick and Place Station

The pick and place station from Festo Didactic is an automated (dis)assemble device that (dis)assembles a work piece insert (clock, thermometer, hygrometer) into a work piece housing [1]. This station consists of five modules: a conveyor belt with a pneumatic separator, a stack for up to six work piece inserts, a pneumatic handling device, and a status light. The mechanical structure of the pick and place station is illustrated in Figure 19.2.

The conveyer belt is responsible for transporting the work piece(s) through the station. It is equipped with two diffuse sensors, one at the entrance of the conveyor belt which detects the work piece housing entering, and one in the middle before the assembly point. A light barrier sensor at the exit of the conveyor belt detects the completed product leaving the belt. The work piece housing enters the station by passing the first diffuse sensor, which causes the start of the conveyer belt. The conveyor belt transports the work piece to the pneumatic separator, which detects the arriving piece by the second diffuse sensor, and stops the part. The pneumatic separator is a lever located in the assembly which stops the work piece housing at the assembly point. The slide stores the work piece to be inserted. The pneumatic manipulator is a two-axis device which picks the work piece to be inserted from the stack

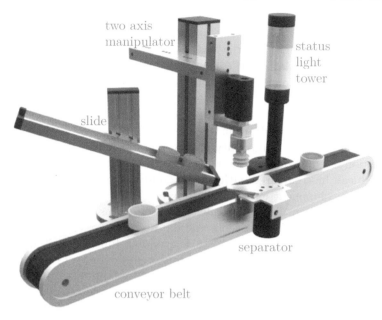

FIGURE 19.2
Structure of pick and place station based on [1].

and places it on the work piece housing to form the completed product. The completed product is released by the separator and transported to the end of the conveyor belt, where it is detected by the light barrier sensor. The green, red, and yellow status lights indicate the mode of the station. An empty stack is indicated by the vacuum sensor of the manipulator and a red blinking status light. The initialization phase is indicated by a orange blinking status light. A running pick and place station is indicated by a green status light.

19.4 Hierarchical Control Application

Initially, following the design principles mentioned above, the mechanical structure of the station is thoroughly analyzed and different components are identified at different levels. The conceptual design of the hierarchically structured control application is then built based on the identified components resembling the overall structure of the the pick and place station, as shown in Figure 19.3. The lowest atomic components which represent digital inputs and outputs (IOs) are placed in the lowest level 0. In layer 1 elementary components, *status light, horizontal axis, vertical axis, gripper, part count, separator, transport* are placed to coordinate the functions of atomic components and perform elementary control operation. The modules *status management*, *manipulator* and *conveyor* are arranged in layer 2, coordinating the underlying elementary components. For controlling the overall process, a component called *picknPlace* stays on top level 3.

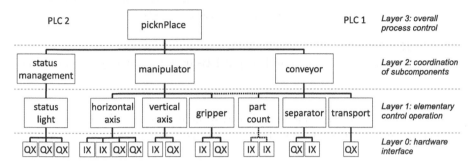

FIGURE 19.3
Hierarchical structure of pick and place station components based on [4].

The hierarchically structured control application for the pick and place station is realized within the open-source framework for distributed industrial automation and control (4DIAC) [5], an implementation of the IEC 61499 standard. The application is developed considering reuse and reconfiguration of the station and its components.

19.4.1 Layer 0: Atomic Components

Atomic components are arranged in the lowest layer 0. These components interact directly with the hardware. Many elements such as motor, valves, and switch type sensors can be modeled directly with digital IO FBs. These FBs get information from any sensor or control any actuator. They provide a hardware interface and need to be modeled as a SIFB within IEC 61499. These SIFBs have to be platform-independent that they can be used on any programmable logic controller (PLC) as well as the upper level FBs. Within 4DIAC IX and QX FBs have been designed, which are generic digital IO FBs. All actuators are modeled with QX FBs and all sensors are modeled with IX FBs, as illustrated by Figure 19.4.

The PARAMS input defines a parameter which supports the specification of the desired input and output numbers, respectively. The parameter setting is hardware-dependent. Upper level components that use these IX and QX FBs for hardware access are dependent on this specification. A direct specification of the IO number prevents any upper level component to be instantiated multiple times within one PLC, since different sensors or actuators cannot use the same IO. This is why a PARAMS input has to be specified by the virtual domain name service (DNS) functionality of 4DIAC. The virtual DNS functionality helps

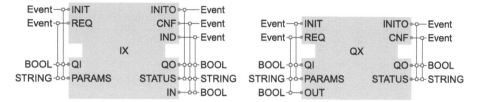

FIGURE 19.4
Platform-independent input and output FBs used for hardware access on layer 0.

create a list of symbolic names and corresponding values. The symbolic names are replaced by the corresponding values during download.

19.4.2   Layer 1: Elementary Components

Elementary components connect atomic components, provide an independent function and can be reused in other applications of the same domain. These components are identified based on the hardware structure of the station and are generalized for better reusability. The elementary components designed for the pick and place station are explained in the following sections according to their corresponding upper layers.

19.4.2.1   Manipulator-Related Components

Considering the two-axis pneumatic handling device, the horizontal axis can be modeled by a double acting cylinder with state information in both directions (feedback from proximity sensors on both sides) and the vertical axis with a single acting cylinder with state information only in one direction (feedback from sensor on a single side). The elementary components for the handling unit are double acting and single acting cylinder and a vacuum gripper.

The double acting cylinder is implemented with a BFB. The interface for the services offered by the component (command from the higher level and state information back to the higher level) are structured with the adapter `ACylinder`, as illustrated in Figure 19.5. In this way the `ACylinder` adapter is an abstraction of the functionality offered by the cylinder.

The cylinder has two input events `Retract` and `Extend` and two confirmation events `Retracted` and `Extended`. If the cylinder reaches the maximum end position after an `Extend` event, it sends the confirmation event `Extended`; if it fails to reach this position, it sends an `ExtendFailed` event. This is similar for retracting. If the cylinder reaches the minimum end position after a `Retract` event, it sends the confirmation event `Retracted`; if it fails to reach the this position, it sends an `RetractFailed` event. Apart from the `ACylinder` adapter, the double acting cylinder is connected to two solenoid valves for ex-

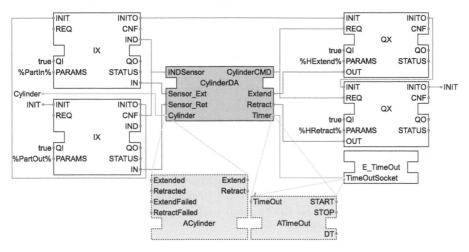

FIGURE 19.5
Layer 1: Subapplication to control the double acting cylinder with double acting cylinder FB for the entire control (gray), adapters providing the moving and the timeout services (light gray), connections to the timeout service provider and the layer 0 FBs for hardware access (white).

tend and retract operations and receives inputs from two proximity sensors. Figure 19.5 shows the double acting cylinder FB to control the horizontal axis and its connections to the layer 0.

The single acting cylinder is also implemented by a BFB. The functionality of the single acting cylinder is similar to that of the double acting cylinder. The services offered by the single acting cylinder for the communication with the upper layer are also modeled with the `ACylinder` adapter. In contrast to the double acting cylinder, there exists only one solenoid valve for extending the cylinder and the retract operation is performed with a spring plunger. This is why the interface only contains one output for the actuator and receives feedback from only one sensor to detect a retracted cylinder.

The gripper is also implemented by a BFB, as illustrated in Figure 19.6. The services offered by the gripper for the communication with the upper layer are modeled with an `AGripperPneumatic` adapter. This adapter achieves the services grip and release by two corresponding events. Feedback is received from the sensor by a `Gripped` event after a `Grip` event has occurred. If no feedback is received within 5 s, a `GripFail` event is sent. For the release service, a similar functionality is realized by the `Release`, the `Released` and the `RelaseFail` events. This FB feeds control input to the suction actuator solenoid valve and receives input from the pressure sensor.

All three components use an `ATimeOut` adapter to model timeout functionality. These components are all connected to an `E_TimeOut` CFB which is responsible for implementing the requested delay.

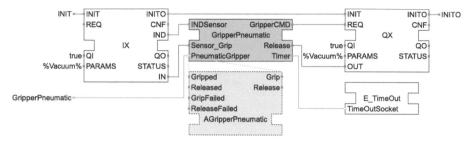

FIGURE 19.6
Layer 1: Subapplication to control the gripper with a pneumatic gripper FB (gray) for the entire control, an adapter providing the gripping service (light gray), and connections to the timeout service provider and the layer 0 FBs for hardware access (white).

19.4.2.2 Conveyor-Related Components

The conveyer belt can be divided into transport, separator and a part counter components. The transport component is realized by a BFB, as illustrated in Figure 19.7. It is the actual conveying unit for the work piece and basically represents a motor which starts and stops the operation of the conveyer belt.

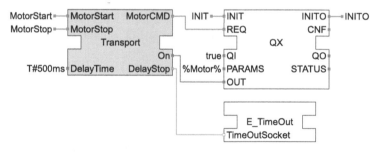

FIGURE 19.7
Layer 1: Subapplication to control the transportation with a transport FB for the entire control (gray) and its connections to the layer 0 FB for hardware access and (white).

This FB operates completely on command and has no status feedback. The transport component has `MotorStart` and `MotorStop` events. The transport component feeds control input to the actuating unit of the conveyer belt. A timer is attached in the interface to support a delay to turn off the motor in case the light barrier that detects completed work pieces cannot be located at the very end of the conveyor.

A separator is a rotating lever used to separate individual components from a group. It detects arriving parts by a diffuse sensor. In the default position, it does not allow any components to pass through. When it rotates

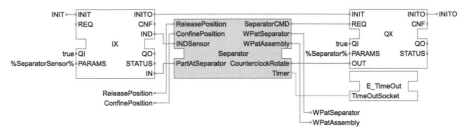

FIGURE 19.8
Layer 1: Subapplication to control the separator with a separator FB for the entire control (gray) and connections to the timeout service provider and the layer 0 FBs for hardware access (white).

anticlockwise, it can grab a work piece and position this work piece for any processing. If it rotates clockwise back to the default position, any grabbed work piece is released. The separator component is also realized by a BFB which has **Release** and **Confine** events to control the separator's position, as shown in Figure 19.8. This block feeds input to the solenoid of the separator and takes input from the diffuse sensor in front of the separator. An external timer is connected in the interface of the separator component to provide a time delay after a release. The time delay prevents the separator from rotating to grab a work piece until the just-released work piece is outside the separator's working space.

The part counter component counts the number of parts on the conveyor belt dependent on the incoming parts which are detected by the proximity

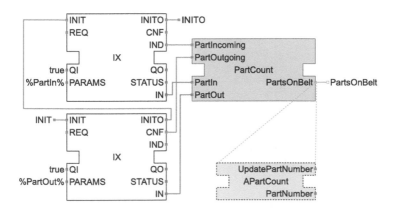

FIGURE 19.9
Layer 1: Subapplication to count the parts on the belt with a part counter FB (gray), an adapter providing the counting service (light gray), and connections to the layer 0 FBs for hardware access (white).

sensor at the beginning of the conveyor belt and the outgoing parts detected by the light barrier at the end of the conveyor belt. This component is also a BFB as shown in Figure 19.9. To prevent multiple detection of work pieces, a debouncing of 100 ms is implemented for the increment and decrement of the number of work pieces. The actual number of parts on the conveyor belt is provided as a service to the next upper layer by an `APartCount` adapter.

19.4.2.3   Status Light Component

The status light provides indications of the station. It consist of a tower of red, orange, and green lights. The status light is realized with a CFB, as illustrated in Figure 19.10. The service offered by this component is provided to the next upper level with an `AStatusLight` adapter which triggers events to control the three lights that glow or blink red, orange and green, with an adjustable blink frequency set to 2 Hz.

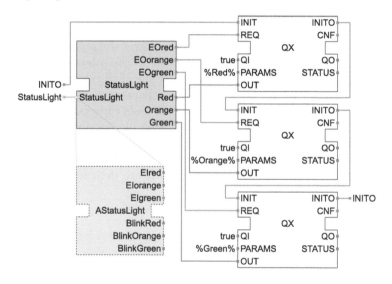

FIGURE 19.10
Layer 1: Subapplication to control the status light with a status light FB for the entire control (gray), an adapter providing the indication service (light gray), and its connections to the layer 0 FBs for hardware access (white).

19.4.3   Layer 2: Coordinating Components

The coordinating components in layer 2 coordinate the operation of the elementary components in layer 1. Considering the station as a whole, four hardware modules were identified: the slide, the manipulator, the conveyer, and the status light.

19.4.3.1 Slide

The slide does not have to be modeled since it is just a physical unit and no sensor or controlling function is associated with it. The slide stores the work piece inserts and transports them to the pick and place unit. The slide function is achieved with the inclination and height variation available in the physical design.

19.4.3.2 Manipulator

The manipulator component is responsible for coordinating the underlying module horizontal axis, vertical axis and pneumatic gripper which are implemented by the subapplications `CylinerDAControl`, the `CylinderSAControl` and the `GripperControl`, respectively. To coordinate the lower layer components, common interface elements of both layers are combined within adapters, as shown in Figure 19.11. The corresponding elementary component communications are coordinated by the adapters `ACylinder`, `AGripperPneumatic` and `APartCount`.

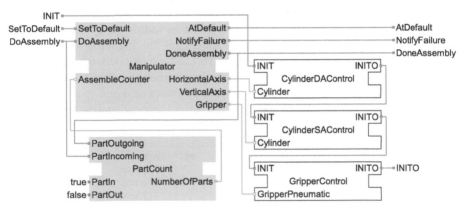

FIGURE 19.11
Layer 2: Subapplication to control the manipulator with the manipulator FB for the entire control, assemble command counting FB (gray) and connections to the layer 1 subapplications (white).

Besides these adapters, the coordinator contains additional input events `SetToDefault` and `DoAssembly` to move the manipulator into the default position and coordinate the assembly process, respectively. To communicate with the next upper layer, the output events `AtDefault`, `DoneAssembly` and `NotifyFailure` inform the upper layer when the default position is reached, the assembly process is performed or a gripping error has occurred. The coordinator component is connected to an instance of the `PartCount` BFB which counts incoming assemble commands of the upper layer to repeat the assemble process.

19.4.3.3   Conveyor

The conveyor component is responsible for coordinating the underlying modules for transportation and separation, which are implemented by the subapplications `TransportControl` and `SeparatorControl`, as illustrated in Figure 19.12. Incoming parts are counted by the `PartCountControl` subapplication and reported to the conveyor component through the `APartCount` adapter connection. The transportation component is activated at any incoming part by sending a `MoveWPtoAssembly` event. It is stopped when there are no parts on the belt by sending a `ConveyorEmpty` event. The separator component informs the conveyor component about parts which arrive at the separator and the assembly position respectively by sending a `WPatSeparator` or a `WPatAssembly` event. The conveyor component decides whether the separator has to be rotated into the release or confine position. The conveyor component also has an output event `ReadyforAssembly` to inform the upper layer that a work piece has arrived at the assembly position and is ready for processing.

The coordination FB for the conveyor is implemented as a BFB. Its ECC is illustrated in Figure 19.12. The ECC consists of two transition loops. States with two outgoing transitions are prioritized as indicated by small numbers at the transitions where the smaller numbers get the higher priority.

The small loop on the left of Figure 19.12 stops the belt when there is no part on the belt. This transition loop is entered when the `APartCount` adapter sends an update event with a part number less than or equal to zero. The second transition loop controls the transportation process and is entered when

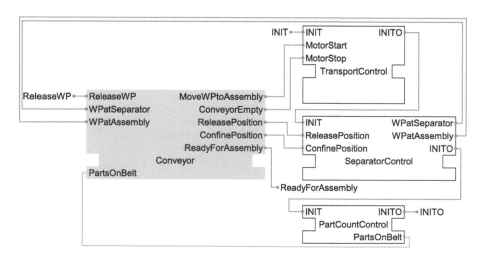

FIGURE 19.12
Layer 2: Subapplication to control the conveyor with the conveyor FB for the entire control (gray) and its connections to the layer 1 subapplications (white).

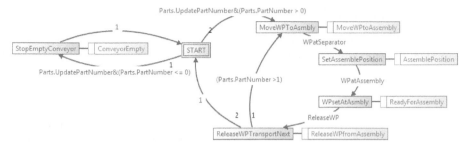

FIGURE 19.13
Layer 2: ECC for coordination of conveyor subcomponents.

the `APartCount` adapter sends an update event with a part number greater than zero. After reaching the assemble position and sending an assemble request to the next upper layer it waits until a release request is returned from the upper layer. As long as the number of parts on the belt is greater than one, the transportation process is repeated. The conveyor coordination BFB is therefore dependent on the values provided by the `PartCountControl` FB.

19.4.3.4 Status Control

The status control component coordinates the operation of the light tower to act as an indication for the station as shown in Figure 19.14. For coordinating the red, orange, and green lights, the `StatusManagement` FB has been developed. During initialization of all hardware modules, the orange light blinks.

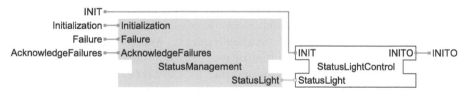

FIGURE 19.14
Layer 2: light management FB (gray) and its connection to the layer 1 FB (white).

After setting the manipulator into default position, the orange light resets and the green light glows, which indicates that the station is running and can process any arriving part. If an error occurs, the red light starts blinking until the error is acknowledged. The lighting commands are sent to the next lower layer through the `AStatusLight` adapter.

19.4.4   Layer 3: Process Control Component

The overall process control of the pick and place station is realized in layer 3. The `PickNPlace` FB coordinates this top layer and functionally resembles the pick and place station, as illustrated in Figure 19.15.

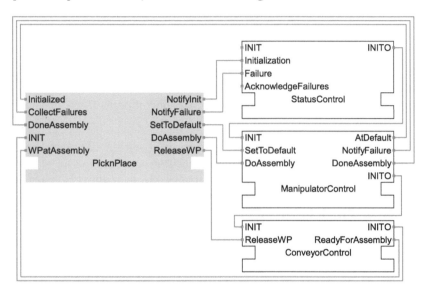

FIGURE 19.15
Layer 3: overall control of pick and place station (gray) and its connections to the layer 2 subapplications (white).

The pick and place component is realized as a BFB and coordinates the subapplications `StatusControl`, `ManipulatorControl`, and `ConveyorControl`. During initialization, this coordination component initializes its hardware modules by sending a `SetToDefault` event. This command initiates the manipulator to move into its default position and inform the pick and place component after completion by sending an `AtDefault` event. After initialization, the `ConveyorControl` subapplication is able to autonomously detect incoming parts, start the conveyor belt and control the separator. When any work piece has reached the assembly position, the `ConveyorControl` subapplication informs the coordination component by sending a `ReadyForAssembly` event. The coordination component then decides whether the work piece is supposed to be released or should be assembled by sending a `ReleaseWP` or a `DoAssembly` event to the conveyor or the manipulator control. In case no work piece insert can be taken from the stack, the manipulator control informs the coordinating component by sending an `NotifyFailure` event. The coordinating component forwards the failure to the status control component so that it can be indicated by a red blinking light until it is acknowledged by an operator. Each component corresponds to a mechatronic unit and can stand on its own from a functional view.

19.5 Conclusion

By hierarchically structuring a control application, it is possible to assign every hardware component one software component to control this component. The presented hierarchically structured control application facilitates the reuse of components and aids easy reconfiguration. The software components implemented resemble the hardware components and this allows a direct hardware-software mapping in principle. Among the FBs developed for the specific control application, the components up to the generic module are completely reusable in principle. The atomic components such as motors and solenoid valves designed with the IX, QX FBs are independent of the applied hardware and can be used with any kind of hardware configuration and in any kind of application irrespective of the domain. The elementary components coordinate the atomic components and can be reused within the same domain for similar hardware components. For example, the double acting cylinder FB can be applied to any type of double acting cylinder for telescoping functions. All elementary components can be considered in the same way. The proposed structure also provides reconfigurability from the atomic level to the process control level. The design of a hierarchical software structure is sometimes difficult since information required for one component may be required for another component of a different branch within the hierarchy tree. The PartCount FB is a typical example of information required within different branches of the hierarchy tree, which led to two instances of this FB: one for the conveyor to count the parts on the belt and one for the manipulator to count incoming assemble commands.

The requirement for creating hierarchically structured applications is hardware independence. Without hardware independence the level 0 components cannot be reused and this influences the interfaces of the higher level components. Therefore IX and QX FB have been introduced to provide a common interface for any type of device and make the application hardware independent. The assignment of IOs to a specific component is sometimes ambiguous since one IO might be needed within more than one component. This concerns the presence sensors on the conveyor belt which are represented by IX FBs and may be assigned to the separator and to the transport elementary component. The introduction of the PartCount FB allowed us to encapsulate the part tracking IOs and transfer the coordination functionality to the upper hierarchy level. The part tracking component also improved the hierarchical structuring since the IOs PartIn and PartOut can now be assigned exactly to one FB, while the ECC of the conveyor FB could be simplified due to a reduced number of transitions.

The configuration of the IX and QX FBs causes a hardware relation, since their parameter value determines the pin on the PLC. Where there are several PLCs, there are also several equal pins. IX and QX FBs with equal pin numbers cannot be mapped onto the same PLC. The pin numbering therefore influences the mapping onto a specific resource. An extended device description which describes the device in terms of provided IOs and network interfaces would allow unambiguous symbolic names for each IO, which in turn could be selected in a dropdown at the IX and QX FBs after mapping.

Distributed applications require a coordinated initialization. The initialization order of the distributed FBs is important to initialize them correctly. This especially concerns communication FBs such as PUBLISH and SUBSCRIBE. SUBSCRIBE FBs need to be initialized first to receive the desired initialization values from corresponding PUBLISH FBs. The initialization order of the FBs therefore determines the download order of the resources. Within the pick and place application, the status light is affected by the initialisation order. Due to the limited number of IOs on a compact PLC, the status light control runs on a different PLC from the rest of the application. The SUBSCRIBE FBs running on the status light control PLC need to get the initialization values from the PUBLISH FBs running on the other PLC.

For collaborative development but also for maintenance of IEC 61499 applications, naming conventions would be helpful. A proper naming of FBs and interface elements help clarify the purpose of an element and indicate how different FBs have to be connected. Useful naming conventions for IEC 61499 applications could be:

- Naming FBs according to their functionality or corresponding hardware components
- Naming any Boolean element according to its purpose, e.g., name an output to extract a cylinder extract or retract but not InOut
- Using equivalent names for events of different FBs that are supposed to be connected

The debugging support in engineering tools of the automation domain is very limited. The IEC 61499 implementation 4DIAC provides monitoring of events as well as data value changes. IEC 61131 engineering tools as logi.CAD and CoDeSys V2.3 provide tracing support to watch value changes over time. To effectively debug an event-driven application under IEC 61499, neither monitoring nor tracing is sufficient. Applications which contain BFBs require a debugging mechanism that follows the execution steps in terms of occurred events and entered states.

Bibliography

[1] F. Ebel and M. Pany. Pick&Place Station Manual. Technical report, Festo Didactic GmbH & Co. KG, 2006.

[2] I. SC65B. IEC 61499-1: Function blocks for industrial process measurement and control systems. Part 1: Architecture, 2005. Geneva.

[3] C. Sünder, A. Zoitl, and C. Dutzler. Functional structure-based modelling of automation systems. *International Journal of Manufacturing Research*, Vol. 1, No. 4, 405–420, 2006.

[4] A. Zoitl and H. Prähofer. Guidelines and patterns for building hierarchical automation solutions in the iec 61499 modeling language. *IEEE Transactions on Industrial Informatics*, Vol. 9, No. 4, 2387–2396, 2013.

[5] A. Zoitl, T. Strasser, and A. Valentini. Open source initiatives as basis for the establishment of new technologies in industrial automation: 4DIAC a case study. In *IEEE International Symposium on Industrial Electronics, Bari*, pages 3817–3819, 2010.

20

Toward Batch Process Domain with IEC 61499

Wilfried Lepuschitz

Practical Robotics Institute Austria

Alois Zoitl

fortiss GmbH

CONTENTS

20.1   Introduction

In the domain of batch process industries, a certain grade of flexibility exists from the process-oriented view due to the usage of modifiable recipes. However, the applied control systems are commonly based on centralized concepts, which makes modifications to the provided equipment functionalities inevitably complex. More flexibility can be achieved by applying a distributed control approach with similar structures as the physical equipment. Thus, entities can be achieved that encapsulate physical equipment with distinct functionality and information. These entities can then be combined for realizing functionality on a more complex level and additional functionality can be added by integrating further entities [1].

IEC 61499 is a well-suited standard for realizing the software aspects of such entities. To apply approaches based on this standard in the industry, compliance with other relevant standards is vital. In regard to batch process automation, the standard ANSI/ISA-88 Batch Control (S88), respectively its counterpart IEC 61512 [4], is significant.

This chapter summarizes work performed in regard to the application of the standard IEC 61499 for a laboratory batch process plant [6, 7, 8]. Three approaches based on IEC 61499 for batch process automation were developed and implemented on a laboratory process plant. These approaches represent evolutionary work on bringing IEC 61499 closer to industrial practice in the batch process domain. Before explaining the implemented approaches, Section 20.2 gives an overview of S88 and its usage in practice. The first approach is described in Section 20.3 and applies a standardized architecture for components on different hierarchical levels in the batch process domain. To be more compliant with industrial batch management systems, the second approach presented in Section 20.4 is based on a state machine as defined in S88 within a function block. Section 20.5 introduces the third approach, which overcomes the limitations of the second approach by introducing a more generic design for the state machine and its application, resulting in an integration with an industrial batch management system. Finally, this chapter is concluded in Section 20.6 with a summary.

20.2   ANSI/ISA-88 Batch Control

Currently applied batch management systems take S88 as the basis for their applied structural models and thereby achieve a certain grade of comparability and interoperability [5]. S88 describes "a framework for the specification of requirements for batch process control, and for their subsequent translation into application software" [9]. Modifications and extensions to the introduced models are explicitly allowed if the consistency of the defined models is assured.

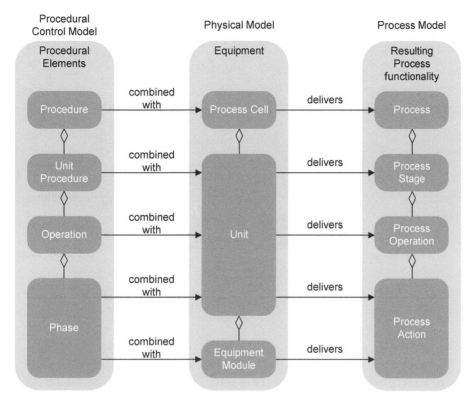

FIGURE 20.1
Structural models of S88 and their relations [4].

20.2.1 Structural Models

The standard introduces a set of structural models for the organization of processes, physical equipment and control software. Generally all these structural models contain four significant levels:

- Process model: process, process stage, process operation, process action

- Physical model: process cell, unit, equipment module, control module

- Procedural control model: procedure, unit procedure, operation, phase

Mapping elements of the procedural control model onto the physical model delivers equipment entities that realize process functionality (see Figure 20.1). For example, mapping a phase onto an equipment module realizes an equipment entity that provides a process action.

Moreover, the concept of recipes is described in S88 as means for the execution of a process. Four types of recipes are distinguished: general recipe, site recipe, master recipe and control recipe. General recipe and site recipe are

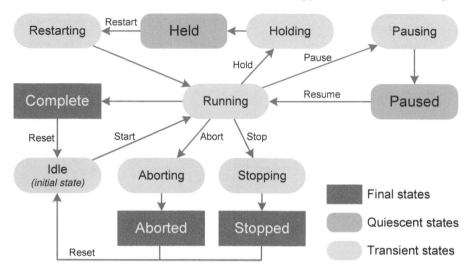

FIGURE 20.2
States and transitions of the S88 state machine [4].

formulated independently of equipment information and structured according to the process model. Conversely, both master recipe and control recipe are structured according to the procedural control model. The master recipe determines the types of needed equipment entities while the control recipe specifies the distinct equipment entities to be used for production. Based on this information in the control recipe, the batch management system is able to activate the according control software of the equipment entities commonly located on programmable logic controllers (PLCs). This interlinking of the batch management system with the control software can be done on any level but most industrially applied systems perform this interlinking on the phase level [9]. A so-called phase logic interface is employed for this interlinking and it translates the sequence of the control recipe into commands for the control software. Status information from the control software is returned to the batch management system using this interface.

20.2.2 State Machine

A set of operational states and their transitions is provided by S88 for equipment entities and procedural elements (see Figure 20.2). Also the commands for state changes in this state machine are introduced in the standard. This state machine can be diminished or extended according to the application requirements and applied on any hierarchical level.

Industrial batch management systems such as FactoryTalk Batch [10] and zenon Batch Control [3] link the control recipe on the phase level with the control software. As a consequence, the state machine only needs to be im-

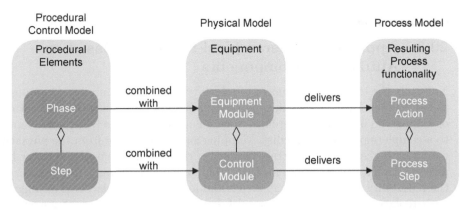

FIGURE 20.3
Region of interest concerning the control software of the presented approaches.

plemented on this level in the PLCs. Higher levels of control software are not required as these levels are controlled by the batch management system.

20.2.3 Amendments and Region of Interest

As can be seen in Figure 20.1, elements of various levels of the procedural control model are mapped onto the physical model to achieve elements of the process level. However, the levels of the three models are not completely equal, as can be seen at the unit level of the physical model. Obviously, procedural elements of three different levels can be mapped onto units to achieve different levels of process functionality. Also a phase of the procedural control model can be mapped onto both units or equipment modules. However, the control module does not appear in the relations between the models.

It is stated in the standard that the structural models can be diminished or extended as long as the consistency of the models is assured. Furthermore, according to the standard, a phase can be segmented into steps and transitions. This further aggregation is emphasized in the presented approaches and the step is introduced as the lowest level of the procedural control model for the remainder of this chapter. Such a step can be mapped onto a control module for achieving process functionality on a very low level. Thus, the process model shall be extended by the process step as the lowest level for increasing consistency throughout the structural models.

As mentioned earlier, industrial batch management systems link the control recipe with the control software on the phase level. As a result, the region of interest for the approaches regarding control software encompasses only the phase level and the newly introduced step level (see Figure 20.3).

20.3 Approach 1: Hierarchical Structure Based on Automation Components

Naturally the technical equipment of a facility can be structured in a compositional manner. Distinct machines can be subdivided into sensors and actuators or aggregated into machine groups. Such an aggregation can be expanded into to a complete production facility. Mapping such a structure from the physical model onto the control structure facilitates the development of the latter by providing a natural structure for functionality encapsulation. This emphasizes the reusability of software components and makes them replaceable as long as their replacements offer the required functionality and interfaces for meeting the necessary specifications.

20.3.1 Concept of Approach

The approach presented in this section represents a hierarchical control system composed of devices with a standardized architecture. Having the same architecture for all components regardless of their position in the system facilitates the usage of this approach. In principle these automation components (ACs) consist of three types of sub-components [12]:

- Sensors and actuators for interfacing the process

- A control device with interfaces to the sensors, actuators, and to the network

- Software components hosted on the control device with data and control logic for carrying out automation functions

According to Sünder et al., each software component can be further segmented into the following elements [11]:

Logic: This software part incorporates the basic control functionality of the AC combining inputs and inner device states for generating according outputs.

Diagnostics: Diagnostic algorithms detect errors.

Universal component interface (UCI): The UCI provides the communication with higher-level and external components.

Human–machine interface (HMI): The HMI is utilized for interactions between human operators and the AC and represents an external component requiring access to the UCI. It is not a central part of the approach and thus not explained in detail.

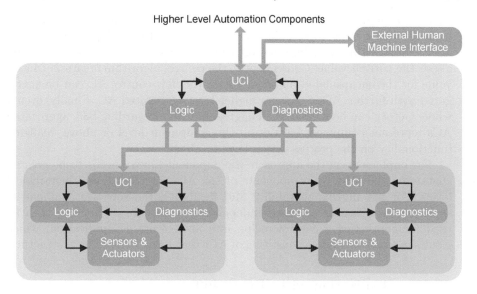

FIGURE 20.4
Architecture of an aggregated automation component incorporating two basic automation components [11].

According to Dutzler, two types of ACs can be distinguished [2]: basic ACs and aggregated ACs (see Figure 20.4). Basic ACs, denoted as intelligent sensor/actuator units, directly interact with the process through their logic and diagnostics parts, which have access to the component's physical equipment. Certain diagnostic algorithms are carried out independently from higher level components. For example, the combination of a ball valve with two sensors for verifying the valve's position delivers a basic AC with diagnostic capabilities. The ball valve is controlled by the logic part for providing the component functionality of enabling or preventing the flow of liquid. When the ball valve is required to change its state, the diagnostics part carries out a check on the valve's position using the sensors. If the sensors do not indicate a position change of the valve as required after a given amount of time, an error event is issued to a higher level entity through the UCI. In regard to S88, the basic ACs represent equipment entities on the control module level that provide functionality on the process step level.

The combination of ACs delivers aggregated ACs. By encapsulating lower level ACs, which can be basic or aggregated ACs, a more abstract and complex functionality can be achieved. The logic and diagnostics parts of such an aggregated AC communicate with the UCI of the encapsulated ACs and provide a UCI for further higher level entities. Besides, an aggregated AC can incorporate additional functionality within its software components. For example a flow control entity realized as an aggregated AC comprises two basic ACs: a pump and a flow sensor. A control algorithm in the flow control's logic receives a required setpoint through the UCI from a higher-level entity and combines this with information from the encapsulated flow sensor AC about the current value. Consequently, the control algorithm creates a control value for the encapsulated pump AC. Also the flow control AC can be aggregated with further ACs to form a higher-level aggregated AC. Finally, even a whole factory can be viewed as an aggregated AC. Regarding S88, aggregated ACs represent equipment entities on the equipment level or above, realizing functionality on the process action level or above.

For implementing the approach, a component library needs to be created containing the software components of the ACs for all required hierarchical levels. For maximizing reusability and flexibility, the software components should be designed as hardware-independent as possible. They can then be used for converting a control recipe into an executable application, which results in a control system consisting of ACs that form a hierarchical structure.

20.3.2   Realization with IEC 61499

Realizing the software components of the ACs in IEC 61499 means implementing the software elements logic, diagnostics and UCI by means of function blocks (FBs). The software components need to be mapped onto physical entities for achieving complete ACs.

In the presented approach, each of the software elements is realized with a composite FB. Having only three FBs in principle delivers a clear layout of each component's FB network (see Figure 20.5). It has to be pointed out that the aggregation of ACs into a higher level AC is only performed on a functional level. This is evident as the UCI of a lower level AC needs to be accessed by both logic and diagnostics of the higher level AC resulting in interfaces within the composite FBs of logic and diagnostics to the lower

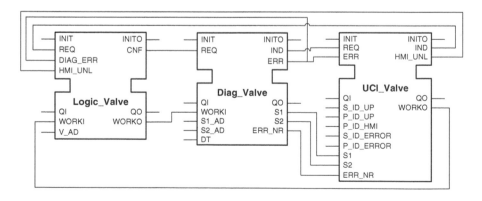

FIGURE 20.5
Function block network of the step for controlling a valve.

level UCIs (see Figure 20.6). Moreover, this design offers the possibility to exchange the lower level AC without the need to adapt its superior AC as long as the corresponding interface FBs within the superior AC do not need to be modified. Hence, replacing for instance a magnetic valve with a ball valve and accordingly its software components does not require adapting the software component of a superior AC as the interface is similar for both types of valves.

The conversion of a control recipe into an application is realized by determining the required ACs according to the procedural elements of the recipe.

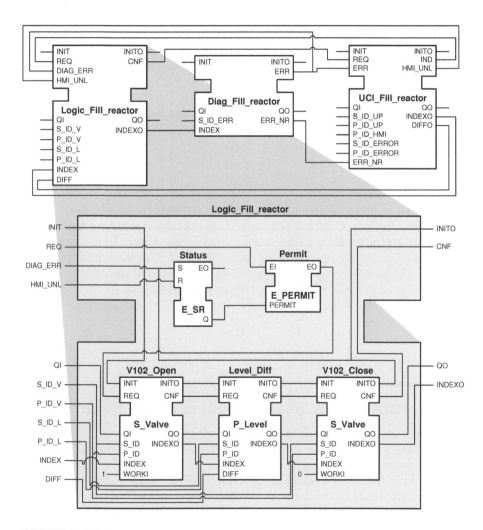

FIGURE 20.6
Function block network of the phase for filling a reactor and detailed view of its logic with the interface function blocks to the step level UCIs.

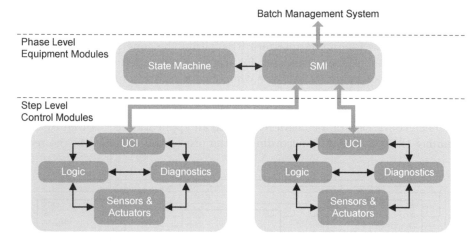

FIGURE 20.7
Control software of an equipment module incorporating control modules.

An application is formed by sequencing their activities, which can theoretically be done on any level of the hierarchical structures. However, creating the application on the lowest level requires a control recipe including every single step to be carried out, which requires extensive efforts from the process engineer. On the contrary, if the conversion is performed on the highest level, a control recipe would only invoke one specific procedure. Consequently, procedural elements on all levels need to be provided, which includes procedures for all imaginable recipes. This is evidently an impossible task. Hence, the conversion of control recipe to application is reasonable on the phase level and on the operation level. Therefore, elements up to the phase or operation level are required.

20.4   Approach 2: Implementation of S88 State Machine

The previously mentioned approach is well suited for converting a control recipe into an executable application. However, it does not take compliance with industrial batch management systems into account. FactoryTalk Batch [10] and zenon Batch Control [3] offer protocols for creating a phase logic interface. According to the name of this interface, the batch management system and actual control software are connected on the phase level. To establish a connection, the control software on the phase level is required to incorporate the S88 state machine. This is clearly missing in the approach presented in the previous section.

20.4.1 Concept of Approach

Figure 20.7 depicts the control software structure of an equipment module. The step level is organized similar to the hierarchical approach described earlier. Control software on the phase level consists of the following two elements:

- State machine: This S88 state machine incorporates operational states and their transitions and algorithms for providing the functionality of an equipment module.

- State machine interface: This interface is used for communicating with the batch management system and with the subjacent control modules.

The control software on this level is responsible for controlling equipment modules that incorporate one or several control modules. Functionality on this level is a combination of the incorporated control modules' functionalities with additional functionality. Integrating the S88 state machine for each equipment module results in compliance with the standard S88.

20.4.2 Realization with IEC 61499

As the software elements on the step level are each realized according to the previously described approach with one distinct FB (for logic, diagnostics and UCI), also the elements on the phase level are realized with one FB each. The approach is explained by taking an equipment module for controlling the temperature in a reactor tank as an example. The task of this module is to heat the liquid in the reactor to a certain temperature and maintain it until a stop event is issued.

As mentioned earlier, the phase level control software incorporates a S88 state machine. It is implemented within a basic FB (see Figure 20.8) and comprises additional algorithms for the equipment module. In the given example, the algorithms encompass temperature control and error handling. Errors can be indicated either by internal algorithms in the state machine or by the subjacent control modules. The state machine is realized as states in the FB's execution control chart (ECC) and comprises the following operational states and algorithms for the given example (see Figure 20.9):

- Initializing: This state sets the internal variables to initial values.

- Idle: The internal variable are reset.

- RunInit: For initializing the control loop, the set value is saved and a first current value from the temperature sensor module is requested.

- Running: Based on the set value and the current temperature value, an event is issued to turn the heater on or off.

- GetTemp: The current temperature value is gathered and compared with

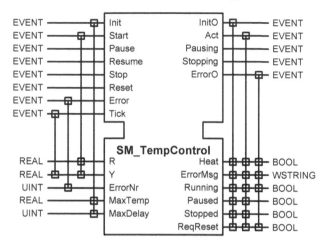

FIGURE 20.8
Basic function block representing the state machine of the equipment module for temperature control.

the previously gathered value. If the heater is active and the new value is lower than the old value, a defect of the heater must have occurred and an error flag is set.

- Paused: The heater is turned off until the process is resumed.

- Stopped: The heater is turned off and the temperature control loop is stopped.

- E1: This error state is entered if the heating module reports the detection of an insufficient amount of liquid in the reactor for a safe heating process.

- E2: This error state is entered if a critically high temperature is reached within the reactor.

- E3: This error state is entered if the heater is damaged.

Transitions to the error states are triggered by internal error flags or by error events from subjacent control modules. For clearing errors, a reset event is required that can be issued by a human operator or the batch management system. Leaving the "Stopped" state requires a reset event for bringing the state machine into the "Idle" state.

The described state machine does not contain all states as defined in the standard S88, which is legitimate as various tasks and components may require different implemented states. As an example, the "Stopping" state, defined in S88 for performing a safe stopping procedure of the equipment, is not implemented for the temperature control module. Turning off the heater is simply achieved by deactivating the according digital output of the controller.

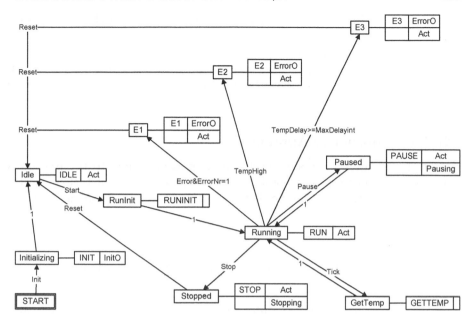

FIGURE 20.9
ECC realizing the state machine of the equipment module for temperature control.

As a consequence, a stop means a direct transition from the "Running" state to the "Stopped" state by turning off the heater.

20.5 Approach 3: Generic S88 Phases Structure and Conjunction with Industrial Batch Management System

The previous approach shows the feasibility of IEC 61499 for implementing control software on the phase level with an integration of the S88 state machine. However, the approach does not take the communication protocols of batch management systems by different vendors into account. Most batch management systems communicate with the control software using tags in the PLC. In FactoryTalk Batch, a command handshake protocol exists and each phase has a command tag for transmitting commands from the batch server to that phase. Commands are represented by numbers and trigger a state change in the state machine of the phase. Furthermore, a failure protocol is used for gathering notifications about failures and a watchdog protocol verifies the liveliness of the communication between the batch server and the phases.

An approach to be integrated with an industrial batch management system, must handle such protocols accordingly. This approach should provide the means for integrating IEC 61499 with an industrial batch management system.

20.5.1   Concept of Approach

Control software on the step level is organized according to the previously mentioned approaches incorporating logic, diagnostics and UCI. However, the phase level control software needs to be structured to perform the following tasks:

- Communication with the batch management system

- Decoding and interpretation of commands from the batch management system

- Encompassing of the S88 state machine to be compliant with the batch management system

- Controlling the subjacent control modules on the step level for realizing actual process functionality

- Handling of errors

These tasks can be easily handled by using a "separation of concerns" principle, which results in a clear structure of the control software application as can be seen in Figure 20.10. Each task is performed by one distinct software element.

Accessing the tags in a PLC is commonly achieved by drivers in the batch management system. However, a general tag table does not exist in a distributed control concept based on IEC 61499 and few suitable drivers for message-based communication are available. As a consequence, communication is realized by using an OPC server (see Figure 20.11). The functionality for accessing the OPC server is provided by the universal phase interface (UPI). It can read and write tags in the OPC server and is therefore able to communicate with the batch management systems.

Batch management systems of different vendors rely on various methods of encoding of information such as commands from the batch management system to the phases. The command handler is responsible for decoding the commands received by the UPI, checks their validity and thus incorporates the command handshake protocol for ensuring a correct communication. In the case of a valid command, an appropriate action such as a state change of the phase is triggered. In case of an invalid command, the failure handler is notified.

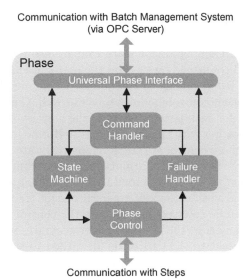

FIGURE 20.10
Control software structure of a phase.

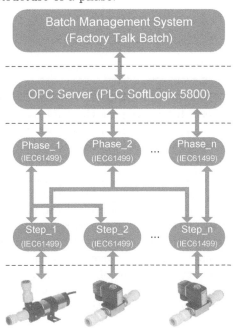

FIGURE 20.11
Architecture of the implementation using an OPC server for communication.

The state machine realizes the S88 state machine similar to the approach mentioned earlier but in a more generic way and is thus strongly compliant with the standard. Commands for state changes are received from the command handler and confirmed by sending the new state value to the UPI to be forwarded to the batch management system. Changing the state commonly requires the subjacent control modules to act as well. Consequently, the phase control is notified for issuing commands to the step level.

The communication with the step level is performed by the phase control. Activities of the phase control are triggered by the state machine on the basis of the momentary state and confirmation of their completion is sent back. Commands are sent to the control modules for inducing actions of their physical equipment and notifications of their successful completion are received. Error messages can be received as well from the control modules.

In the case of an error, the failure handler generates an appropriate failure message, which is sent to the UPI to be forwarded to the batch management system. Errors can be invalid commands or failures of the physical equipment detected by the control modules.

20.5.2   Realization with IEC 61499

Each of the software elements of the phase level control software is realized with one FB (see Figure 20.12). The UPI is a composite FB containing a set of service interface FBs for realizing the communication with the OPC server. Upon initialization, server name and item addresses for accessing the tags in the server need to be provided as parameters. When a new command is read from the server, its value is sent to the command handler together with an according event. By exchanging the implementation of the UPI with a different one, other communications means than OPC may be realized.

The command handler is a basic FB incorporating two main algorithms. They are executed sequentially each time a new command is received from the UPI. The command is checked for validity and based on its value an appropriate event is generated, which is sent to the state machine to trigger a state change. In case of an invalid command, the failure handler is notified. Subsequently a value is determined to be written on the server by the UPI for notifying the batch management system that the command was read successfully. Thus, the command handshake protocol is realized. This FB represents a driver for processing the encoding of states and commands of a distinct batch management system. As a consequence, different implementations of the command handler allow the conjunction with batch management systems of other vendors.

A further basic FB represents the state machine. Its ECC contains states and transitions according to the S88 state machine. Transitions are triggered by events received from the command handler or by the phase control when a control module completes an activity. After reaching a new state in the ECC,

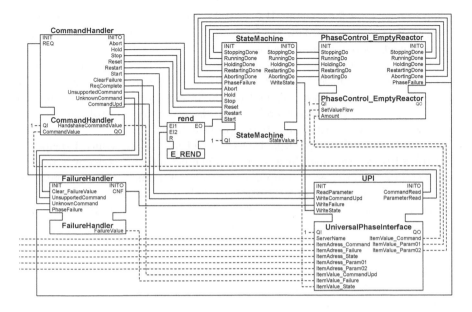

FIGURE 20.12
Function block network for a phase (initialization event lines are not displayed for a clearer visibility).

an event is issued to the phase control and the state value is sent to the UPI for being written on the OPC server.

The phase control is a composite FB containing service interface FBs for communicating with the subjacent control modules and further FBs for incorporating coordination and support algorithms (e.g., an algorithm for a closed-control loop). Thus, the actual functionality of the equipment module is realized. Commands to the step level are triggered by events received from the state machine and relevant events are returned upon notification of activity completion. If a control module reports an error, an event is generated and sent to the failure handler.

The failure handler is a basic FB, which receives failure notifications and generates the appropriate failure value. This is then sent to the UCI to be written on the OPC server.

20.6 Conclusion

This chapter presented evolutionary work on the integration of IEC 61499 for the batch process domain. The standard S88 is significant and has to be taken into account for bringing IEC 61499 closer to industrial practice. In this context, it is relevant to show the integrability of IEC 61499 with higher level systems such as batch management.

Three approaches were described in this chapter. The first focused on the usage of a standardized architecture for components at various levels in the process domain. While not sufficient for the integration with a batch management system, this component architecture is usable for control modules, i.e., the lowest physical entities in a batch process plant.

The phase level was identified as the level for linking control recipe and control software. S88 introduces a state machine and this concept is required on the phase level to assure compliance with batch management systems. As a consequence, a further approach introduced the state machine on this level in the form of a basic FB. This approach lacked the implementation of required protocols for communicating with such a system.

Finally, the third approach introduced a concept for control software on the phase level with more flexibility of its software elements. However, the approach can be adapted according to the communication means as well as the encoding of states and transitions of a batch management system. As an example, the integration of IEC 61499 with FactoryTalk Batch of Rockwell Automation was presented. In order to communicate with this batch management system, an OPC server was employed.

The last approach showed the feasibility of IEC 61499 for integration with an industrial batch management system. The approach is generic on the phase level while maintaining flexibility on the step level due to its foundation on entities with encapsulated functionality.

Further research could improve the integration with other batch management systems. Consequently, the tag-based communication of current batch management systems with phases could be avoided by implementing according drivers for a message-based communication.

Bibliography

[1] R. W. Brennan. Toward real-time distributed intelligent control: A survey of research themes and applications. *IEEE Transactions on Systems, Man, and Cybernetics, Part C: Applications and Reviews*, Vol. 37, No. 5, 744–765, 2007.

[2] C. Dutzler. *A Modular Distributed Control Architecture for Industrial Automation Systems – Designed for Holonic Systems and Vertical Integration*. PhD thesis, Vienna University of Technology, Austria, 2003.

[3] Ing. Punzenberger COPA-DATA GmbH. zenon Batch Control. http://www.copadata.com/en/products/food-beverage/isa-88-compatible-batch-control-in-zenon.html.

[4] International Electrotechnical Commission. *IEC 61512 Batch Control. Parts 1-4*. Geneva, 2000–2010.

[5] S. Kuikka. *A Batch Process Management Framework: Domain-Specific, Design Pattern and Software Component Based Approach*. PhD thesis, Helsinki University of Technology, Finnland, 1999.

[6] W. Lepuschitz, F. Königseder, and A. Zoitl. Conjunction of a distributed control system based on IEC 61499 with a commercial batch management system. In *Proceedings of 14th IEEE International Conference on Emerging Technologies and Factory Automation*, pages 1–8, 2009.

[7] W. Lepuschitz and A. Zoitl. An engineering method for batch process automation using a component oriented design based on IEC 61499. In *Proceedings of the 13th IEEE International Conference on Emerging Technologies and Factory Automation*, pages 207–214, 2008.

[8] W. Lepuschitz and A. Zoitl. Integration of a DCS based on IEC 61499 with industrial batch management systems. In *Proceedings of 13th IFAC Symposium on Information Control Problems in Manufacturing*, pages 690–695, 2009.

[9] M. Parker and J. Rawtani. *Practical Batch Process Management*. Newnes, The Netherlands, 2005.

[10] Rockwell Automation. FactoryTalk Batch. http://www.rockwellautomation.com/rockwellsoftware/products/factorytalk-batch.page.

[11] C. Sünder, A. Zoitl, and C. Dutzler. Functional structure-based modelling of automation systems. *International Journal of Manufacturing Research*, Vol. 1, No. 4, 405–420, 2006.

[12] V. Vyatkin. Intelligent mechatronic components: control system engineering using an open distributed architecture. In *Proceedings of 10th IEEE International Conference on Emerging Technologies and Factory Automation*, pages 277–284, 2003.

21

Smart Grid Laboratory Automation Approach Using IEC 61499

Filip Andrén

AIT Austrian Institute of Technology GmbH

Georg Lauss

AIT Austrian Institute of Technology GmbH

Roland Bründlinger

AIT Austrian Institute of Technology GmbH

Philipp Svec

AIT Austrian Institute of Technology GmbH

Christian Seitl

AIT Austrian Institute of Technology GmbH

Thomas Strasser

AIT Austrian Institute of Technology GmbH

CONTENTS

21.1 Introduction and Motivation

Renewable energy sources are seen as key enablers to decrease greenhouse gas emissions and cope with the anthropogenic global warming trend. Their intermittent behaviors and limited storage capabilities present new challenges to power system operators in maintaining power quality and reliability [18].

However, the availability of information and communication technologies (ICT), advanced automation approaches, and smart devices provides new and intelligent solutions to cope with these challenges. As a consequence of these developments the traditional power system is transformed into a cyber-physical system, the smart grid [8, 10, 15]. This vision of integrating power systems with support of sophisticated ICT technology has the potential to significantly contribute to a more stable and secure electricity supply [10, 11, 22, 26].

Along technical, organizational, and educational progress [24], a proper infrastructure validation for smart grid systems and components is required. The Smart Electricity Systems and Technologies Laboratory (SmartEST lab) established by the AIT Austrian Institute of Technology provides a research and testing infrastructure to analyze the interactions between smart grid components and the power grid under realistic conditions. Furthermore, it is also used for development of grid components. Candidates for testing in the SmartEST lab include grid-connected and off-grid inverter systems for distributed energy resources (DER), energy storage systems, algorithms and strategies for grid control, communication infrastructure and corresponding energy management systems and services, and charging infrastructure for electric vehicles [6].

Operating such a multi-functional research, development and testing infrastructure in the domain of smart grids requires a flexible and scalable automation environment. The aim of this chapter is to discuss the automation and control application for the SmartEST lab based on the IEC 61499 reference model [25] using open source tools. This automation approach provides the necessary basis for the realization of a flexible and configurable laboratory automation supporting the design, development and validation of new approaches, methods and concepts in the domain of smart grids.

This chapter summarizes previous work performed using the IEC 61499 standard for the automation of a smart grid laboratory [2, 6, 17]. The main focus in this chapter is related to the developed automation architecture and the modeling of the automation application for the laboratory environment.

Section 21.2 briefly discusses the needs for smart grid laboratories followed by an introduction of the SmartEST multi-functional laboratory in Section 21.3. The core part of this chapter—the IEC 61499-based automation systems and the corresponding automation application—is presented in Section 21.4. Section 21.5 concludes this chapter with the main findings and the planned future work.

21.2   Smart Grid Laboratories: Needs and Requirements

The design, development, and implementation of smart grid technologies and components and their installation in the field continue to become more complex. One main reason is the cyber-physical nature of smart grid concepts and components. In addition, the validation and testing of various configurations (i.e., hardware and software) will play a major role in future research and technology development to ensure that reliable and error-free components and applications are installed.

Several demonstration and field test projects in the domain have proven successful in recent years in various countries. A large scale roll-out of new and improved products, services, and solutions is expected in the near future. Public authorities have designed smart grid research and development infrastructures as major priorities [5, 9, 16, 19].

To validate and test different smart grid configurations in an integrated (cyber-physical) manner, the power system and the corresponding information, communication and control components have to be covered equally [24]. The complexity of such a validation environment for future, intelligent power systems requires virtual and real-world based methods addressing the following points [2, 13, 14, 17, 24]:

- A multi-domain, cyber-physical approach for analyzing, validating and testing smart grid configurations on the system level is necessary.

- Methods, concepts, and procedures for analysing, developing, and testing smart grid components (power system, ICT, automation and control) are required.

- Well-educated researchers and engineers understanding smart grid configurations in a cyber-physical manner need to be trained on a broad scale.

These points require a proper and integrated research infrastructure taking simulated and laboratory-based design, development, validation, and testing methods for smart grids and related components (including a combination of both methods) into account. Testing new approaches and components on a broad scale in the field is often not a suitable option since the availability and security of supply of the power system must be guaranteed. Hence, a research infrastructure using simulated systems with real laboratory hardware provides a reasonable environment.

21.3   Brief Overview of SmartEST Lab Environment

In the following sections a brief overview of the multi-functional SmartEST laboratory environment and corresponding case studies are provided. In addition, important requirements and needs for a flexible and scalable laboratory automation system are introduced [2, 17].

21.3.1   Purpose and Usage of Laboratory

The SmartEST lab is an infrastructure for testing smart grid components and their integration into the existing power system. It also offers a flexible and multi-functional environment for validating different smart grid configurations (i.e., hardware and software). It plays a central role in Austrian smart grid research and development. The lab specializes in inverter and (sub)system testing, system tests with multiple components and environmental tests. The SmartEST lab features advanced possibilities to combine simulation and experimental validation. Its applications range from research, design and validation for smart grid systems and their corresponding component development to the analysis of automation and communication concepts [6].

The laboratory infrastructure accommodates DER components, energy storage systems, voltage regulators and controllers, and other related electrical equipment. Figure 21.1 provides an overview of the topology and the main elements of the SmartEST lab. Large-scale, free programmable, high-bandwidth AC and DC sources allow full-power testing capability up to 1 MVA (AC), including a high-performance PV Array (DC) simulation up to 1 MW (DC). Additional equipment for simulating control and communication interfaces and the possibility of operating the equipment under defined (extreme) temperature and humidity conditions offer additional testing capabilities. Furthermore, the lab features highly flexible possibilities of different low-voltage network configurations that exert a huge impact on the research of complex power grid problems [6].

Additionally, the SmartEST lab offers the opportunity to simulate complex power systems in real-time and interconnect them to the laboratory equipment in closed loop operation mode. This so-called controller-hardware-in-the-loop (CHIL) and/or power-hardware-in-the-loop (PHIL) configuration allows real components to be integrated into a virtual power grid environment and tested under realistic conditions (i.e., closed loop real-time simulation with time steps down to $10 \, \mu s$, depending on the emulated network and real-time computing power [21]. It also allows connecting offline co-simulation of power and ICT solutions [11].

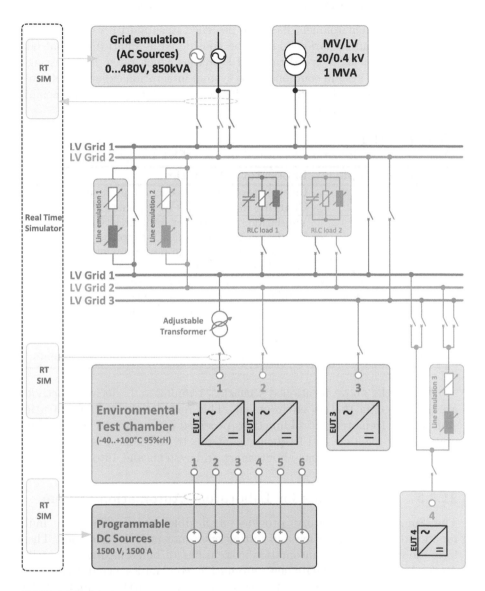

FIGURE 21.1
Overview of SmartEST laboratory environment [2, 6, 17].
Legend:
MV/LV: Medium voltage/low voltage transformer
LV Grid: Low voltage grid
RT SIM: Digital real-time simulator
EUT: Equipment-under-test

21.3.2 Laboratory Use Cases and Scenarios

The SmartEST environment has been designed to support a variety of different use cases and scenarios and can be considered as multi-functional laboratory environment. A broad range of activities which range from testing of DER and power system components to system integration experiments can be performed. Table 21.1 briefly summaries the validation and testing scenarios performed in this laboratory [2].

TABLE 21.1
Use Cases and Scenarios Performed in SmartEST Lab [2]

Research Topic	Laboratory Activities
DER component *validation*	Inverter tests (component, integration)
	Charging devices tests (component, integration)
ICT/controller *development &* *validation*	Validation of controller implementation
	Validation of communication protocols for DER
	Test of SCADA system developments
Development of new *network components* *(power electronics)*	Test of new topologies
	Test of materials for power electronic devices
	Validation of component control methods
Co-simulation *based validation*	Grid control & communication behavior testing
	Component control & communication testing
Real-time *simulation & HIL*	Integration test for DER
	Validation of new power electronic components

Besides the above mentioned scenarios, training and education activities are performed in the lab environment. This includes the training of utility staff on new (remote-)control functions provided by inverter-based DER components (e.g., voltage and frequency control). In addition, it is also used for teaching university students about smart grid systems and automation issues.

21.3.3 Requirements for Laboratory Automation System

The SmartEST laboratory provides a flexible and configurable multi-functional design, validation and testing environment for smart grids. Therefore, the corresponding automation system has to be as flexible as possible. To develop, test and validate new automation, communication and safety concepts for smart grids applications, especially in the SmartEST lab, the following requirements can be derived from the above collected use cases [2]:

Hardware environment:

- *Flexibility:* The lab evaluation and test environment should provide sufficient flexibility to incorporate various communication protocols and interfaces as well as controller platforms. The support of different simulation environments for power distribution systems, ICT, and automation concepts should be possible.

- *Scalability:* An easy enlargement of the laboratory evaluation and test environment should be possible. If the limits of scalability are reached, simulation methods must be provided and developed to achieve scalability in power and flexibility (e.g., real-time simulation, CHIL/PHIL).

Software application environment:

- *Configurability:* The configuration of heterogeneous controllers by various software tools should be possible.

- *Portability:* The tested automation and control programs should be easily portable to real controller devices.

- *Hardware-independent:* Control applications should be implemented in a hardware-independent manner in order to execute them on different hardware resources.

- *Distribution:* The distribution of programs to controller hardware should be supported to be compliant with the distributed nature of smart grids applications.

Simulation environment:

- *Offline simulation:* Offline simulation of power distribution models (i.e., physical and control) and ICT concepts need to be supported.

- *Real-time simulation:* Real-time simulation of power distribution systems and their corresponding components (i.e., physical hardware and emulated models in software) along with ICT concepts should be possible.

Open system architecture and standard compliance:

- *Interoperability:* The interoperability of simulation models and environments, real network components, and controller hardware and software should be supported.

- *Standard-compliant implementation:* The lab environment should be based on well-known standards from the communication (i.e., Industrial Ethernet, Ethernet, fieldbuses), automation and control (i.e., IEC 61499, OPC UA) and power systems domain (i.e., IEC 61850, IEC 60870) domains [20, 23].

- *Open communication interfaces:* Various open communication networks and interfaces (e.g., IEC 61850, OPC UA, Modbus RTU/TCP, Industrial Ethernet) should allow the testing and validation of a wide range of applications and components in the smart grid domain [20, 23].

- *Free and open source approaches:* The usage of free and open source software components and modules should help enhance the laboratory evaluation and test environment which is not always possible by closed source approaches.

21.4 IEC 61499-Based Laboratory Automation System

In order to fulfill the needs for a flexible and scalable smart grid laboratory, an automation system using the IEC 61499 reference model [25] for distributed automation and control has been developed. In the following sections the automation architecture, the automation application and the resulting open source-based automation environment are presented.

21.4.1 Architecture and Concept

A major requirement for SmartEST automation system is a flexible and rapidly adaptable system architecture which takes scalability into account. A fast adaptation to new lab equipment and devices (i.e., hardware and software) without changing the core elements of the automation application is a key point for its realization.

The general concept—a four layer approach—is motivated by the smart grid architecture model (SGAM) provided by the International Electrotechnical Commission (IEC) [20]. For the implementation of the automation and control logic, the IEC 61499 standard is used since it is a promising candidate approach as suggested by DKE[1] and IEC [7, 20]. The different layers have the following meanings [3]:

- *Supervision, Visualization and Interaction Layer:* The user interaction (i.e., with the laboratory operator) and the visualization are covered by this layer. Typically, a human-machine interface (HMI) is provided to show the status of the lab configuration by representing measurement and monitoring results in a human-readable format. This layer can be used to configure the laboratory setup. It is also called supervisory control and data acquisition (SCADA) layer.

- *Control and Monitoring Layer:* In this layer, control and automation functions which are usually time critical or independent from user interaction are carried out. Measurements from sensors in the system components layer are processed and filtered in this layer before they are distributed

---

[1]German Commission for Electrical, Electronic and Information Technology of DIN Deutsches Institut für Normung e. V. and VDE Verband der Elektrotechnik Elektronik Informationstechnik e.V.

to the SCADA layer. Furthermore, safety related functions (e.g., lockout, emergency signals) protect the lab against faults and wrong configurations. Because of the critical nature of the functions in this layer in terms of computer performance, they are usually executed on embedded devices.

- *System Components Layer:* This layer contains the electrical components used for interaction with the technical process. Examples are sensors (i.e., measurement devices, feedback sensors), actuators (i.e., power switches and breakers) and inputs and outputs (I/Os) (i.e., digital and analogue distributed I/Os). These components are used to interact with and influence the technical process, i.e., the system components can be controlled and monitored by the functions in the control and monitoring layer.

- *Technical Process Layer:* The laboratory grids and corresponding components (power switches, AC and DC simulator, transformer, loads, etc.) as shown in Figure 21.1 excluding the system components in the previous layer are located in this layer.

The resulting architecture is shown in Figure 21.2. The control and monitoring layer is realized using the IEC 61499 reference model for distributed automation and control (see Section 21.4.2). With this concept in hand, the above mentioned requirements and needs for a flexible laboratory environment can be fulfilled to a large extent.

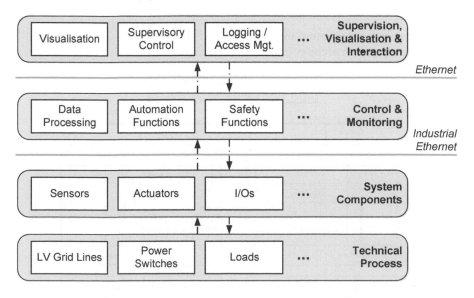

FIGURE 21.2
Automation architecture allowing a flexible setup of smart grid lab projects.

21.4.2 Automation Application Model Based on IEC 61499

The automation application model described below contains, *(i)* the system configuration and the *(ii)* automation application. The whole setup is modeled and implemented using the IEC 61499 reference model [25] for distributed automation and control.

21.4.2.1 System Configuration

The main tasks of the SmartEST automation system are the secure controlling of the laboratory configurations (i.e., different setups for the lab grids and components) and their visualization for interaction with the operator. The whole system must be flexible and adaptable to new hardware without changing the core automation and control software.

Figure 21.3 illustrates the IEC 61499 system configuration which allowed us to achieve this challenging goal. It contains two IEC 61499-compliant embedded control devices for the execution of laboratory automation and control application. Both control devices are connected through the communication interface to the communication network and finally to the laboratory SCADA. In addition, the process interface connects these devices with the laboratory equipment (i.e., lab grids, switches, loads, AC and DC simulator as depicted in Figure 21.1). The interaction with offline and/or real-time simulated components is possible. The control devices are responsible for executing different control and monitoring functions (denoted as *automation application* in Figure 21.3).

As a result, the requirements described in Section 21.3.3 can be fulfilled with this proposed architecture.

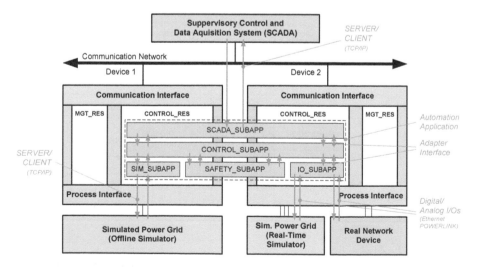

FIGURE 21.3
IEC 61499 system configuration for the SmartEST lab [2].

21.4.2.2 Automation Application

For the development of the automation application the focus was placed on the reusability of the software components (i.e., IEC 61499 function blocks) and clarity of the control code since the SmartEST lab contains a large number of different devices. Such an issue has already been discussed in several IEC 61499 publications. According to [1, 12, 28], an IEC 61499 control application should be modeled using the subapplications on different layers to form a hierarchical structure. It is suggested to provide on the top-most layer an overview of the different components contributing to the application. Each subapplication representing a specific automation function should describe its corresponding behavior.

In the SmartEST context, the automation application is divided into the following application parts using IEC 61499 subapplications as main modeling elements on the top layer and underlying layers:

- SCADA interaction (i.e., `SCADA_SUBAPP`)

- Control functions, monitoring and data processing (i.e., `CONTROL_SUBAPP`)

- Safety functions (i.e., `SAFETY_SUBAPP`)

- I/O integration (i.e., `IO_SUBAPP`), and

- Simulation integration (i.e., `SIM_SUBAPP`)

The interaction of the subapplications is realized using adapter interfaces and generic communication interfaces patterns (i.e., `Server/Client`) as proposed by [1, 12, 28]. The adapter interfaces are sketched in Figure 21.4, which also provides the top level view of the automation application with the main parts as listed above.

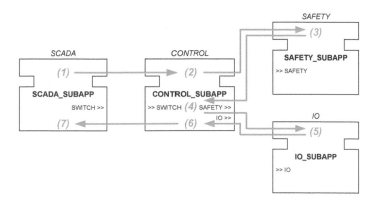

FIGURE 21.4
Sketch of the automation application and the corresponding signal flow.

With the decoupling of the automation application a largely hardware independent realization of the SmartEST automation system is possible (except the IO_SUBAPP and the SIM_SUBAPP). The following paragraphs provide more details about the realization of the different subapplications.

SCADA Interaction

The SCADA_SUBAPP provides the interface between the control layer and the SCADA layer. Control signals from the SCADA system are transmitted via this subapplication to invoke control functions and to provide feedback signals (steps *(1)* and *(7)* in Figure 21.4). Moreover, status information from the sensors and measurement equipment are also communicated to the SCADA system to provide an overview of the laboratory state to the operator. The generic IEC 61499 communication service interface functions block (SIFB) patterns CLIENT/SERVER are the main parts of this subapplication. The control device represents the server functionality whereas the SCADA system is the corresponding client.

Control Functions, Monitoring and Data Processing

This application part is used to configure, control and monitor the laboratory configuration (AC/DC simulator, adjustable transformer, earthing system, topology of the laboratory grids—radial, meshed, island, etc.). It provides automation functionality to control the power switches manually via the SCADA system or via an automatic switching sequence (e.g., adjusting the reference value for the active and/or reactive power of the RLC-load, testing sequence for configuring the laboratory grids).

Figure 21.5 shows the corresponding IEC 61499 function block (FB) for controlling a power switch. Digital signals, provided via the decentralized I/Os (see I/O integration below), are necessary to turn a power switch on or off. This FB is responsible to check whether a switch commando can be executed and generate the appropriate I/O signals (steps *(2)* and *(4)* in Figure 21.4).

Furthermore, the monitoring of the laboratory state (e.g., switch position,

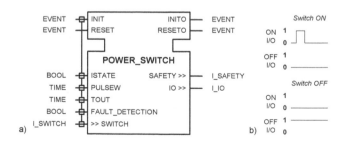

FIGURE 21.5
Power switch FB: a) FB interface and b) generated switching signal.

emergency signals) is also performed and reported to the SCADA (step *(6)* in Figure 21.4). For this task the generic IEC 61499 communication SIFB CLIENT is used to interact with sensors and measurement equipment via the IEC 61499 process interface. Modbus/TCP is mainly used as communication protocol to interact with the measurement devices.

Safety Functions

The SmartEST lab grids contain multiple switches and components and many of them have specific safety conditions. For example there exist several cases where two or more switches cannot be on at the same time. Some of the most critical safety conditions are covered using hardware functionality; all other conditions must be covered in the software.

Safety conditions can be covered in the control or SCADA layer. Also, where the safety conditions are checked by hardware, proper user interactions and visualizations are needed. However, in practice a SCADA system is not used for safety critical checks, mostly because of the low task priority of the SCADA software. To keep the safety applications as reliable as possible, they were implemented in the control layer using IEC 61499 elements (i.e., FBs).

The safety subapplication (SAFETY_SUBAPP) contains functions which check different safety conditions in the lab. Before a control action (e.g., switch request initiated either from the operator via SCADA or via control functions) can be executed using the I/Os connected to an actor (e.g., power switch) it has to pass the safety subapplication (step *(3)* in Figure 21.4). This prevents any control actions to circumvent the safety subapplication and connect directly to the I/Os. The two main safety functions implemented for the lab environment are:

- One common safety function is a so-called lock-out functionality. This was implemented using an IEC 61499 FB, which is shown in Figure 21.6. The main functionality of this FB is to prevent switches from being turned on at the same time. The FB receives from two switches one request signal and one feedback signal with the current state of the switch. The allowed states are forwarded to the outputs, as also described in this figure.

- A more comprehensive safety function is the inrush current delimitation. Because of the high inrush current which may occur when the voltage is switched on, some of the grid components in the RLC load may be damaged (see Figure 21.1). To prevent this, a defined order is used in which the switches are allowed to be turned on, i.e., the AC voltage must be switched on before the RLC load.

I/O Integration

To allow the IEC 61499 control devices to interact with the lab equipment, decentralized I/Os were used (step *(5)* in Figure 21.4). The I/Os were connected with an Industrial Ethernet interface using Ethernet POWERLINK

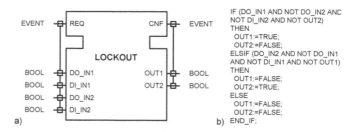

FIGURE 21.6
Implementation of a lockout functionality using an IEC 61499 FB: a) FB
interface and b) encapsulated algorithm.

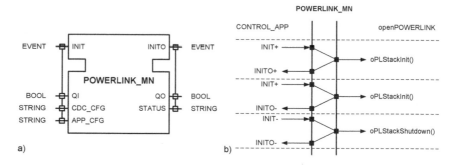

FIGURE 21.7
EPL master node SIFB: a) FB interface and b) service sequence [4].

(EPL), encapsulated into IEC 61499 FBs [4]. The EPL specification defines
master nodes (MN) and controlled nodes (CN). Usually, programmable logic
controllers (PLC) or industrial PCs (IPC) act as MNs whereas CNs (i.e., I/O
devices, drives, etc.) are connectable via Ethernet. As shown in Figure 21.3,
an IEC 61499 device represents the MN services (i.e., IO_SUBAPP). To encapsulate the MN and CN functions, special IEC 61499 FBs—service interface
functions blocks (SIFBs)—are necessary to define the IO_SUBAPP. The following two SIFBs have been developed [4]:

- *EPL Master Node SIFB:* This SIFB represents the MN. It has two data
  inputs: CDC_CFG contains the mapping of the I/Os to the process image
  of the EPL configuration as a binary file and APP_CFG presents the network configuration as an XML file. With this SIFB, the EPL driver can
  be started and stopped. During the start-up phase, the MN deploys the
  network configuration (i.e., CDC_CFG) to the CNs. Figure 21.7 shows the
  interface and the service sequences of the MN SIFB.

- *EPL Client Node I/O SIFB:* The access to EPL I/O devices is encapsulated into this SIFB. It provides services to read inputs and write outputs.

Since several CNs are supported in an EPL network, the identification of a specific CN has to be provided to this IEC 61499 SIFB (i.e., `CNID` data input). Decentralized I/Os have usually several I/Os. Therefore, it is necessary to specify the right I/O channel. This is covered by the `PARAM` input of this SIFB. `OUTDATAm` and `INDATAn` represent the corresponding inputs and outputs of the decentralized I/Os encapsulated into an IEC 61499 SIFB. Figure 21.8 shows the SIFB interface and the service sequences of the CN representing the access to decentralized I/Os via EPL.

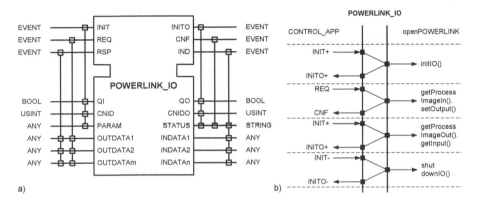

FIGURE 21.8
EPL controlled node SIFB (i.e., I/O node): a) FB interface and b) service sequence [4].

Since the EPL specification uses cyclic execution behavior and IEC 61499 event-based execution behavior, a mapping between both worlds is necessary. The following procedure was chosen: The POWERLINK_IO SIFB notifies the EPL driver whether outputs have to be updated. During the next EPL cycle, the corresponding outputs are updated. The inputs are normally read at every EPL cycle but an output event of the POWERLINK_IO SIFB (i.e., `IND`) is generated only if this SIFB detects changes in the values of the inputs.

Simulation Integration

To fulfill the requirements for smart grid labs as discussed in Section 21.2, the integration of simulated components is also of importance [24]. This issue is addressed by the proposed architecture and system configuration depicted in Figure 21.3 by the interaction with offline power system simulators (MATLAB®/Simulink SimPowerSystems, PSAT, PowerFactory, etc.) or digital real-time simulators (OPAL-RT, Typhoon HIL, etc.). `CLIENT/SERVER` SIFBs using TCP/IP are mainly used to connect to offline simulators whereas digital I/Os controlled by EPL are available to communicate with the real-time simulators. With this possibility in mind, simulated power system components can be integrated into the whole laboratory design.

21.4.3   Open-Source-Based Automation Environment

To realize the above introduced concept and corresponding IEC 61499 system configuration open source software and tools are used [17]. The open source and IEC 61499-compliant distributed control environment 4DIAC is used for the implementation of the control functions [27, 29]. It provides an engineering environment for IEC 61499 systems and the corresponding execution environment for embedded devices. This environment can execute different control and safety functions and interact with the SCADA layer and the system components layer.

For interacting with the lab hardware, about 1,300 digital I/Os are installed using 16 distributed I/O nodes controlled via EPL. Their main tasks are the controlling of the actuators (power switches and relays) and providing feedback and status information from sensors and measurement devices.

The communication and data exchange between the 4DIAC control environment executes on two Linux-operated IPC and the distributed I/O system is carried out using the open source EPL driver openPOWERLINK encapsulated into SIFBs (see Section 21.4.2.2) [4]. Along with the distributed digital I/O system, a large amount of measurement data (about 700 channels covering voltages, currents, active and reactive power, frequency, etc.) are observed by SmartEST. In total 22, power measurement devices are integrated and utilized for both operation and control purposes. In order to transfer the measured values from the lab equipment to the control and SCADA layers, Modbus/TCP is used. Similar to the openPOWERLINK integration, an open source approach has been integrated into IEC 61499 communication SIFBS (i.e., `CLIENT/SERVER`). The libmodbus library is used for this purpose.

The SCADA layer for supervisory control and monitoring is also very important for operating the SmartEST lab, as it provides a HMI to the operator. Since 4DIAC covers the control functionality without any visualization possibilities the open source implementation ScadaBR is used [17]. It visualizes the available laboratory equipment and provides status information as depicted in Figure 21.9. Also high-level supervisory and control functions are executed in ScadaBR. An overview of the chosen configuration of the lab grids is provided to the operator. The communication and data exchange with the control layer (i.e., 4DIAC environment) is carried out using a TCP-based protocol suggested by IEC 61499 which is implemented in ScadaBR and 4DIAC.

21.5   Summary and Conclusions

For the development and validation of smart grid solutions, a flexible and multi-functional environment is necessary. It should facilitate testing of hardware and software components and their integration into the power system.

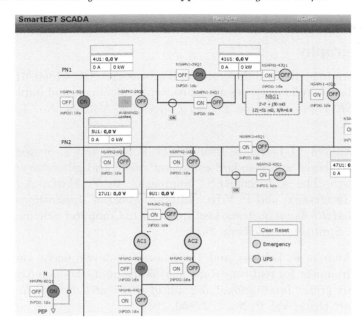

FIGURE 21.9
Extract of implemented HMI for operating the SmartEST laboratory.

This can be achieved by laboratory hardware, or proper simulation models or a combination of both. The SmartEST laboratory environment provide such a setup for the power system and smart grid domain.

For the automation and control of this research infrastructure, various challenging hardware and software requirements have to be fulfilled. To achieve this, an automation architecture and corresponding concept using the IEC 61499 reference model is proposed. For its realization the open source tools 4DIAC, ScadaBR, openPOWERLINK, and libmodbus are used. This time-consuming integration of freely available software was necessary to achieve a very flexible and highly configurable software environment. The IEC 61499-compliant 4DIAC approach serves as an integrating middleware between the laboratory components and the SCADA. It provides the core of the proposed automation system.

Future work focuses on the further development of automation functions required by specific testing methods. Other communication approaches like IEC 61850, OPC UA, and smart metering protocols need to be integrated into the SmartEST automation system. The open source-based approach allows the required integration of new concepts, methods and protocols instead of proprietary and closed source solutions from commercial automation vendors.

Bibliography

[1] F. Andrén, R. Bründlinger, and T. Strasser. IEC 61850/61499 control of distributed energy resources: concept, guidelines, and implementation. *IEEE Transactions on Energy Conversion*, Vol. 29, No. 4, 1008–1017, 2014.

[2] F. Andrén, G. Lauss, R. Bründlinger, P. Svec, and T. Strasser. An open source-based and standard-compliant smart grid laboratory automation system: The AIT SmartEST approach. In V. Marík, A. Shirrmann, D. Trentesaux, and P. Vrba, editors, *Industrial Applications of Holonic and Multi-Agent Systems*, Lecture Notes in Computer Science, pages 195–205. Springer, Heidelberg, 2015.

[3] F. Andrén, F. Lehfuss, and T. Strasser. A development and validation environment for real-time controller-hardware-in-the-loop experiments in smart grids. *International Journal of Distributed Energy Resources and Smart Grids*, Vol. 9, No. 1, 27–50, 2013.

[4] F. Andrén and T. Strasser. Distributed open source control with Industrial Ethernet I/O devices. In *Proceedings of 16th IEEE Conference on Emerging Technologies Factory Automation*, pages 1–4, 2011.

[5] R. Bründlinger, D. Craciun, W. Heckmann, V. Helmbrecht, B. Klebow, P. Kotsampopoulos, L. Martini, S. Numminen, I. Papaioannou, T. Strasser, M. Sosnina, and P. Vaessen. European Distributed Energy Resources Laboratories Activity Report 2012/2013, 2012.

[6] R. Bründlinger, T. Strasser, G. Lauss, A. Hoke, S. Chakraborty, G. Martin, B. Kroposki, J. Johnson, and E. de Jong. Lab tests: verifying that smart grid power converters are truly smart. *IEEE Power and Energy Magazine*, Vol. 13, No. 2, 30–42, 2015.

[7] DKE. The German Standardisation Roadmap E-Energy/Smart Grid, German Commission for Electrical, Electronic and Information Technologies of DIN and VDE, 2010.

[8] EC. European SmartGrids Technology Platform: Vision and Strategy for Europe's Electricity Networks of the Future, European Commission, 2006.

[9] EC. Smart Grids: From Innovation to Deployment, Communication from the Commission to the European Parliament, the Council, the European Economic and Social Committee and the Committee of the Regions, European Commission, 2011.

[10] H. Farhangi. The path of the smart grid. *IEEE Power and Energy Magazine*, Vol. 8, No. 1, 18–28, 2010.

[11] V. Gungor, D. Sahin, T. Kocak, S. Ergut, C. Buccella, C. Cecati, and G. Hancke. Smart grid technologies: communication technologies and standards. *IEEE Transactions on Industrial Informatics*, Vol. 7, No. 4, 529–539, 2011.

[12] I. Hegny, T. Strasser, M. Melik-Merkumians, M. Wenger, and A. Zoitl. Towards an increased reusability of distributed control applications modeled in IEC 61499. In *Proceedings of 17th IEEE Conference on Emerging Technologies Factory Automation*, pages 1–8, 2012.

[13] K. Heussen and O. Gehrke. Lab Survey: State of the Art Smart Grid Laboratories, Department of Electrical Engineering, DTU, Kongens Lyngby, 2014.

[14] K. Heussen, A. Thavlov, and A. Kosek. Use Cases for Laboratory Software Infrastructure: Outline of Smart Grid Lab Software Requirements, Department of Electrical Engineering, DTU, Kongens Lyngby, 2014.

[15] IEA. Smart grid insights, International Energy Agency, 2011.

[16] IEA. Smart Grid International Research Facility Network (SIRFN), International Energy Agency, 2015.

[17] G. Lauss, F. Andrén, M. Stifter, R. Bründlinger, T. Strasser, K. Knöbl, and H. Fechner. Smart grid research infrastructures in Austria: examples of available laboratories and their possibilities. In *Proceedings of 13th IEEE International Conference on Industrial Informatics*, pages 1539–1545, 2015.

[18] M. Liserre, T. Sauter, and J. Hung. Future energy systems: integrating renewable energy sources into the smart power grid through industrial electronics. *IEEE Industrial Electronics Magazine*, Vol. 4, No. 1, 18–37, 2010.

[19] SGA. Roadmap Smart Grids Austria: Pathway to the Future of Electrical Power Grids!, Smart Grids Austria, 2015.

[20] SMB Smart Grid Strategic Group (SG3). IEC Smart Grid Standardization Roadmap, International Electrotechnical Commission, 2010.

[21] M. Steurer, C. Edrington, M. Sloderbeck, W. Ren, and J. Langston. A megawatt-scale power hardware-in-the-loop simulation setup for motor drives. *IEEE Transactions on Industrial Electronics*, Vol. 57, No. 4, 1254–1260, 2010.

[22] T. Strasser, F. Andren, J. Kathan, C. Cecati, C. Buccella, P. Siano, P. Leitao, G. Zhabelova, V. Vyatkin, P. Vrba, and V. Marik. A review of architectures and concepts for intelligence in future electric energy systems. *IEEE Transactions on Industrial Electronics*, Vol. 62, No. 4, 2424–2438, 2015.

[23] T. Strasser, F. Andrén, M. Merdan, and A. Prostejovsky. Review of trends and challenges in smart grids: an automation point of view. In V. Marík, J. Lastra, and P. Skobelev, editors, *Industrial Applications of Holonic and Multi-Agent Systems*, volume 8062 of Lecture Notes in Computer Science, pages 1–12. Springer, Heidelberg, 2013.

[24] T. Strasser, M. Stifter, F. Andrén, and P. Palensky. Co-simulation training platform for smart grids. *IEEE Transactions on Power Systems*, Vol. 29, No. 4, 1989–1997, 2014.

[25] TC 65/SC 65B. IEC 61499: Function blocks. International Electrotechnical Commission, Geneva, 2nd ed., 2012.

[26] P. Vrba, V. Marik, P. Siano, P. Leitao, G. Zhabelova, V. Vyatkin, and T. Strasser. A review of agent and service-oriented concepts applied to intelligent energy systems. *IEEE Transactions on Industrial Informatics*, Vol. 10, No. 3, 1890–1903, 2014.

[27] V. Vyatkin. IEC 61499 as enabler of distributed and intelligent automation: state-of-the-art review. *IEEE Transactions on Industrial Informatics*, Vol. 7, No. 4, 768–781, 2011.

[28] A. Zoitl and H. Prähofer. Guidelines and patterns for building hierarchical automation solutions in the IEC 61499 modeling language. *IEEE Transactions on Industrial Informatics*, Vol. 9, No. 4, 2387–2396, 2013.

[29] A. Zoitl, T. Strasser, and A. Valentini. Open source initiatives as basis for the establishment of new technologies in industrial automation: 4DIAC a case study. In *Proceedings of 2010 IEEE International Symposium on Industrial Electronics*, pages 3817–3819, 2010.

Index